Arctic and Alpine Plants: Ecology, Adaptations and Conservation Biology

Arctic and Alpine Plants: Ecology, Adaptations and Conservation Biology

Editor

Gregor Kozlowski

MDPI

Basel • Beijing • Wuhan • Barcelona • Belgrade • Novi Sad • Cluj • Manchester

Editor
Gregor Kozlowski
University of Fribourg
Fribourg
Switzerland

Editorial Office
MDPI AG
Grosspeteranlage 5
4052 Basel, Switzerland

This is a reprint of articles from the Special Issue published online in the open access journal *Plants* (ISSN 2223-7747) (available at: https://www.mdpi.com/journal/plants/special_issues/Conservation_Arctic_Alpine_Plants).

For citation purposes, cite each article independently as indicated on the article page online and as indicated below:

Lastname, A.A.; Lastname, B.B. Article Title. *Journal Name* **Year**, *Volume Number*, Page Range.

ISBN 978-3-7258-1553-1 (Hbk)
ISBN 978-3-7258-1554-8 (PDF)
doi.org/10.3390/books978-3-7258-1554-8

Cover image courtesy of Gregor Kozlowski

Contents

About the Editor

Gregor Kozlowski

Prof. Dr Gregor Kozlowski is a director of the Botanic Garden and group leader at the Department of Biology of the University of Fribourg (Switzerland). He is also a conservator of botany at the Natural History Museum in Fribourg (Switzerland). Furthermore, his group is tightly associated with the Shanghai Chenshan Botanic Garden in China at the Eastern China Conservation Centre for Wild Endangered Plant Resources of the Chinese Academy of Sciences, where he leads a research group. His main research interests are the biogeography and conservation biology of relict, endemic, and threatened species. The trademark of his research group is intensive field work, both locally (e.g., Switzerland, Central Europe, and the Alps) and globally (e.g., the Mediterranean, Arctic, Transcaucasia, and Eastern Asia).

One of the most important study topics of his group is the biogeography, ecology, and evolution of arctic-alpine and boreo-alpine taxa such as *Calamagrostis* (Poaceae), *Nuphar* (Nymphaeaceae), *Papaver* (Papaveraceae), *Arenaria* (Caryophyllaceae), as well as selected members of monilophytes (ferns and allies). Their second very important research topic covers woody species. For the past 15 years, their major model organisms have been relict trees, mainly from the following families: Ulmaceae, Juglandaceae, Fagaceae, Betulaceae, Pinaceae, Rosaceae, and Fabaceae. More recently, his group also investigated the ecology and conservation biology of an atypical woody shrub *Ptilostemon greuteri* (Asteraceae) endemic to Sicily, and the woody liana *Clematis alpina* (Ranunculaceae) across the European continent. One of the most important turning points of his career was the foundation in 2010 of the international and interdisciplinary Project *Zelkova* and its enlargement in 2015 with the parallel Project *Pterocarya*. Originally, as the names indicate, the main focus was on the relict tree genera *Zelkova* (Ulmaceae) and *Pterocarya* (Juglandaceae). However, over the past few years and in collaboration with associated researchers, this focus was enlarged, and today, the projects investigate many other rare and threatened woody plants in various geographical regions and carry out a large spectrum of research activities (e.g., global change biology, phylogeny, phylogeography, population genetics, population structure, dendrochronology, taxonomy, morphometry, etc.). More than 30 researchers representing universities and research institutes from 15 countries worldwide are now forming the *Zelkova/Pterocarya* network (e.g., those from Switzerland, China, Italy, Greece, Germany, Poland, United Kingdom, USA, Colombia, Georgia, Azerbaijan, Iran, Japan, and Vietnam), organizing joint workshops, conferences, symposia, scientific expeditions, capacity-building activities, and publishing together. The key of these successful scientific activities is the intense and reciprocal collaboration and exchange of data and knowledge with local scientists, local administration, stakeholders, and foresters.

plants MDPI

Editorial

Understanding Arctic–Alpine Plants from Ecological and Evolutionary Perspectives

Gregor Kozlowski [1,2,3]

1 Department of Biology and Botanical Garden, University of Fribourg, Chemin du Musée 10, CH-1700 Fribourg, Switzerland; gregor.kozlowski@unifr.ch
2 Natural History Museum Fribourg, Chemin du Musée 6, CH-1700 Fribourg, Switzerland
3 Eastern China Conservation Centre for Wild Endangered Plant Resources, Shanghai Chenshan Botanical Garden, 3888 Chenhua Road, Songjiang, Shanghai 202602, China

Conditions in arctic and alpine ecosystems impose great challenges on the plants and other organisms that live there [1]. Despite this, thousands of plant species worldwide survive or even prosper under extreme climatic, edaphic, and ecological conditions in the High North (or South) and in the high mountains [2]. Arctic and alpine plants continue to surprise researchers with their ingenious strategies and adaptations [3]. Today, global warming, the ever-increasing demand for resources, and the expansion of tourism are growing threats to arctic and alpine plant life, even in the most remote regions of the world. This applies not only to glacial relics and endemics in isolated mountain refugia but also to tundra areas that were intact until recently and are now under increasing pressure from man-made global changes [4,5]. In such a context, this Special Issue of *Plants*, entitled *Arctic and Alpine Plants: Ecology, Adaptations and Conservation Biology*, has been launched to improve our understanding of this unique group of plants from ecological, biogeographical, and evolutionary perspectives.

A publication by Körner [6] provides an excellent opportunity to investigate the complexity and specificity of high mountain plants and their communities, which are often found at the edge of their fundamental niches. This review paper presents no less than 12 concepts in alpine plant ecology, starting with life forms and aspects of the physical environment, such as topography, moving through physiology and reproductive biology, and finishing with global change drivers and their influence on plants growing in the alpine ecological zone. Despite the enormous complexity of such factors, Körner [6] states in his concluding comments that "nothing make sense in alpine plant biology unless one accounts for micro-climate". This is further developed upon in concept 11 of his review, entitled "To be or not to be—the edge of the fundamental niche". The author postulates that any projections of future species distributions, especially in the alpine and arctic context, need to rest on a very fine-grained representation of the landscape and stresses the essential role of microtopography in any modeling study.

The second review paper, by Hörandl [7], presents so-called geographical parthenogenesis and explores its importance for alpine and arctic plants. Geographical parthenogenesis describes the phenomenon of asexual organisms usually having larger distribution areas than their sexually reproducing relatives and colonizing more frequently at higher latitudes and in previously glaciated or otherwise devastated areas. However, the causal factors behind this phenomenon are still unclear. Polyploidy, epigenetic flexibility, and phenotypic plasticity are often seen as important factors fostering geographical parthenogenesis patterns. Additional large-scale studies are needed that include information on reproduction and that integrate ecological, experimental, and molecular research involving many genera and large-scale screening from large geographical areas. High arctic regions are of great importance since they are not easily accessible and, thus, are not well explored.

Polyploidy is an important phenomenon in understanding the biogeography and taxonomy of disjunct arctic–alpine species complexes. Wagner et al. [8] provided detailed

Citation: Kozlowski, G. Understanding Arctic–Alpine Plants from Ecological and Evolutionary Perspectives. *Plants* **2024**, *13*, 1560. https://doi.org/10.3390/plants13111560

Received: 22 May 2024
Accepted: 3 June 2024
Published: 5 June 2024

insights into taxonomically challenging hexaploid alpine shrub willows (*Salix* sp., *Salicaceae*). In this study, RAD sequencing data, infrared spectroscopy, and morphometric data were used to analyze the phylogenetic relationships of the species from the *Phylicifoliae* and *Nigricantes* sections in a framework of 45 Eurasian *Salix* species. The results reveal, among other insights, that both sections are polyphyletic and need to be redefined. Additionally, *S. bicolor* should be included in *S. phyllicifolia* s.l., and the alpine endemic *S. hegetschweileri* is a distinct lineage showing close relationships to the *Nigricantes* clade.

The importance of large-scale geographical sampling when studying disjunct and polyploid arctic–alpine plant complexes was confirmed by two studies performed by Kozlowski et al. [9,10]. These two studies explored the genome sizes and ploidy levels of the *Arenaria ciliata* species complex (*Caryophyllaceae*). Their papers enabled the first synthesis of the ploidy level variability within this group, concluding that three taxa are predominantly tetraploid (2n = 4x = 40): the arctic *A. pseudofrigida* and the two alpine taxa *A. multicaulis* and *A. ciliata* subsp. *ciliata*. Higher ploidy levels were detected in *A. norvegica* (2n = 8x = 80) and *A. gothica* (2n = 10x = 100), and the highest detected ploidy level was for the narrow endemic of the Western Alps *A. ciliata* subsp. *bernensis* (2n = 20x = 200). These results confirm that there are no diploid taxa within the *A. ciliata* species complex and that this group is an example of a so-called *mature polyploid complex*, a common phenomenon among arctic–alpine taxa. The *A. ciliata* species group is, thus, an example of an arctic–alpine species complex characterized by reticulate evolution, polyploidization, and hybridization, most likely associated with rapid latitudinal and altitudinal migration from the Pleistocene through to the Holocene.

Due to their adaptations to cold environments, arctic and alpine plants are very sensitive to climate change. Two papers published in this Special Issue explore the influence of changing climate conditions on the past and future distributions of plants growing at high latitudes and altitudes. Both studies use traditional species distribution models (SDMs) based on data extracted from the Global Biodiversity Information Facility (GBIF) and, thus, do not use fine-grained representations of the landscape, as proposed by Körner [6]. Walas et al. [11] assessed and compared the current and future potential niche areas of *Kalmia procumbens* (*Ericaceae*) in the Pyrenees and Carpathians. Both mountain ranges represent the southernmost localities of this circumpolar, arctic–alpine species in Europe. The results demonstrated, as expected, that the species covered a larger distribution area during the Last Glacial Maximum; due to recent, accelerated climate warming, a reduction in the potential distribution area is expected by 2100. However, the reduction will be substantial in the Carpathians, where only a few South Carpathian populations would persist, and would be rather moderate in the Pyrenees.

Zhang and Wang [12] presented another study using SDMs, aiming to predict the potential distribution of endangered *Meconopsis punicea* (*Papaveraceae*), which is endemic to China and grows in high-altitude grasslands and shrublands between 2800 and 4300 m a.s.l. Four types of species distribution models were applied: generalized linear model (GLM), generalized boosted model (GBM), random forests (RF) and flexible discriminants analysis (FDA). Under future climate change, the potential distribution of *M. punicea* will expand from southeastern to northwestern China. However, there were significant differences in the distributions predicted by the different SDMs. Therefore, the authors propose the use of consistent results from different SDMs as the basis for developing conservation strategies.

In addition to the frequently covered effects of climate warming, many other human-made changes have important impacts on the survival of arctic and alpine plants. Chardon et al. [13] explored the response of arctic shrubs and graminoids to human trampling, an integral component of global change; however, a comprehensive understanding of the effects of trampling on alpine and arctic ecosystems is lacking. The authors surveyed trail-side (disturbed) and off-trail (undisturbed) transects along altitudinal gradients in the Garibaldi Provincial Park in Canada. The main focus was placed on dominant shrubs (*Phyllodoce empetriformis*, *Cassiope mertensiana*, and *Vaccinium ovalifolium*, all three of which are *Ericaceae*) and on graminoids (*Carex* species, *Cyperaceae*). The study demonstrated that trampling reduces plant cover in all species and that growth traits are more sensitive than reproductive traits. However, the disturbance response is species-specific. For example, a reduction in size was observed only for *P. empetriformis* and *V. ovalifolium*, whereas in *C. mertensiana*, human trampling mainly reduced reproductive output. Interestingly, in *Carex* spp., height, but not diameter, was sensitive to trampling, indicating the need to measure the traits of multiple species to better understand the effects of disturbance on arctic and alpine plants. The results of this study are of great conservation importance, highlighting that specific management approaches may need to be tailored toward the protection of particular species.

Another aspect linked to land-use changes and climate warming is the spread of shrubs in arctic and alpine environments. Oberhuber et al. [14] focused on the most rapidly expanding shrub species in the European Alps, the green alder (*Alnus alnobetula, Betulaceae*). This study, using dendrochronological methods, was carried out in alder populations spread above 2100 m a.s.l. within the alpine treeline ecotone on Mt. Patscherkofer (Tyrol, Austria). The main problem in such high-altitude individuals is their asymmetric radial growth and anomalous growth ring patterns, which make representative measurements very challenging. Consequently, it is very difficult to differentiate between the influences of climate and microsite conditions on radial growth. In topographically heterogeneous environments, site-specific variability in the climate–growth relationships of *A. alnobetula* must be considered. Therefore, the authors provide the following sampling recommendations when undertaking investigations on site-specific variability: measuring one radius per shoot, three shoots per individual, and a minimum of ten individuals within the same stand, which results in ca. ten radii per study plot. However, for the determination of absolute growth rates and growth trends with respect to recent climate warming, the following recommendations are made: measuring 4 radii per shoot, $\geq$5 shoots per individual, and $\geq$10 individuals within one stand (total ca. 200 radii per study plot).

Another important phenomenon in alpine ecosystems is the decrease in the diversity of life forms with altitude. López et al. [15] investigated the changes of trees to shrubs and other life forms along an elevational gradient. The authors used *Polylepis tarapacana* (*Rosaceae*), a small, cold-tolerant, evergreen tree species growing in the Andes in the high-elevation Altiplano as a model taxon. After Tibet, the Altiplano Plateau, which lies between Bolivia, Peru, and Chile, is the second most extensive high plateau on Earth. In northwestern Argentina, where the study was carried out, *P. tarapacana* forms monospecific stands in a treeline ecotone. The authors propose a new classification of life forms for this species: arborescent, shrub, dwarf shrub, and *brousse tigrée* (tiger bush). The justification for this new classification lies not only in the marked biometric characteristics but also in the different influences of topographic, climatic, and human-use factors on their frequency. Therefore, it is important to consider these different life forms of *P. tarapacana* when conducting studies and developing management plans, since each life form occupies a particular environmental situation. The authors concluded that the conservation of *P. tarapacana* forests without differentiating between their life forms is achieved through the loss of part of their ecological niche.

In summarizing the subject matter presented in this Special Issue of *Plants*, entitled *Arctic and Alpine Plants: Ecology, Adaptations and Conservation Biology*, it is evident that, despite the long tradition of alpine and arctic research, there is still much to be discovered. The future of these highly specialized organisms is uncertain. All the authors of this Special Issue agree that much more intensive field work and field research should be carried out, as well as more experimental work. Arctic–alpine plants and vegetation can serve as a barometer of global change. Therefore, research on this group of plants is highly valuable.

Acknowledgments: I would like to thank Alain Müller, Yann Fragnière, and Benoît Clément from the Botanical Garden of the University of Fribourg (Switzerland) and Peter Wandeler from the Natural History Museum in Fribourg (NHMF, Switzerland) for their long-standing support.

Conflicts of Interest: The author declares no conflicts of interest.

References

1. Körner, C. *Alpine Plant Life. Functional Plant Ecology of High Mountain Ecosystems;* Springer: Berlin/Heidelberg, Germany, 1999.
2. Bliss, L.C. Adaptations of arctic and alpine plants to environmental conditions. *Arctic* **1962**, *15*, 117–144. [CrossRef]
3. Crawford, R.M.M. *Plants at the Margin. Ecological Limits and Climate Change;* Cambridge University Press: Cambridge, UK, 2008.
4. Niskanen, A.K.J.; Niittynen, P.; Aalto, J.; Väre, H.; Luoto, M. Lost at high latitudes: Arctic and endemic plants under threat as climate warms. *Divers. Distrib.* **2018**, *25*, 809–821. [CrossRef]
5. Monz, C.A. The response of two arctic tundra plant communities to human trampling disturbance. *J. Environ. Manag.* **2002**, *2*, 207–217. [CrossRef] [PubMed]
6. Körner, C. Concepts in alpine plant ecology. *Plants* **2023**, *12*, 2666. [CrossRef]
7. Hörandl, E. Geographical parthenogenesis in alpine and arctic plants. *Plants* **2023**, *12*, 844. [CrossRef]
8. Wagner, N.D.; Marinček, P.; Pittet, L.; Hörandl, E. Insights into the taxonomically challenging hexaploidy alpine shrub willows of *Salix* section *Phylicifoliae* and *Nigricantes* (Salicaceae). *Plants* **2023**, *12*, 1144. [CrossRef]
9. Kozlowski, G.; Fragnière, Y.; Clément, B.; Meade, C. Genome size in the *Arenaria ciliata* species complex (Caryophyllaceae), with special focus on *A. ciliata* subsp. *bernensis*, a narrow endemic of the Swiss Northern Alps. *Plants* **2022**, *11*, 3489. [PubMed]
10. Kozlowski, G.; Fragnière, Y.; Clément, B.; Gilg, O.; Sittler, B.; Lang, J.; Eidsen, P.B.; Lang, S.I.; Wasowicz, P.; Meade, C. Genome size in the *Arenaria ciliata* species complex (Caryophyllaceae), with special focus on Northern Europe and the Arctic. *Plants* **2024**, *13*, 635. [CrossRef] [PubMed]
11. Walas, Ł.; Pietras, M.; Mazur, M.; Romo, Á.; Tasenkevich, L.; Didukh, Y.; Boratyński, A. The perspective of arctic-alpine species in southernmost localities. The example of *Kalmia procumbens* in the Pyrenees and Carpathians. *Plants* **2023**, *12*, 3399. [CrossRef] [PubMed]
12. Zhang, H.-T.; Wang, W.-T. Prediction of the potential distribution of the endangered species *Meconopsis punicea* Maxim under future climate change based on four species distribution models. *Plants* **2023**, *12*, 1376. [CrossRef] [PubMed]
13. Chardon, N.I.; Ston, P.; Hilbert, C.; Maclachlan, T.; Ragsdal, B.; Zhao, A.; Goodwin, K.; Collins, C.G.; Hewitt, N.; Elphinstone, C. Species-specific response to human trampling indicate alpine plant size is more sensitive than reproduction to disturbance. *Plants* **2023**, *12*, 3040. [CrossRef] [PubMed]
14. Oberhuber, W.; Dobler, A.-L.; Heinzle, T.; Scandurra, F.; Gruber, A.; Wieser, G. Climate overrides the influence of microsite conditions on radial growth of the tall multi-stemmed shrub *Alnus alnobetula* at treeline. *Plants* **2023**, *12*, 1708. [CrossRef] [PubMed]
15. López, V.L.; Bottan, L.; Pastur, G.M.; Lencinas, M.V.; Cuyckens, G.A.E.; Vellini, J.M. Characterization of *Polylepis tarapacana* life forms in the highest-elevation Altiplano in South America: Influence of the topography, climate and human uses. *Plants* **2023**, *12*, 1806. [CrossRef] [PubMed]

plants

MDPI

Perspective

Concepts in Alpine Plant Ecology

Christian Körner

Department of Environmental Sciences, University of Basel, Botany, 4056 Basel, Switzerland;
ch.koerner@unibas.ch

Abstract: The alpine life zone is perhaps the only biome that occurs globally where mountains are high enough. At latitudinally varying elevation, the alpine belt hosts small stature plants that vary greatly in morphology, anatomy and physiology. In this contribution, I summarize a number of principles that govern life in what is often considered a cold and hostile environment. The 12 conceptual frameworks depicted include the key role of aerodynamic decoupling from free atmospheric climatic conditions, the problematic concepts of limitation and stress in an evolutionary context, and the role of developmental flexibility and functional diversity. With its topography driven habitat diversity, alpine plant diversity is buffered against environmental change, and the multitude of microclimatic gradients offers 'experiments by nature', the power of which awaits multidisciplinary exploration.

Keywords: climate; biodiversity; ecological theory; high elevation; mountains; niche concept; productivity; reproduction; stress; topography

Citation: Körner, C. Concepts in Alpine Plant Ecology. *Plants* **2023**, *12*, 2666. https://doi.org/10.3390/plants12142666

Academic Editor: Gregor Kozlowski

Received: 15 June 2023
Revised: 4 July 2023
Accepted: 10 July 2023
Published: 17 July 2023

Why Care for Concepts?

As any other field of scientific research, alpine plant ecology is embedded in certain concepts, theoretical frameworks, or paradigms. These emerge from research history, belong to certain research communities, and often simply reflect the fact that some aspects of alpine plant life are understood better than others because methods were and are available. Indeed, the availability of tools often shapes scientific paradigms [1]. It seems useful to revisit some of the existing concepts and ask whether they just reflect legacies or available methods or rest on facts and robust theoretical backing. In this contribution, I depict a number of such concepts, explain their implications, reflect on their acknowledgment in the research community, and discuss future research needs to sharpen their theoretical foundation.

Not a concept, but rather a matter of definition or convention, 'alpine' represents the naturally treeless belt above the montane belt, both separated by the high elevation, climatic treeline [2]. Note the difference between this biogeographic definition of alpine and the common-language meaning that often substitutes mountain by alpine (e.g., alpine skiing, alpine cities). There are no cities at alpine elevations (that is, above the treeline) anywhere on the globe, although it often may look like there are, when the montane forest had been destroyed by land use, with trees absent from treeline, a source of confusion. Since the potential position of the treeline can be predicted by climatic data [3], the biogeographic delineation of 'alpine' also rests on climatology, and comprises 2.6% of the global land area outside Antarctica [4,5].

Across the treeless alpine world from arctic–alpine to tropical–alpine conditions, plant life is ruled by a number of common drivers, each belonging to different scientific domains and thus rarely considered jointly, given the disciplinary barriers. The sequence of the 12 concepts depicted has no particular meaning, but I start with plant morphology and the physical environment, then move to more physiological aspects, and close with population- and community-related concepts. For the sake of coherence, this overview is restricted to angiosperms, thereby not neglecting the rich alpine cryptogam flora (bryophytes, ferns, lichens) that would deserve a separate assessment. This selection of concepts and their meaning leans on earlier works as summarized in Körner [6,7].

1. 'Small Is Beautiful'—The Life Form Concept

All plant species belong to certain growth forms (the genotype) which become shaped to certain life forms by the environment (the phenotype). While the genotype–phenotype contrast can be large in the case of trees, which could be forced to small shrubs by the environment above treeline (krummholz), the growth form–life form difference becomes increasingly smaller with elevation above treeline. In a first approximation, the life form exhibited by most alpine plants is quite close to their growth form, meaning that transplantation to a warmer, lower elevation location does not fundamentally change appearance, but depending on species, plants may become slightly taller, without losing the architectural principles that control their stature (morphology). Because the life form is what we actually see, I use this term instead of growth form in the following text, with growth form implicitly included (definitions in the tradition of A. von Humboldt, C. Troll and W. Rauh, in [7].

As a unifying principle, plants above the climatic treeline are small 'by design' (genotypically) and in a large part belong to the four life forms: dwarf shrub, mostly tussock-forming graminoids, mostly rosette-forming herbs and cushion plants [7–9] (Figure 1). The restriction of alpine plant life to these low-stature architectures is the outcome of evolutionary selection of taxa that must meet several requirements: (a) decoupling aerodynamically from ambient air conditions as much as possible, and thus engineering a warmer micro-environment close to the ground, (b) restricting apical meristems (buds) to the buffered atmospheric conditions near the soil surface or to below the soil surface (escaping severe freezing), (c) ensuring snow cover in winter (at extratropical latitudes) in species in which winter temperatures above snow would be fatal, (d) guaranteeing mechanical robustness to disturbances by strong below-ground (clonal) structures that are ensuring long life and represent a buffer against the failure of sexual reproduction, (e) enabling rapid seasonal development (leaf dynamics, flowering, seed maturation). At all latitudes, these requirements become increasingly critical with increasing elevation, hence plant species become smaller the higher they are located, and the clonal life strategy becomes increasingly important as sexual reproduction becomes riskier. The life form concept and its implications unify the alpine flora globally. Neglect of the 'small by design' principle leads to substantial misconceptions such as the idea that alpine plants are small because they are unable to grow taller phenotypes under the given life conditions [10]. There are numerous other environments that opt for small stature life forms for different reasons, such as pastures, ruderal habitats, steppe, deserts, coastal habitats, salt marshes, fens and mires.

Figure 1. Various alpine plant life forms assembled around a rock in the Swiss central Alps at 2440 m elevation. The graminoid life form is represented by *Nardus stricta* and *Juncus trifidus*, cushions by *Silene acaulis* ssp. *exscapa*, dwarf shrubs by the creeping *Salix herbacea* and herbs by *Geum montanum*, *Leontodon helveticus* and *Ligusticum mutellina*.

2. Habitat Mosaics That Matter—The Role of Topography and Relief

In addition to life form, the diversity of microhabitats resulting from exposure to sun and wind, slope inclination, landscape fragmentation by relief (geologically, by periglacial processes and by erosion) creates a multitude of life conditions at very close proximity (often less than a meter). At extratropical latitudes, these land surface properties produce rather diverse snow distribution and thus snowmelt patterns (e.g., [11,12] (Figure 2). At all latitudes, moisture, nutrients and soil organic matter differ substantially across these habitat mosaics, with each of these micro-habitats selecting for assemblages of certain plant species [13–15].

Figure 2. Topography shapes alpine habitat diversity, exemplified here by snow distribution patterns in spring. After snowmelt, this slope exhibited a range of seasonal mean plant temperature of 5 to 14 °C [16]. Swiss central Alps, 2500 m.

A central implication of the concept of habitat diversification at alpine elevations is the unsuitability of elevation-related explanations of current and projections of future plant distributions. Topographic diversity by far outranges elevation-specific changes in environmental conditions [9,15]. For instance, air temperature declines with altitude by c. 0.55 K per 100 m. In contrast, seasonal mean temperatures of alpine plant meristems on a single slope have been shown to vary between 5 and 14 °C at the same elevation (Figure 2) and thus the same mean air temperature in temperate, arctic and high-arctic landscapes [16].

3. A Decoupled World—The Overarching Role of Microclimate

As the result of small plant size and often compact morphologies (life forms) on one hand and topographic shelter on the other hand, the actual microclimate alpine plants experience has very little in common with what a weather station or a climate data base would deliver for a given elevation (Figure 3). The substantial physical, that is, aerodynamic decoupling of alpine plants from atmospheric conditions is the central factor in alpine plant life [7,17–19]. It comes as a surprise to many when they realize that alpine plants at supposedly cold locations show tropical temperatures in their leaf canopy under bright weather conditions. The interaction of solar radiation with a high aerodynamic boundary layer resistance to heat transfer does indeed cause heat stress problems in some alpine plants. Not surprisingly, the heat resistance of compact alpine plants is quite high (commonly >50 °C, up to 60 °C; [7,20–22].

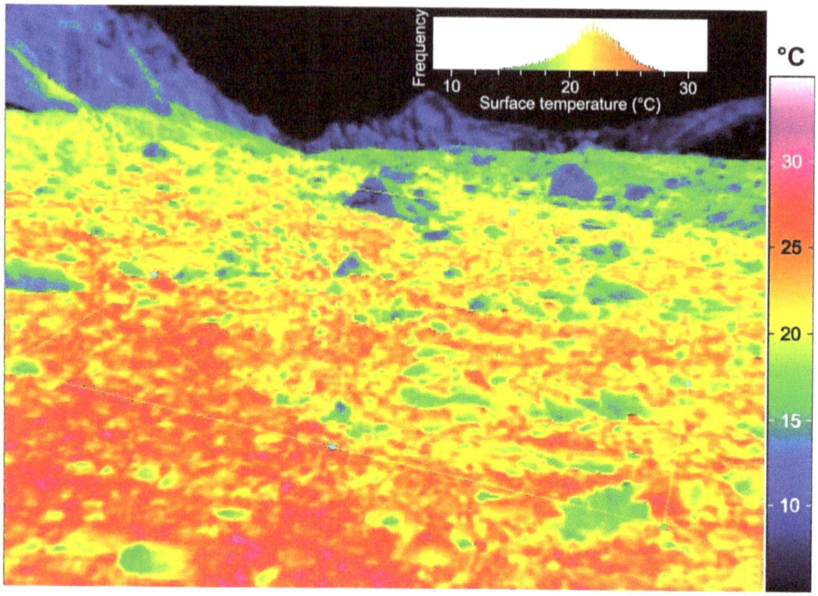

Figure 3. The microclimate in alpine plant communities varies strongly over very short distances, as exemplified by an infra-red thermal image taken in midsummer at 2460 m elevation in the Swiss Alps (Furkapass). Note that the frequency distribution shows a temperature range from 12 to 28 °C, a range that diminishes if averaged over an entire sunny day. However, thermal microhabitat diversity, corresponding to several hundred meters of altitude-related air temperature differences, is retained, and thus disqualifies air temperature data from weather stations for making a case for life conditions of alpine plants (see [16]; photograph by C. Mullis).

The life-form- and topography-driven microclimate is the gateway to understand alpine plant life. It explains why the photosynthetic temperature optimum of alpine plants (including plants from most extreme summit habitats such as *Ranunculus glacialis* above a 3000 m elevation in the Alps) does not differ from that of their low-elevation relatives [23]. It also explains why >80% of all roots and meristems are located in the top 10 cm of the soil profile, where the heat accumulates. Large inflorescences heat up by 10 K or more above air temperature when the sun is shining [24]. The idea that alpine plants are living in a critically low-temperature world is simply biased by weather station data and hiker experiences [25] and is not based on ground-truth data [7,16]. This divergence of plant and air temperature complicates interpretations of alpine plant performance without in situ temperature records.

The good news is that it was never as simple as it is today to obtain accurate data about the actual life conditions of plants in most remote places [26,27]. Miniature data loggers provide year-round information about the thermal regime experienced. Such signals also indicate presence or absence of snow. It is hard to understand why so much speculation about alpine thermal life conditions becomes published (assuming air temperature to reflect plant temperature) while it is so simple to assess the truth. A functional alpine plant ecology without ground truth data is ill-founded, and the use of climate data obtained from weather services is a no-go in light of what we know (for a summary, see [7,25]. Therefore, conceptualizing alpine plant life in a microclimate framework is imperative. In the absence of on-site data, an empirical variance term of likely plant and soil temperature for a given mean air temperature can be applied to scale from air temperature to a proxy of plant temperature [16]. Spaceborne remote-sensing can also provide proxy data, although such data commonly have limited temporal resolution and are affected by clouds.

A note on facilitation: The microclimate effects explained above rest on aerodynamic boundary layer phenomena related to structural density (leaf area density, LAD; [7] near the ground, irrespective of species diversity. Also, single clonal species such as grass tussocks ([28], mat-forming dwarf shrubs [29], or cushion plants [7] show dramatic microclimate benefits (a sort of 'self-facilitation'). In fact, all late successional plant communities, including tropical forests, exhibit mutual shelter benefits of closed canopies, because all, except pioneer species, evolved embedded in communities. Thus, it is a truism that most alpine plants need neighbors or sheltered niches in order to thrive—classical wisdom that can also be framed in a 'facilitation' concept [30]. Hosting of non-cushion plants in cushion plants is a facilitation that goes beyond traditional micro-climatology [31] in highly eroded and seasonally dry alpine habitats, and this can rise local species diversity.

4. From Opportunism to Internal Clock—The Key Role of Developmental Controls

Growth and development are two different categories of life processes, with the latter controlling the former and exerting overarching functions. Development is the genetically controlled transition between plant states such as dormancy, bud break, greening, flowering seed release and senescence. Growth can occur within the 'windows' opened by developmental cues transmitted by hormones. The visible part of development is called phenology. These developmental controls are deeply rooted in evolutionary adaptation. Evolution has selected for seasonal dynamics of development that ensure long-term survival under rather unpredictable weather conditions. Whether and how much the day-to-day or year-to-year variation in actual climatic conditions exerts additional influences on plant performance depends on species and life strategies. Species belonging to later successional stages and species of high longevity exhibit stronger genetic control over seasonal activity (phenology) than short-lived pioneer species. Certain environments (e.g., wind edges with insecure snow cover) opt for rather conservative developmental controls, while others (e.g., snow beds) opt for a more opportunistic phenology. The important point is that phenology reflects the long-term mean of season length so that the reproductive cycle can be completed at least in some years, with clonal persistence securing life in other years.

The concept of evolutionarily selected developmental control comes in conflict with ideas that seasonal plant performance is largely driven by concurrent environmental conditions. Therefore, it was predicted that warmer climates will reduce snow duration and thus allow plants to grow for a longer period, with a higher biomass production at community scale. When explored experimentally, this is not what was found in late successional alpine grassland. The most important species of the alpine grassland belt in the Alps, *Carex curvula*, has an internal clock that causes senescence (leaf browning) whenever 6–7 favorable weeks are over, irrespective of the date of release from snow [32]. Therefore, this dominant species makes no use of any extra time. Another form of internal clock is used by the snow-bed species *Soldanella pusilla*. This species grows its complete 5–6 cm flowering stalk and flower from a 1–2 mm bud at snowing-in to the time when it emerges through thin spring snow, ready to welcome bumblebees [11]. Even 3 m under snow, *Soldanella* starts growing in early January (Figure 4). Given the unpredictable day of release from snow, the inflorescence must be ready to perform by late May, even though snow may disappear in early August only, after a snow-rich winter; this is an opportunistic life strategy.

The important point with developmental controls is that short season climates (often at high elevation) opt for rapid development, which requires high rates of metabolism over a rather short period of time, a clear advantage for herbaceous plants, and an obvious disadvantage for woody plants that require far more time to complete seasonal maturation. This is an analogy to the response of hot desert ephemerals to occasional rain events, with the difference that alpine plants can rely on a certain regularity of seasonality and thus have been selected for long life (perennial below-ground structures).

Figure 4. The alpine snow bed plant *Soldanella pusilla* emerges through melting snow in July with fully grown flowering stalks. An internal clock initiates growth in early January under meters of snow pack. Note the spaces around dark plant structures are created by thermal re-radiation (see [11]).

5. Persistence Is More Important Than Vigor—A Life Insurance Principle

As mentioned several times above, the clonal life strategy is one of the common denominators among the alpine floras. Just as the life form of a tree, clones of small-stature plants are selected for long life, but in the case of clones, theoretically, for eternal life, year-to-year success of sexual reproduction is not very important. In fact, clonal plants may 'hold position' for centuries without sexual offspring. Some genets of late successional alpine clonal taxa have been found to have a life history of several millennia [33,34].

Commonly, such most successful clonal taxa (the majority) grow comparatively slowly on a year-to-year basis, reminding us that evolution does not select for productivity but for fitness, that is, for retaining genes in space over time. In fact, lush (or even uncontrolled) growth is the anti-thesis for persistent life in harsh environments. In other words, the alpine flora makes no exception from the vigor–stress tolerance trade-off that has been described for trees [35]. While seasonal development and short-term above-ground growth may be rater rapid, the controls over its timing (see 4), irrespective of concurrent weather conditions, are part of the survival concept. The exceptionally high investment into below-ground structures may constrain above-ground productivity, but it ensures buffering the effect of adverse conditions [36], including 'missing a summer', when snow cover does not disappear in an unfavorable year [11].

Importantly, slow overall growth is not to be confused with low metabolic capacity. It rather reflects a compromise related to longevity, robustness against rapidly changing life conditions, extreme events, dependency in sturdy below-ground (heterotrophic) structures and a need to rapidly complete the seasonal life cycle. With this 'design', the majority of alpine plant species is unlikely to track rapid environmental change, with microevolutionary selection of more flexible genotypes likely to take quite some time. I wish to recall that the *Carex curvula* clones examined near our alpine research station at Furka Pass in the Swiss central Alps occupied the same piece of land during the medieval warm period when the Wikings discovered Labrador and people grew grapes in Scotland [33], and during the little ice age, when expanding glaciers endangered mountain villages and trees at treeline hardly grew [37]. The persistence syndrome in late successional alpine plant communities also explains the low sensitivity to ongoing climatic warming in high alpine grassland [38].

6. The Many Solutions to The Same Problem—Plant Functional Diversity

There is no 'archetype' alpine plant. The morphological, anatomical and physiological diversity is overwhelming even under extremely high elevation conditions [7,39]. In herbaceous plants at >3000 m elevation in the Alps, dry matter allocation (how much biomass is invested in roots, storage organs, shoots and foliage) covers the entire possible spectrum [40,41] (Figure 5), with similar patterns across the globe [7,42]. On the very same square meter, one may find herbs that have 10% or 90% of their total biomass below ground. Similarly, photosynthetic capacity per unit leaf area or the discrimination of the heavy ^{13}C isotope (a measure of carboxylation efficiency) cover ranges as wide as can be found in any other environment [23,43,44]. Also, leaf anatomy varies greatly across the globe's alpine world [45].

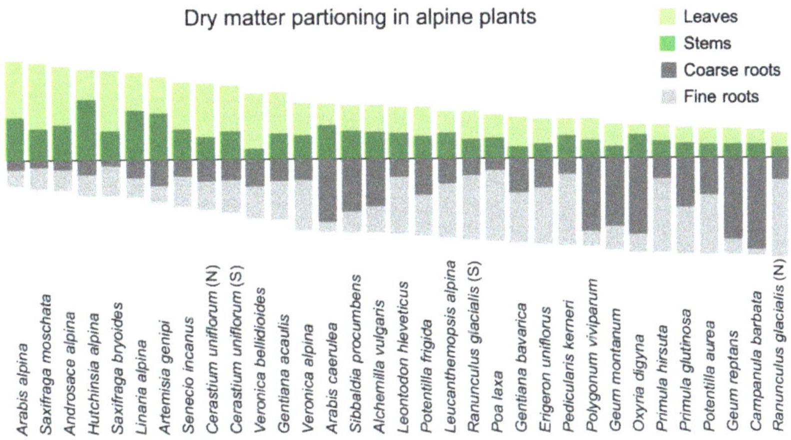

Figure 5. Dry matter allocation in the nival flora of the Alps (data from the central Alps in Tyrol, [40]. The range of allocation patterns covers almost the full known range for angiosperms at sites c. 1000 m above treeline (nival habitats).

In other words, the conditions at high alpine or nival elevation select for traits that are not finding expression in certain common carbon investment patterns, except for reducing plant height, and thus investment in upright stems. Within the life form spectrum present, neither anatomical nor tissue quality traits belong to a common high-elevation syndrome. A delicate small herb may grow next to sturdy sedge tussocks, a dwarf shrub or a semi-woody cushion plant. These patterns clearly falsify (except for small stature) the concept of a common pattern in alpine plant architecture. Different micro-topographies select for higher or lower abundance of different morpho- and physiotypes (e.g., [7,13,46–50], the explanation of which deserves more research. Microtopography-related 'experiments by nature' also hold promise for explaining species range limits [51] (see 11). As an example, the graminoid life form is less successful in snowbeds than the life form herb [52], and contrasting snow duration regimes select for certain pheno-rhythmotypes. Persistent clonal structures are abundant across all alpine habitats. It is a truism that alpine plants must be able to cope with the thermal extremes in their habitats (see 8). Therefore, adequate freezing and heat tolerance are a 'filter' through which all alpine plants species have to pass, with the consequence that what the successful plant species experience is not stress but part of normal life [9,53]. Yet, their habitat conditions would be rather stressful for non-adapted taxa.

Trait selection may well operate at the earliest life stage (seedling establishment), and not at the life stage at which the commonly measured functional traits of plants are obtained from (see 10). Since it is obvious that alpine plants most commonly require a decoupling from atmospheric conditions (see 3), the type of neighborhood matters as well [54], shifting habitat selection from the individual to the community level. Given the exceptional diversity of the soil microbiome (with thousands of bacterial and fungal organismic taxonomic units, OTUs, obtained from alpine soil samples; M. Grube in [52]), it seems that the microbial diversity covers the requirements of any type of habitat. Even the plants at the coldest known place where plants can live exhibit a rich fungal microbiome [55].

7. The Species Diversity–Productivity Relationship Is Confounded

Over the years, the idea became popular that plant species diversity correlates with biomass production. This rationale roots in the valid assumption that species have different resource requirements, and thus a functionally diverse community may extract more from a given resource (e.g., nutrients, light) through the complimentary utilization of a growth-limiting resource (e.g., [56]). Many experiments with designed grassland diversity on common substrates ('common gardens') at low elevation supported the concept of the diversity–productivity connection (e.g., [57]). However, when explored in plant communities where species presence was the outcome of natural selection, no such patterns were found [58].

Under alpine conditions, habitat diversity (and thus very local life conditions) challenges exploring such diversity–productivity relationships. Extremely species-poor communities at harsh locations (wind edges, eroded soils, center of snow beds) exhibit very poor productivity, and highest productivity is found in lush, species-poor or even monospecific stands (e.g., *Carex* fens, *Rumex alpinus* stands on ungulate resting places), with the remaining, closed-cover communities filling the middle ground. When tested within comparable types of grassland, no species diversity–productivity relationship can be found, but a weakly positive trend (r^2 0.12–0.19, $p < 0.01$, n = 67) occurs across all non-woody vegetation types in a larger alpine area, provided the above-mentioned extreme cases are excluded [59] (Figure 6). It seems that moderately favorable soil conditions with a moderate productivity facilitate the co-existence of a higher number of species, with a functional interaction between the two rather uncertain. Species diversity–productivity relationships also do not exhibit uniform patterns, and species identity was found to exert a strong influence on productivity in the Tibetian plateau [60]. Thereofre, species diversity and biomass production reflect habitat diversity (soil fertility) while their potential interaction does not match results from common garden experiments where habitat diversity is eliminated and the presence of species is enforced by experimental design.

Figure 6. Extremely productive (*Rumex alpinus*) and extremely unproductive (*Sempervivum montanum* with *Leucanthemopsis* alpina and *Leontodon* sp.) communities are species poor, and moderately productive communities are species rich (an alpine grassland with *Carex curvula*, *Nardus stricta*, *Arnica montana* and *Trifolium alpinum*). Swiss central Alps 2440 m (Furkapass).

The functional benefit of high species diversity rather seems to come into play when stochastic external effects or novel threats occur, similar to the role of 'portfolio diversity' for buffering shareholder value against financial crises. The persistent plant presence and thus the stabilizing of diverse microhabitats [50] often depend on key-stone species, the functional importance of which becomes obvious under disturbance only. For instance, high species diversity ensures persistent land cover under erosion pressure, as was shown for the Alps and the High Caucasus. A single, common tussock grass species out of many other grassland species became the engineer of erosion edges in both regions. Quite unexpectedly, this extreme situation made this species a keystone species [61,62]. Below-ground diversity was also found key in maintaining grassland integrity after disturbance [63]. Diverse seasonal phenologies are also likely to buffer effects of climatic change (such as changing snow duration) on the integrity and productivity of alpine plant communities [64].

Therefore, alpine species diversity secures continued ecosystem integrity (e.g., preventing erosion) and is rather confounded with productivity than driving it. Habitats with very high species richness fall in the moderately productive category, that is, life conditions that permit co-existence, with no single species overgrowing others. There is a risk that nutrient input (e.g., by ongoing nitrogen deposition) will act as a game changer, favoring a few responsive species at the loss of others (see 12). These few examples illustrate that 'response traits' are likely to be ecologically more important than 'static traits' (such as leaf size, specific leaf area, tissue nutrient concentration) that might be obtained, for instance, from an herbarium.

8. The Concept of Limitation and Stress—Utmost Confusion

It is rather misleading to apply limitation concepts as they were developed in yield-oriented agronomy (Liebig's law of the minimum) to ecosystem ecology [6,9,53]. Representing the most frequently employed term in ecology, the yield-oriented concept of limitation has no place in ecology. Natural vegetation such as alpine or arctic reflects nature's answer to the local life conditions. As species assemblages mirror life conditions, any change in these conditions causes the assembly to change. A so-called nutrient-limited alpine grassland turns into a 'fat' meadow if fertilizer is added, with all the species believed to have suffered from nutrient limitation becoming locally extinct once relieved from that 'limitation'.

The concept of plant stress, a severe form of limitation, is a similarly misleading concept [9,53]. A certain degree of stress keeps vigorous neighbors away and ensures persistence for those selected for their stress tolerance. There are lots of anthropocentric interpretations of life conditions that seem hostile for humans. Such conditions are essential for many species that do not tolerate competition but can cope with what by an observer's rating might be hostile. This does not mean that obviously stress-dominated habitats are providing luxury for those inhabiting it. What might appear like rather supportive physiological life conditions for a species based on its abundance patterns in reality might represent marginal life conditions from a growth physiological perspective, with the abundant species simply coping with these conditions better than others. Often addressed as a discrepancy between the ecological and the physiological 'optimum', the first is rated by fitness, the second is rated by biomass accumulation. The latter is unsuitable for making a case for survival and abundance limitations in alpine environments (but possibly also in most other environments). What matters is retaining a species' gene pool in space over time rather than biomass yield.

Hence, a certain degree of limitation or stress by resources or climate, respectively, is vital for plant existence in the wild. In their natural environment, most plant species operate far away from what might be rated as 'optimal' from a growth-physiological point of view. Therefore, the agronomic concept of optimality in terms of productivity needs to be abandoned from plant ecology in general, and alpine plant ecology in particular. The challenge ahead is to identify the degrees of environmental limitations of growth and physiological stress that are essential for keeping a plant's aggressive neighbors away while at the same time affecting survival and the range limits of species (see 11).

9. Alpine Productivity—A Matter of Timing and Ground Cover

Given the low mean ambient temperature and the increasingly shorter alpine growing season at extratropical latitudes, the productivity (biomass yield) per unit land area can be expected to decline with elevation. However, all such comparisons rest on the year as a reference period (the concept of annual net plant biomass production, NPP). What if the actual period of plant growth is accounted for? And what if one accounts for uncovered ground fractions because of rocks or disturbances and refers to productivity per unit of plant covered ground area only?

Such a recalculation of alpine productivity yields a surprise. Per unit of time (per day of growing season) and calculated for full cover areas only, NPP of alpine vegetation does not differ from that of humid low-elevation vegetation (2.2 g d.m. m^{-2} d^{-1} or a mean of 185 g m^{-2} dry matter production over a mean length of the growing season of 84 days across temperate alpine sites; [7,9]. Since the delineation of the actual growing season is very difficult [65], and most of that above-ground biomass production does actually occur during the first 4–6 weeks of the season, the productivity would be twice as high on a 'per day of actual growth' basis and thus approaching that of a humid tropical forest. Explanations for this high productivity per unit of time with measurable growth and per unit of fully covered ground include the fact that actual life conditions are not necessarily as bad during the period of active growth as one might expect from elevation (see 3). Physiological adjustments can buffer the remaining thermal limitations, and an efficient below-ground microbiome may compensate for effects of cool soil. Further, in extratropical, high-latitude mountains, nutrients recycled over an entire 12-month period (including continuous heterotrophic respiration under snow [66]) become available during a few weeks of vigorous growth after snowmelt, and stored growth in the form of carbohydrate reserves may also contribute to seasonal above-ground biomass accumulation [36]. The below-ground productivity is likely to be as high as or higher than that above the ground, but it is hard to measure because of the multi-year longevity of all below-ground organs. In the temperate zone, a reasonable estimate of the mean root duration in alpine grassland might be 3–5 years, including roots that recycle within a few weeks of their production and roots that live more than 50 years (e.g., some tap roots). Such estimates of productivity that account for the duration of the growing season and full ground cover falsify the concept of low alpine productivity, provided moisture shortage does not come into play, as is the case in semi-arid subtropical or continental temperate mountains.

Tropical alpine productivity, using a 'per year concept' of NPP, is a special and largely unknown case because, similar to root growth in extratropical regions, the continuous growth makes it difficult to identify which biomass was produced in a given year. Seasonally dry tropical alpine settings were thought to represent an exception with a clear 'rainy season', but it turned out that wide plant spacing with large root spheres and soil moisture reserves buffer the climatic seasonality, so that growth continues year-round (data from the Bolivian Altiplano by [67] (Figure 7). The only way such above-ground data can be obtained when there is no clear-cut start and end of the season is by assessing shoot (tiller and leaf) turnover by labelling and revisiting. While still neglecting cover effects, these tall altiplano tussocks were estimated to produce c. 1200 g m^{-2} in 12 months [7], which is roughly six times the productivity of grassland in the Alps in a 2-month period of active growth. Repeated harvests are not helpful when growth is continuous. Cutting off biomass may actually stimulate re-growth (as does herbivory) and thus lead to an overestimation of sustainable productivity (compensatory growth; [68–71]. Therefore, it remains largely unknown whether closed humid tropical alpine vegetation can compensate the cool climate (often very little sunshine, but fog) by year-round growth. Such data are urgently needed to validate the idea that the degree of ground cover and the period of active growth rather than the climatic conditions during this period are the unifying factors that determine alpine productivity.

Figure 7. Alpine productivity data must account for ground cover and actual periods of growth. Here is an example of wide spacing of tussock grasses in the Bolivian Altiplano at 4200 m elevation.

10. The Cradle Is the Bottleneck—Alpine Plant Reproduction

What is it that constrains alpine plant reproduction, that is, the production of reproductively successful offspring (evolutionary fitness)? As explained in Section 5, the overwhelming abundance of clonal life strategies across alpine floras can be interpreted as a safeguard against periodic failure of sexual reproduction. There is also a trade-off between investments in clonal growth and sexual reproduction (e.g., [72]). The high abundance of clonal alpine plants suggests higher risks of sexual reproduction than elsewhere. Yet, the likely risks are manifold, including the loss of flowers by freezing events during the warm season, poor pollination, excessively slow embryogenesis, insufficient seed maturation, seed dispersal, germination and seedling establishment. There is rich literature on each of these steps (summarized in Chapter 16 in [7]). The starting hypothesis is always a concept of serious limitations of each of the various steps till seed dispersal given the assumed hostile life conditions. Surprisingly, all these steps were found to exert no or little risk that would place alpine flowering plants in an inferior position compared to plants from lower elevation, where successful reproduction is not self-evident, either. Even nival plants were found to exhibit no particular pollen limitation [73]. The delicate process of embryogenesis was found to perfectly match even the most extreme nival life conditions in the Alps (e.g., [74] and references to earlier works therein). Whoever explored this was surprised that seeds of most alpine species germinate well, if properly stratified. What is left is the last and much less studied step: the establishment of a new plant from viable seed.

Although not specific to the alpine world, but a crucial step in any environment, the first and second of often very short seasons with a harsh winter, including the risk of low snow cover in early winter and dehydration in summer, provide alpine-specific challenges of seedling survival. It is surprising how little attention this step of the reproductive cycle received compared to pollination and seed viability (but see [75–79]. It may come as a surprise that heat stress is one of the major obstacles that small alpine seedlings have to cope with in bare seed beds (e.g., [22]). Perhaps seedling establishment received less attention because it is a rather tedious fieldwork to track the fate of seedlings under alpine field conditions. However, if we are to understand plant species radiation and species range limits, seed bed ecology needs to be brought to the forefront. Seeds can be found almost anywhere, including very high, exposed summits [27,80,81], but seedling establishment is likely to select for certain response traits and thus species. While summit floras are commonly assembled from species that can grow in isolation or are confined to micro-shelters and tolerate raw substrates, the novel establishment of grassland depends on the success of late-successional species and soil development [82].

11. To Be or Not to Be—The Edge of the Fundamental Niche

In my rating, the number one question in plant ecology is why species occur where they do and why they are absent from certain habitats. Once we can explain species distribution, we can attempt mechanism-based projections where species might be occurring when life conditions change. This question boils down to defining the fundamental niche of a species and further to defining and explaining the cold edge of that niche. Given the central role of that question, its neglect in past alpine plant research is surprising (with one first step discussed below). My explanation is the immense task in terms of micro-habitat mapping, microclimatology, reproduction biology, stress physiology, life history traits, etc. The edge of the fundamental niche has, for long, not even been assessed for common tree species at low elevation, with a recent attempt ending up in a multidisciplinary mega-project [83], employing a design that also holds promises for alpine plant species range limits. Importantly, the edge of the fundamental niche cannot be assessed by geostatistics or mapping of occurrences, but requires a mechanistic approach, including identifying and explaining 'extreme' life conditions.

For explaining alpine plant species distribution, it is essential to know the species' niche preferences, ideally though difficult, those of the fundamental niche [84–86]. Identifying potential range limits of species and the traits that determine that potential edge of their life will open a novel arena of functional ecology. Note that similar to the treeline concept [2,7], this concept is built upon potential performance, with the realized niche (the actual distribution) reflecting distribution history, stochastic events, biotic interactions, disturbances, lack of soil, plant–animal interactions, etc. Since these influences vary from place to place, it seems near impossible to formulate hypotheses and predictions. However, in contrast to the edge of the fundamental niche (representing 'extreme' life conditions for a species), the edge of the realized niche can be seen in the field and mapped (Figure 8). We do not know in which alpine species the realized edge matches the fundamental edge, nor do we know for any alpine species how far from its potential range limit it is currently operating. The rapid arrival of new species in previously almost empty summit floras (see 12) points at a substantial leeway for spreading under climate warming conditions, and it also shows which species are more likely to track the spatial shift of their fundamental niche edge.

To the best of my knowledge, the only case in which the range limits of alpine species was nailed down to a mechanistic explanation is that by Von Büren et al. [51]. I think this study adopted a most promising conceptual framework that includes detailed spatial mapping of occurrences, very detailed micro-climatology and a professional stress-physiological assessment, all three possibly depending on a field station very close to the study area. This work revealed that the spatial segregation of the two dominant alpine graminoids in the central Alps, *Carex curvula* and *Nardus stricta*, is controlled by snow distribution in winter, with periodic low or absent snow cover (thus, micro-topography) and the maximum freezing tolerance of the tissues surrounding the apical meristems co-explaining their distribution.

Figure 8. At the same elevation (here, in the Austrian Hohe Tauern National Park at c. 2250 m elevation), topography and associated microclimate and soil conditions shape plant community distribution, challenging the niche concept of species ranges. What we see and what can be mapped is the edge of communities and associated realized niche edges of species (1–11). What is needed for predicting species occurrences for specific abiotic environmental conditions is the edge of their fundamental niche, because it is species rather than communities that shift (though modulated by biotic interactions, to varying degrees).

Given the fragmentation of the edge of the fundamental niche to micro-habitats [15], any projections of future species distributions need to rest on a rather fine-grained representation of the alpine landscape. Randin et al., [85] showed that even a 25 × 25 m grid is too coarse to reliably represent the (statistical) spatial distribution of alpine species across two separate test regions. The predicted 'space-for-time' loss of species ranged from 20% to 96% for a 3 K warming scenario depending on whether one adopts a 1 m or a 100 m grid [87]. Thus, fine-grain representation of the alpine landscape (thereby accounting for potential micro-topography effects) is essential in modelling alpine plant distribution [88]. The overarching role of micro-habitats as micro-refugia during adverse periods had been illustrated for rock specialists by [89].

12. Environmental Change Is 'Normal' in Alpine Life

The previous points illustrated that variability in space and time is a unifying factor of alpine environments. This is the matrix in which long-term changes in atmospheric conditions and other anthropogenic influences need to be rated. The term 'global' means that the associated changes are considered to apply over large regions, hence they are not region specific. Among these changes, those in atmospheric chemistry (concentrations of CO_2, reactive N-compounds, ozone) and climate (temperature, humidity, precipitation, snow cover) act globally. Land-use changes, either by intensification or abandonment (e.g., pastoralism, mining, hydrology, tourism), do occur globally, but the intensity of these changes varies greatly among different regions. For earlier assessments of such impacts on the alpine flora, see ([7], Chapter 17; [90,91]). Here, I depict a few aspects of atmospheric changes which are nested in traditional concepts: the concept of carbon limitation and of nitrogen 'limitation' (see 8), with a few words on the role of climate warming (for recent statistics of the plant literature on alpine global change, see [92].

As atmospheric CO_2 rises, plants living in 'thin' air (low partial pressure) might be expected to draw particular benefits from that rise, because leaf photosynthesis in alpine plants increases when supplied with extra CO_2 [23,93], and thus plants might be expected to grow faster (the concept of carbon limitation). By all what we know today, the idea of carbon limitation has been falsified empirically, in situ, for both late successional and pioneer alpine vegetation [94,95]. A complementary test of the C limitation hypothesis is reducing light. When seasonal light consumption was experimentally reduced in half, this also did not affect peak season biomass ([7], Figures 11 and 14; [96]), thus validating the results of CO_2-enrichment experiments. Not unexpectedly, a 90% reduction in light does, however, exert dramatic effects [97]. However, an important caveat to these bulk biomass responses is that in each case, species-specific responses were observed. In the Alps, one grass species that currently contributes little to bulk biomass profited from CO_2 enrichment and suffered from shading (*Helictrotrichon versicolor*), pointing at the possibility of long-term community adjustments. To date, these results suggest that alpine vegetation is carbon saturated, but long-term biodiversity effects cannot be ruled out. Once more, responsiveness (response traits) plays a key role. The conventional static traits do not show anything special for *Helictotrichon versicolor* compared to, for instance, *Poa alpina*. This similarly looking pair of species might offer an explanation for what causes a species' growth to become CO_2-responsive under alpine life conditions.

Given that photosynthetic capacity does not limit growth (see above) and that temperature hardly affects seasonal CO_2 assimilation in alpine plants (see 3, [7]), any mitigation of low temperature limitations in a warmer world must act via direct influences on growth and development (including night time effects) and indirectly via controlling season length in extratropical regions. For low-temperature-related growth limitation to come into play, the meristematic tissues must be critically cold. For apical shoot meristems and leaf primordia, this is mostly not the case (see 3), but for deep roots and roots on permafrost, such constraints are possible. It was shown that roots grow very slowly below 5 °C and not at all at 0 °C. The strongest stimulation of alpine root growth by rising temperature occurs between 5 and 10 °C, with little additional benefits at temperatures above 10 °C [98,99]. Overall, soil heating under solar radiation diminishes such limitations, with the remaining influences of temperature indirect, via season length and snow duration.

A wide-spread assumption is that a longer season accelerates development and productivity. Both these assumptions have not been supported empirically in late successional alpine vegetation so far [32,38] (see 8), but a thermophilization of summit floras is evident (e.g., [100,101]. Developmental responses to climatic warming that translate into changes in phenology are rather mixed so far, with some species flowering earlier, and others responding little (see discussion in [7,64]. The concept of a simple tracking of temperature by development has been falsified, and photoperiod controls or internal clocks are important constraints. For climate warming to change development (phenology), micro-evolutionary selection from the existing gene pool would be required, which is another promising field of future research.

Finally, I wish to draw attention to an underestimated global change driver: soluble N-deposition. While all alpine ecosystem (except for cattle resting places) would become more productive (while losing species at the same time; see 8), the ongoing rates of reactive N-deposition are a multiple of pre-industrial background deposition (in essence, lightning effects) and reach 7 kg N $ha^{-1}a^{-1}$ in the Swiss Alps [102], with very high rates also reported for the Rocky Mountains and parts of China (for references, see [7], Chapter 17). The effects on plant species are clearly different, with vigorous species, and some sedges in particular, taking more advantage than slow-growing species (e.g., small herbs), leading to a trivialization of the flora ([7] Chapter 17; [103]). Most field experiments with reactive nitrogen addition apply excessively high rates, which might reveal fast responses and a ranking of species responsiveness but are unsuitable to infer long-term effects of nitrogen deposition. This aspect of global change also underpins the misleading nature of the concept of nutrient limitation (see 8).

Concluding Comments

Extending the famous statement by Dobzhansky [104], 'Nothing in biology makes sense except in the light of evolution', one could add for alpine plants: 'Nothing makes sense in alpine plant biology unless one accounts for micro-climate'. The 12 conceptual frameworks addressed in this paper are certainly not covering all aspects of alpine plant biology, but each of them deserves re-thinking some of the mind models that have been driving much of the research we have seen published over the years.

In my view, research on species range limits (the edge of their fundamental niche) and on genetic variants of existing response traits within species (illuminating where microevolution can select from, under climatic change) deserve high priority. From all what we know about life cycles in alpine plants, seedling establishment appears to be the most critical step in plant species distribution. In light of the many structural and metabolic solutions to cope with the same environmental conditions, static plant traits hold little promise as predictive tools. What is needed are response traits, that is, the way plant species react to local changes in their environment, either naturally or by manipulation experiments.

Among these response traits, the consequences of varying snow cover and the consequences of long-term nitrogen deposition, the differential species responses in particular, deserve far more attention, with existing response indicators for the flora of the Alps awaiting more applications. Expert-knowledge-based 'indicator values' proved to be valuable proxies for plant responses to environmental conditions [105–112]. At the community level, such response proxies are often as strong or stronger as on-site physico-chemical assessments of growth conditions.

Given the importance of existing genotypic variation of plant responses to environmental change as well as phenotypic plasticity within genets, experiments that permit assessing both are required. Common gardens for experimentally exploring these fields are one established method, but they bear a risk of bias, because any given common garden provides (a) asymmetric climatic life conditions for species originating from contrasting 'home' climates and (b) the common soil represents an asymmetric treatment as well, given the known soil preferences of plant species. Hence, what is considered 'common' is in fact an unbalanced treatment. A species from a warm and a species from a cold origin brought to an intermediate location will experience contrasting directions of shift. There is no perfect solution, but replicated common gardens at contrasting conditions can improve the situation, with the central obstacle being the substrate influence. Neither a common (inevitably artificial) substrate nor substrate translocation solve the problem. Substrate fertility may affect all climate response traits. For genotype tests, a most robust approach might be to sample replicated clonal fragments of different genets of the same species and replant them across the species' home range [113–115]. Reciprocal transplantations of entire sods or monoliths can overcome the soil bias, but target species remain tied to the given neighborhood in the plant community. Alexander et al. [54] showed that as species move upslope, effects of species–species interactions can exceed the direct effect of climatic warming on species. Acknowledging and critically applying these concepts in the study of alpine plant life will require more tedious, more complex field work, avoiding over-simplified, seemingly 'standardized' experiments in greenhouses, and will bring us closer toward answering 'big' questions such as why species grow where they do.

Funding: This research received no external funding.

Institutional Review Board Statement: Not applicable.

Informed Consent Statement: Not applicable.

Data Availability Statement: Not applicable.

Acknowledgments: I thank S. Riedl for adjusting the Figures. The ALPFOR research station at Furka Pass, Switzerland facilitated the research that contributed to the development of these concepts. I also thank E. Hiltbrunner for continued support and fruitful discussions.

Conflicts of Interest: The author declares no conflict of interest.

References

1. Körner, C. Tools Shape Paradigms of Plant-Environment Interactions. *Prog. Bot.* **2021**, *82*, 1–42.
2. Körner, C.; Hoch, G. Not every high-latitude or high-elevation forest edge is a treeline. *J. Biogeogr.* **2023**, *in press*. [CrossRef]
3. Paulsen, J.; Körner, C. A climate-based model to predict potential treeline position around the globe. *Alp. Bot.* **2014**, *124*, 1–12. [CrossRef]
4. Körner, C.; Paulsen, J.; Spehn, E.M. A definition of mountains and their bioclimatic belts for global comparison of biodiversity data. *Alp. Bot.* **2011**, *121*, 73–78. [CrossRef]
5. Körner, C.; Jetz, W.; Paulsen, J.; Payne, D.; Rudmann-Maurer, K.; Spehn, E.M. A global inventory of mountains for bio-geographical applications. *Alp. Bot.* **2017**, *127*, 1–15. [CrossRef]
6. Körner, C. Concepts in empirical plant ecology. *Plant. Ecol. Div.* **2018**, *11*, 405–428. [CrossRef]
7. Körner, C. *Alpine Plant Life*, 3rd ed.; Springer: Cham, Switzerland, 2021.
8. Billings, W.D. Arctic and alpine vegetations: Similarities, differences, and susceptibility to disturbance. *BioSci* **1973**, *23*, 697–704. [CrossRef]
9. Körner, C. Alpine plants: Stressed or adapted? In *Physiological Plant Ecology*; Press, M.C., Scholes, J.D., Barker, M.G., Eds.; Blackwell: Oxford, UK, 1999; pp. 297–311.
10. Crivellaro, A.; Büntgen, U. New evidence of thermally constrained plant cell wall lignification. *Trends Plant. Sci.* **2020**, *25*, 322–324. [CrossRef] [PubMed]
11. Körner, C.; Riedl, S.; Keplinger, T.; Richter, A.; Wiesenbauer, J.; Schweingruber, F.; Hiltbrunner, E. Life at 0 °C: The biology of the alpine snowbed plant *Soldanella pusilla*. *Alp. Bot.* **2019**, *129*, 63–80. [CrossRef]
12. Verrall, B.; Green, K.; Pickering, C.M. Temporal dynamics in alpine snowpatch plants along a snowmelt gradient explained by functional traits and strategies. *Oecologia* **2023**, *201*, 155–171. [CrossRef]
13. Billings, W.D.; Bliss, L.C. An alpine snowbank environment and its effects on vegetation, plant development, and productivity. *Ecology* **1959**, *2*, 388–397. [CrossRef]
14. Pauli, H.; Gottfried, M.; Grabherr, G. Vascular plant distribution patterns at the low-temperature limits of plant life—The alpine-nival ecotone of Mount Schrankogel (Tyrol, Austria). *Phytocoenologia* **1999**, *29*, 297–325. [CrossRef]
15. Körner, C.; Hiltbrunner, E. Why is the alpine flora comparatively robust against climatic warming? *Diversity* **2021**, *13*, 383. [CrossRef]
16. Scherrer, D.; Körner, C. Infra-red thermometry of alpine landscapes challenges climatic warming projections. *Glob. Chang. Biol.* **2009**, *16*, 2602–2613. [CrossRef]
17. Billings, W.D. Physiological ecology. *Ann. Rev. Plant. Physiol.* **1957**, *8*, 375–391. [CrossRef]
18. Larcher, W. Bioclimatic temperatures in the high Alps. In *Plants in Alpine Regions*; Lütz, C., Ed.; Springer Wien: New York, NY, USA, 2012; pp. 21–27.
19. Oldfather, M.F.; Ackerly, D.D. Microclimate and demography interact to shape stable population dynamics across the range of an alpine plant. *New Phytol.* **2019**, *222*, 193–205. [CrossRef] [PubMed]
20. Gauslaa, Y. Heat resistance and energy budget in different Scandinavian plants. *Holarct. Ecol.* **1984**, *7*, 1–78. [CrossRef]
21. Larcher, W.; Kainmüller, C.; Wagner, J. Survival types of high mountain plants under extreme temperatures. *Flora* **2010**, *205*, 3–18. [CrossRef]
22. Marcante, S.; Erschbamer, B.; Buchner, O.; Neuner, G. Heat tolerance of early developmental stages of glacier foreland species in the growth chamber and in the field. *Plant. Ecol.* **2014**, *215*, 747–758. [CrossRef]
23. Körner, C.; Diemer, M. In situ photosynthetic responses to light, temperature and carbon dioxide in herbaceous plants from low and high altitude. *Funct. Ecol.* **1987**, *1*, 179–194. [CrossRef]
24. Dietrich, L.; Körner, C. Thermal imaging reveals massive heat accumulation in flowers across a broad spectrum of alpine taxa. *Alp. Bot.* **2014**, *124*, 27–35. [CrossRef]
25. Körner, C.; Hiltbrunner, E. The 90 ways to describe plant temperature. *Perspect. Plant. Ecol. Evol. Syst.* **2018**, *30*, 16–21.
26. Klimes, L.; Dolezal, J. An experimental assessment of the upper elevational limit of flowering plants in the western Himalayas. *Ecography* **2010**, *33*, 590–596. [CrossRef]
27. Körner, C. Coldest places on earth with angiosperm plant life. *Alp. Bot.* **2011**, *121*, 11–22. [CrossRef]
28. Monteiro, J.A.F.; Hiltbrunner, E.; Körner, C. Functional morphology and microclimate of *Festuca orthophylla*, the dominant tall tussock grass in the Andean Altiplanto. *Flora* **2011**, *206*, 387–396. [CrossRef]

29. Cernusca, A. Energie- und Wasserhaushalt eines alpinen Zwergstrauchbestandes während einer Föhnperiode. *Arch. Met. Geoph Biokl Ser. B* **1976**, *24*, 219–241. [CrossRef]
30. Callaway, R.M.; Brooker, R.W.; Choler, P.; Kikvidze, Z.; Lortie, C.J.; Michalet, R.; Paolini, L.; Pugnaire, F.L.; Newingham, B.; Aschehoug, E.T.; et al. Positive interactions among alpine plants increase with stress. *Nature* **2002**, *417*, 844–848. [CrossRef] [PubMed]
31. Cavieres, L.A.; Badano, E.I.; Sierra-Almeida, A.; Gomez-Gonzalez, S.; Molina-Montenegro, M.A. Positive interactions between alpine plant species and the nurse cushion plant *Laretia acaulis* do not increase with elevation in the Andes of central Chile. *New Phytol.* **2006**, *169*, 59–69. [CrossRef]
32. Möhl, P.; Von Büren, R.S.; Hiltbrunner, E. Growth of alpine grassland will start and stop earlier under climate warming. *Nat. Commun.* **2022**, *13*, 7398. [CrossRef]
33. Steinger, T.; Körner, C.; Schmid, B. Long-term persistence in a changing climate: DNA analysis suggests very old ages of clones of alpine *Carex curvula*. *Oecologia* **1996**, *105*, 94–99. [CrossRef]
34. De Witte, L.C.; Armbruster, G.F.J.; Gielly, L.; Taberlet, P.; Stöcklin, J.A.F.L.P. Markers reveal high clonal diversity and extreme longevity in four key arctic-alpine species. *Mol. Ecol.* **2012**, *21*, 1081–1097. [CrossRef] [PubMed]
35. Loehle, C. Height growth rate tradeoffs determine northern and southern range limits for trees. *J. Biogeogr.* **1998**, *25*, 735–742. [CrossRef]
36. Hiltbrunner, E.; Arnaiz, J.; Körner, C. Biomass allocation and seasonal non-structural carbohydrate dynamics do not explain the success of tall forbs in short alpine grassland. *Oecologia* **2021**, *197*, 1063–1077. [CrossRef] [PubMed]
37. Paulsen, J.; Weber, U.M.; Körner, C. Tree growth near treeline: Abrupt or gradual reduction with altitude? *Arct. Antarct. Alp. Res.* **2000**, *32*, 14–20. [CrossRef]
38. Vittoz, P.; Randin, C.; Dutoit, A.; Bonnet, F.; Hegg, O. Low impact of climate change on subalpine grasslands in the Swiss Northern Alps. *Glob. Chang. Biol.* **2009**, *15*, 209–220. [CrossRef]
39. Scott, D.; Billings, W.D. Effects of environmental factors on standing crop and productivity of an alpine tundra. *Ecol. Monogr.* **1964**, *34*, 243–270. [CrossRef]
40. Körner, C.; Renhardt, U. Dry matter partitioning and root length/leaf area ratios in herbaceous perennial plants with diverse altitudinal distribution. *Oecologia* **1987**, *74*, 411–418. [CrossRef]
41. Körner, C. Some often overlooked plant characteristics as determinants of plant growth: A reconsideration. *Funct. Ecol.* **1991**, *5*, 162–173. [CrossRef]
42. Patty, L.; Halloy, S.R.P.; Hiltbrunner, E.; Körner, C. Biomass allocation in herbaceous plants under grazing impact in the high semi-arid Andes. *Flora* **2010**, *205*, 695–703. [CrossRef]
43. Yang, Y.; Siegwolf, R.T.W.; Körner, C. Species specific and environment induced variation of delta-13C and delta-15N in alpine plants. *Front. Plant. Sci.* **2015**, *6*, 423. [CrossRef]
44. Körner, C.; Leuzinger, S.; Riedl, S.; Siegwolf, R.T.; Streule, L. Carbon and nitrogen stable isotope signals for an entire alpine flora, based on herbarium samples. *Alp. Bot.* **2016**, *126*, 153–166. [CrossRef]
45. Körner, C.; Neumayer, M.; Pelaez Menendez-Riedl, S. Smeets-Scheel A Functional morphology of mountain plants. *Flora* **1989**, *182*, 353–383. [CrossRef]
46. Kerner, A. *Das Pflanzenleben der Donauländer*; Wagner: Innsbruck, Austria, 1863.
47. Kerner, A. *Die Abhängigkeit der Pflanzengestalt von Klima und Boden. Festschr 43*; Jahresvers dt Naturf Ärzte 29–45; Wagner: Innsbruck, Austria, 1869.
48. Friedel, H. Schneedeckendauer und Vegetationsverteilungen im Gelände. In: Ökologische Untersuchungen in der subalpinen Stufe zum Zwecke der Hochlagenaufforstung. *Mitt. Bundes-Vers. Mariabrunn* **1961**, *59*, 317–369.
49. Braun-Blanquet, J. Pflanzensoziologie. In *Grundzüge der Vegetationskunde*, 3rd ed.; Springer: Wien, Austria, 1964.
50. Onipchenko, V.G.; Semenova, G.V.; van der Maarel, E. Population strategies in severe environments: Alpine plants in the northwestern Caucasus. *J. Veg. Sci.* **1998**, *9*, 27–40. [CrossRef]
51. Von Büren, R.S.; Hiltbrunner, E. Low winter temperatures and divergent freezing resistance set the cold range limit of widespread alpine graminoids. *J. Biogeogr.* **2022**, *49*, 1562–1575. [CrossRef]
52. Körner, C.; Berninger, U.G.; Daim, A.; Eberl, T.; Fernandez Mendoza, F.; Fureder, L.; Grube, M.; Hainzer, E.; Kaiser, R.; Meyer, E.; et al. Long-term monitoring of high-elevation terrestrial and aquatic ecosystems in the Alps—A five-year synthesis. *Eco. Mont.* **2022**, *14*, 48–69. [CrossRef]
53. Körner, C. Limitation or stress—Always or never? *J. Veg. Sci.* **2003**, *14*, 141–143.
54. Alexander, J.M.; Diez, J.M.; Levine, J.M. Novel competitors shape species' responses to climate change. *Nature* **2015**, *525*, 515–518. [CrossRef]
55. Oehl, F.; Körner, C. Multiple mycorrhization at the coldest place known for Angiosperm plant life. *Alp. Bot.* **2014**, *124*, 193–198. [CrossRef]
56. Spehn, E.M.; Joshi, J.; Schmid, B.; Diemer, M.; Körner, C. Above-ground resource use increases with plant species richness in experimental grassland ecosystems. *Funct. Ecol.* **2000**, *14*, 326–337.
57. Weisser, W.W.; Roscher, C.; Meyer, S.T.; Ebeling, A.; Luo, G.; Allan, E.; Beßler, H.; Barnard, R.L.; Buchmann, N.; Buscot, F.; et al. Biodiversity effects on ecosystem functioning in a 15-year grassland experiment: Patterns, mechanisms, and open questions. *Basic. Appl. Ecol.* **2017**, *23*, 1–73.

58. Kahmen, A.; Perner, J.; Audorff, V.; Weisser, W.; Buchmann, N. Effects of plant diversity, community composition and environmental parameters on productivity in montane European grasslands. *Oecologia* **2005**, *142*, 606–615. [CrossRef]
59. Von Riedmatten, L.; Körner, C.; Hiltbrunner, E.; Hefel, C.; Stöcklin, J. Produktivität und Biodiversität auf alpinen Rasen. Master's Thesis, University of Basel, Basel, Switzerland, 2007, *unpublished*.
60. Wu, J.S.; Shen, Z.X.; Zhang, X.Z. Precipitation and species composition primarily determine the diversity-productivity relationship of alpine grasslands on the Northern Tibetan Plateau. *Alp. Bot.* **2014**, *124*, 13–25. [CrossRef]
61. Caprez, R.; Spehn, E.; Nakhutsrishvili, G.; Körner, C. Drought at erosion edges selects for a 'hidden' keystone species. *Plant. Ecol. Divers.* **2012**, *4*, 303–311. [CrossRef]
62. Huck, C.; Körner, C.; Hiltbrunner, E. Plant species dominance shifts across erosion edge-meadow transects in the Swiss Alps. *Oecologia* **2013**, *171*, 693–703. [CrossRef] [PubMed]
63. Kahmen, A.; Perner, J.; Buchmann, N. Diversity-dependent productivity in semi-natural grasslands following climate perturbations. *Funct. Ecol.* **2005**, *19*, 594–601. [CrossRef]
64. Vorkauf, M.; Kahmen, A.; Körner, C.; Hiltbrunner, E. Flowering phenology in alpine grassland strongly responds to shifts in snowmelt but weakly to summer drought. *Alp. Bot.* **2021**, *131*, 73–88.
65. Körner, C.; Möhl, P.; Hiltbrunner, E. Four ways to define growing season. *Ecol. Lett.* **2023**, *in press*. [CrossRef]
66. Scholz, K.; Hammerle, A.; Hiltbrunner, E.; Wohlfahrt, G. Analyzing the effects of growing season length on the net ecosystem production of an Aapine Ggassland using model–data fusion. *Ecosystems* **2017**, *21*, 982–999. [CrossRef]
67. Monteiro, J.A.F.; Körner, C. Leaf turnover and herbivory in the tall tussock grass *Festuca orthophylla* in the Andean Altiplano. *Alp. Bot.* **2013**, *123*, 13–20. [CrossRef]
68. McNaughton, S.J. On plants and herbivores. *Am. Nat.* **1986**, *128*, 765–770. [CrossRef]
69. Briske, D.D. Developmental morphology and physiology of grasses. In *Grazing Management. An Ecological Perspective*; Heitschmidt, R.K., Stuth, J.W., Eds.; Timber Press: Portland, OR, USA, 1991; pp. 85–108.
70. Skarpe, C.; Hester, A. Plant traits, browsing and grazing herbivores, and vegetation dynamics. In *The Ecology of Browsing and Grazing*; Gordon, I.J., Prins, H.H.T., Eds.; Springer: Berlin, Germany, 2008; pp. 217–261.
71. Frank, D.A.; McNaughton, S.J. Evidence for the promotion of aboveground grassland production by native large herbivore in Yellowstone National Park. *Oecologia* **1993**, *96*, 157–161. [CrossRef] [PubMed]
72. Weppler, T.; Stoll, P.; Stocklin, J. The relative importance of sexual and clonal reproduction for population growth in the long-lived alpine plant *Geum reptans*. *J. Ecol.* **2006**, *94*, 869–879. [CrossRef]
73. Wagner, J.; Lechleitner, M.; Hosp, D. Pollen limitation is not the rule in nival plants: A study from the European Central Alps. *Am. J. Bot.* **2016**, *103*, 375–387. [CrossRef]
74. Steinacher, G.; Wagner, J. The progamic phase in high-mountain plants: From pollination to fertilization in the cold. *Plants* **2013**, *2*, 354–370. [CrossRef]
75. Bliss, L.C. Arctic and alpine plant life cycles. *Ann. Rev. Ecol. Syst.* **1971**, *2*, 405–438. [CrossRef]
76. Billings, D.W. Arctic and alpine vegetation: Plant adaptations to cold summer climates. In *Arctic and Alpine Environments*; Ives, J.D., Barry, R.G., Eds.; Routledge: Methuen, MA, USA, 1974; pp. 403–443.
77. Urbanska, C.M.; Schutz, M. Reproduction by seed in alpine plants and revegetation research above timberline. *Bot. Helv.* **1986**, *96*, 43–61.
78. Diemer, M. Population dynamics and spatial arrangement of *Ranunculus glacialis* L., an alpine perennial herb, in permanent plots. *Vegetatio* **1992**, *103*, 159–166. [CrossRef]
79. Stöcklin, J.; Bäumler, E. Seed rain, seedling establishment and clonal growth strategies on a glacier foreland. *J. Veg. Sci.* **1996**, *7*, 45–56. [CrossRef]
80. Noroozi, J.; Pauli, H.; Grabherr, G.; Breckle, S.W. The subnival-nival vascular plant species of Iran: A unique high-mountain flora and its threat from climate warming. *Biodivers. Conserv.* **2011**, *20*, 1319–1338. [CrossRef]
81. Winkler, M.; Lamprecht, A.; Steinbauer, K.; Hülber, K.; Theurillat, J.P.; Breiner, F.; Choler, P.; Ertl, S.; Gutiérrez Girón, A.; Rossi, G.; et al. The rich sides of mountain summits—A pan-European view on aspect preferences of alpine plants. *J. Biogeogr.* **2016**, *43*, 2261–2273. [CrossRef]
82. Bürli, S.; Theurillat, J.P.; Winkler, M.; Lamprecht, A.; Pauli, H.; Rixen, C.; Steinbauer, K.; Wipf, S.; Abdaladze, O.; Andrews, C.; et al. A common soil temperature threshold for the upper limit of alpine grasslands in European mountains. *Alp. Bot.* **2021**, *131*, 41–52. [CrossRef]
83. Körner, C.; Basler, D.; Hoch, G.; Kollas, C.; Lenz, A.; Randin, C.F.; Vitasse, Y.; Zimmermann, N.E. Where, why and how? Explaining the low-temperature range limits of temperate tree species. *J. Ecol.* **2016**, *104*, 1076–1088. [CrossRef]
84. Vetaas, O.R. Realized and potential climate niches: A comparison of four *Rhododendron* tree species. *J. Biogeogr.* **2002**, *29*, 545–554. [CrossRef]
85. Randin, C.F.; Dirnbock, T.; Dullinger, S.; Zimmermann, N.E.; Zappa, M.; Guisan, A. Are niche-based species distribution models transferable in space? *J. Biogeogr.* **2006**, *33*, 1689–1703. [CrossRef]
86. Guisan, A.; Petitpierre, B.; Broennimann, O.; Daehler, C.; Kueffer, C. Unifying niche shift studies: Insights from biological invasions. *Trends Ecol. Evol.* **2014**, *29*, 260–269. [CrossRef]
87. Scherrer, D.; Schmid, S.; Körner, C. Elevational species shifts in a warmer climate are overestimated when based on weather station data. *Int. J. Biometeorol.* **2011**, *55*, 645–654. [CrossRef]

88. Randin, C.F.; Engler, R.; Normand, S.; Zappa, M.; Zimmermann, N.E.; Pearman, P.B.; Vittoz, P.; Thuiller, W.; Guisan, A. Climate change and plant distribution: Local models predict high-elevation persistence. *Glob. Chang. Biol.* **2009**, *15*, 1557–1569. [CrossRef]
89. Patsiou, T.S.; Conti, E.; Zimmermann, N.E.; Theodoridis, S.; Randin, C.F. Topo-climatic microrefugia explain the persistence of a rare endemic plant in the Alps during the last 21 millennia. *Glob. Change Biol.* **2014**, *20*, 2286–2300. [CrossRef]
90. Körner, C.; Ohsawa, M. Mountain Systems. In *Ecosystems and Human Well-Being: Current State and Trends*; Hassan, R., Scholes, R., Ash, N., Eds.; Island Press: Washington, DC, USA, 2005; Volume 1, pp. 681–716.
91. Körner, C. The alpine life zone under global change. *Gayana Bot.* **2000**, *57*, 1–17. [CrossRef]
92. Verrall, B.; Pickering, C.M. Alpine vegetation in the context of climate change: A global review of past research and future directions. *Sci. Total. Environ* **2020**, *748*, 141344. [CrossRef]
93. Körner, C.; Diemer, M. Evidence that plants from high altitudes retain their greater photosynthetic efficiency under elevated CO_2. *Funct. Ecol.* **1994**, *8*, 58–68. [CrossRef]
94. Körner, C.; Diemer, M.; Schäppi, B.; Niklaus, P.; Arnone, J. The responses of alpine grassland to four seasons of CO_2 enrichment: A synthesis. *Acta Oecologica* **1997**, *18*, 165–175. [CrossRef]
95. Inauen, N.; Körner, C.; Hiltbrunner, E. No growth stimulation by CO_2 enrichment in alpine glacier forefield plants. *Glob. Change Biol.* **2012**, *18*, 985–999. [CrossRef]
96. Möhl, P.; Hiltbrunner, E.; Körner, C. Halving sunlight reveals no carbon limitation of aboveground biomass production in alpine grassland. *Glob. Change Biol.* **2020**, *26*, 1857–1872. [CrossRef] [PubMed]
97. Onipchenko, V.G.; Blinnikov, M.S.; Aksenova, A.A. Experimental evaluation of shading effects in seasonal dynamics of four alpine communities in northwestern Caucasus, Russia. *Arct. Antarct. Alp. Res.* **2001**, *33*, 330–339. [CrossRef]
98. Nagelmüller, S.; Hiltbrunner, E.; Körner, C. Critically low soil temperatures for root growth and root morphology in three alpine plant species. *Alp. Bot.* **2016**, *126*, 11–21.
99. Nagelmüller, S.; Hiltbrunner, E.; Körner, C. Low temperature limits for root growth in alpine species are set by cell differentiation. *AoB Plants* **2017**, *2*, plx054. [CrossRef]
100. Gottfried, M.; Pauli, H.; Futschik, A.; Akhalkatsi, M.; Barančok, P.; Alonso, J.L.B.; Coldea, G.; Dick, J.; Erschbamer, B.; Fernández Calzado, M.A.R.; et al. Continent-wide response of mountain vegetation to climate change. *Nat. Clim. Change* **2012**, *2*, 111–115. [CrossRef]
101. Matteodo, M.; Wipf, S.; Stöckli, V.; Rixen, C.; Vittoz, P. Elevation gradient of successful plant traits for colonizing alpine summits under climate change. *Environ. Res. Lett.* **2013**, *8*, 024043. [CrossRef]
102. Kosonen, Z.; Schnyder, E.; Hiltbrunner, E.; Thimonier, A.; Schmitt, M.; Seitler, E.; Thöni, L. Current atmospheric nitrogen deposition still exceeds critical loads for sensitive, semi-natural ecosystems in Switzerland. *Atmos. Environ.* **2019**, *211*, 214–225. [CrossRef]
103. Roth, T.; Kohli, L.; Rihm, B.; Achermann, B. Nitrogen deposition is negatively related to species richness and species composition of vascular plants and bryophytes in Swiss mountain grassland. *Agric. Ecosyst. Environ.* **2013**, *178*, 121–126. [CrossRef]
104. Dobzhansky, T. Nothing in biology makes sense except in the light of evolution. *Am. Biol. Teach.* **1973**, *35*, 125–129. [CrossRef]
105. Diekmann, M. Species indicator values as an important tool in applied plant ecology—A review. *Basic. Appl. Ecol.* **2003**, *4*, 493–506. [CrossRef]
106. Landolt, E.; Bäumler, B.; Erhardt, A.; Hegg, O.; Klötzli, F.; Lämmler, W.; Nobis, M.; Rudmann-Maurer, K.; Schweingruber, F.H.; Theurillat, J.-P.; et al. *Flora Indicativa*; Haupt: Bern, Switzerland, 2010.
107. Bartelheimer, M.; Poschlod, P. Ellenberg-type indicator values for European vascular plant species. *Funct. Ecol.* **2016**, *30*, 506–516. [CrossRef]
108. Nakhutsrishvili, G.; Batsatsashvili, K.; Rudmann-Maurer, K.; Körner, C.; Spehn, E. New indicator values for Central Caucasus flora. In *Plant Diversity in the Central Great Caucasus: A Quantitative Assessment*; Nakhutsrishvili, G., Abdaladze, O., Batsatsashvili, K., Spehn, E., Körner, C., Eds.; Springer International Publishing AG: Cham, Switzerland, 2017; pp. 145–170.
109. Rumpf, S.B.; Hülber, K.; Klonner, G.; Moser, D.; Schütz, M.; Wessely, J.; Willner, W.; Zimmermann, N.E.; Dullinger, S. Range dynamics of mountain plants decrease with elevation. *Proc. Natl. Acad. Sci. USA* **2018**, *115*, 18481853. [CrossRef]
110. Dalle Fratte, M.; Brusa, G.; Pierce, S.; Zanzottera, M.; Cerabolini, B.E.L. Plant trait variation along environmental indicators to infer global change impacts. *Flora* **2019**, *254*, 113–121. [CrossRef]
111. Birks, H.J.B. Reflections on the Use of Ecological Attributes and Traits in Quaternary Botany. *Front. Ecol. Evol.* **2020**, *40*, 166. [CrossRef]
112. Tichý, L.; Axmanová, I.; Dengler, J.; Guarino, R.; Jansen, F.; Midolo, G.; Nobis, M.P.; Van Meerbeek, K.; Aćić, S.; Attorre, F.; et al. Ellenberg-type indicator values for European vascular plant species. *J. Veg. Sci.* **2023**, *34*, e13168. [CrossRef]
113. Stöcklin, J.; Armbruster, G.F.J. Environmental filtering, not local adaptation of established plants, determines the occurrence of seed- and bulbil-producing *Poa alpina* in a local flora. *Basic. Appl. Ecol.* **2016**, *17*, 586–595. [CrossRef]

114. Stöcklin, J.; Kuss, P.; Pluess, A.R. Genetic diversity, phenotypic variation and local adaption in the alpine landscape: Case studies with alpine plant species. *Bot. Helv.* **2009**, *119*, 125–133. [CrossRef]
115. Hautier, Y.; Randin, C.F.; Stöcklin, J.; Guisan, A. Changes in reproductive investment with altitude in an alpine plant. *J. Plant. Ecol.* **2009**, *2*, 125–134. [CrossRef]

plants

MDPI

Review

Geographical Parthenogenesis in Alpine and Arctic Plants

Elvira Hörandl

Department of Systematics, Biodiversity and Evolution of Plants (with Herbarium), University of Goettingen, 37073 Göttingen, Germany; elvira.hoerandl@biologie.uni-goettingen.de

Abstract: The term "Geographical parthenogenesis" describes the phenomenon that asexual organisms usually occupy larger and more northern distribution areas than their sexual relatives, and tend to colonize previously glaciated areas. Several case studies on alpine and arctic plants confirm the geographical pattern, but the causal factors behind the phenomenon are still unclear. Research of the last decade in several plant families has shed light on the question and evaluated some of the classical evolutionary theories. Results confirmed, in general, that the advantages of uniparental reproduction enable apomictic plants to re-colonize faster in larger and more northern distribution areas. Associated factors like polyploidy seem to contribute mainly to the spatial separation of sexual and asexual cytotypes. Ecological studies suggest a better tolerance of apomicts to colder climates and temperate extremes, whereby epigenetic flexibility and phenotypic plasticity play an important role in occupying ecological niches under harsh conditions. Genotypic diversity appears to be of lesser importance for the distributional success of asexual plants. Classical evolutionary theories like a reduced pressure of biotic interactions in colder climates and hence an advantage to asexuals (Red Queen hypothesis) did not gain support from studies on plants. However, it is also still enigmatic why sexual outcrossing remains the predominant mode of reproduction also in alpine floras. Constraints for the origin of apomixis might play a role. Interestingly, some studies suggest an association of sexuality with abiotic stresses. Light stress in high elevations might explain why most alpine plants retain sexual reproduction despite other environmental factors that would favor apomixis. Directions for future research will be given.

Keywords: apomixis; biogeography; DNA methylation; ecology; hybridization; polyploidy; stress response

Citation: Hörandl, E. Geographical Parthenogenesis in Alpine and Arctic Plants. *Plants* **2023**, *12*, 844. https://doi.org/10.3390/plants12040844

Academic Editor: Gregor Kozlowski

Received: 19 January 2023
Revised: 8 February 2023
Accepted: 9 February 2023
Published: 13 February 2023

1. Introduction

The term Geographical Parthenogenesis (GP) was originally introduced by Vandel (1928) [1] and described the phenomenon that sexual and asexual organisms have different geographical distributions. Later on, several authors described geographical parthenogenesis for animals and plants, and refined the pattern: asexual organisms tend to have larger distribution areas than their sexual relatives, and they colonize more frequently in higher latitudes, and in previously glaciated (or otherwise devastated) areas, whereby these features can occur alone or in combination [2–7]. Here I will update the known case studies, and discuss the causality with a focus on arctic and alpine plants of the Northern hemisphere, where biogeographical and climatic pre-conditions favor GP scenarios: firstly, these areas provide large landmasses in the circumarctic-circumboreal zone, secondly, several high mountain systems with elevations and vegetation above timberline exist, and thirdly, these areas were covered by ice-shields during cold periods of the Pleistocene, and left after postglacial glacier retreat huge areas for re-colonization. The climatic oscillations of the Pleistocene resulted in range fluctuations of plant and animal species, thus causing opportunities for secondary contact hybridization [8–10]. Hybridization, in turn, is regarded as the most important factor for the evolutionary origin of asexual lineages [11]. Most examples of arctic and alpine GP patterns are documented in Europe and North America [2,4,12]. An update of case studies indeed suggests that a majority of taxa or complexes (71%) do have great proportions of their geographical distributions in alpine and/or arctic regions (Supplementary Table S1). A GP pattern appeared also in functionally asexual plants reproducing via permanent translocation heterozygosity in the genus *Oenothera* [13]. GP patterns were observed also in vegetatively propagating plants (e.g., in *Allium*, [14]). However, also apomictic-sexual taxon pairs without typical GP patterns have been detected (e.g., *Hieracium intybaceum* in the European Alps [15], or *Boechera* species in North America [16]).

From its discovery onwards, geographical parthenogenesis has been enigmatic, and the causal factors have been under dispute [2,5,17]. Classical theories emphasize the advantages of asexuality per se, such as the advantage of uniparental reproduction for colonization as single individuals can act as founders of populations, a principle known as "Bakers' law" [18]. Other authors stressed interactions of cytotypes and postulated introgression of asexuality into sexual populations, resulting in the reduction or extinction of the latter [19,20]. A set of ecological theories regard the interplay of niche occupation and the population genetic structure of asexuals: Asexual hybrids could have advantages as general-purpose genotypes (GPG) that can survive in several niches [21], or by producing broad arrays of clonal lineages that would better explore the resource space than sexual species (Frozen Niche Variation model, FNV, [22,23]). For plants, already previous authors stressed that the predominant polyploidy in asexual plants could have positive effects on asexuals and provide a better adaptivity to cold climatic conditions [2,5,24]. Biotic interactions were also thought to have indirect effects: asexual clones could perform better in colder climates where the pressure of parasites and pathogens would be expected to be lower than in warmer regions (Red Queen model, [17]).

None of these classical theories exclusively explained the patterns, and hence some authors postulated combinations of factors [5]. Research of the last decade added several new case studies on GP occurrences or re-evaluated some older records of candidates of GP (Supplementary Table S1). Associations between main features as given in the literature will be discussed in the respective sections to shed some light on the "classical" hypotheses, like reproduction modes (Section 2), polyploidy (Section 3), and ecological factors (Section 4). However, recent research revealed that several other factors have to be considered: stress response, phenotypic plasticity, and epigenetic flexibility to acclimate to environmental conditions [25–27], see Section 5. Here I will provide an overview of the findings of the last decade to shed light on old and new questions about GP in plants.

2. Factors Related to Mode of Reproduction

2.1. Uniparental Reproduction (Baker's Law): Autonomous Apomixis Versus Pseudogamy

According to Baker's law, uniparental reproduction is advantageous for colonization because a single individual can found a population, which applies both to selfing and apomixis [18]. Apomixis is in flowering plants defined as asexual reproduction via seeds [11] and occurs in many different developmental pathways (reviewed by [11]). Gametophytic apomixis involves the formation of an unreduced embryo sac out of a somatic nucellus cell (apospory) or an unreduced megaspore (diplospory); the egg cell develops parthenogenetically into an embryo. Sporophytic apomixis involved the development of embryos out of a nucellar or integumental cell. Seeds carry one (or more) embryo(s) that are clones of the mother plant. Both modes of reproduction can be associated with GP patterns, whereby gametophytic apomixis appears to be more frequent (Table 1). Different effectivity in colonization, however, can be assumed from the mode of pollination (pollen-dependent pseudogamy versus pollen-independent autonomous apomixis). Indeed, autonomous apomixis occurs in more than 60% of reported cases (Figure 1).

Table 1. Overview of published studies on GP in plants. Arctic/alpine means position of the whole taxon or group; GA = gametophytic apomixis, SA = sporophytic apomixis; A = autonomous apomixis (pollen-independent), P = pseudogamous apomixis (pollen-dependent); S = any geographical separation of sexuals/apomicts; L = apomicts have larger distribution areas than sexuals; N = apomicts have a more polar distribution than sexuals (towards poles in the northern or southern hemisphere, respectively); G = only apomicts occur in previously glaciated areas, sexuals not. Associations between the main factors (Table 1) were calculated using Past vs. 4.03 [28] (see presence/absence matrix and details in Supplementary Table S1).

Taxon	Family	Arctic/Alpine	Type of Apomixis	Type of Pollination *	Ploidy Difference of Apomicts/Sexuals	GP Pattern **	References
Antennaria friesiana	Asteraceae	yes	GA	A	yes	S, L, N	[29]
Antennaria monocephala	Asteraceae	yes	GA	A	yes	S, L, N	[29]
Antennaria parlinii	Asteraceae	no	GA	A	no	S, G	[30]
Antennaria rosea s.l.	Asteraceae	yes	GA	A	yes	S, L, N	[31,32]
Arnica alpina	Asteraceae	yes	GA	A	yes	S, L, N, G	[33]
Arnica amplexicaulis	Asteraceae	yes	GA	A	yes	S, L, N	[33]
Arnica angustifolia	Asteraceae	yes	GA	A	yes	S, G	[33]
Arnica chamissonis	Asteraceae	yes	GA	A	yes	S, L, N	[33]
Arnica lessingii	Asteraceae	yes	GA	A	yes	S, L	[2]
Arnica lonchophylla	Asteraceae	yes	GA	A	yes	S, L, N, G	[33]
Arnica louiseana	Asteraceae	yes	GA	A	yes	S, L, N	[33]
Arnica mollis	Asteraceae	yes	GA	A	yes	S, L, N	[33]
Chondrilla juncea	Asteraceae	no	GA	A	yes	S, L	[7]
Crepis acuminata	Asteraceae	no	GA	A	yes	no	[2,34]
Crepis bakeri	Asteraceae	yes	GA	A	yes	S, L, N	[2]
Crepis exilis	Asteraceae	yes	GA	A	yes	S, L, N	[2]
Crepis modocensis	Asteraceae	yes	GA	A	yes	S, L, N, G	[2]
Crepis monticola	Asteraceae	yes	GA	A	yes	S, L, N	[2]
Crepis occidentalis	Asteraceae	yes	GA	A	yes	S, L, N, G	[2]
Crepis pleurocarpa	Asteraceae	yes	GA	A	yes	S, L, N	[2]
Erigeron strigosus	Asteraceae	no	GA	A	yes	S, L, N	[35]
Eupatorium altissimum	Asteraceae	no	GA	A	yes	S, L, N, G	[36]
Eupatorium cuneifolium	Asteraceae	no	GA	A	yes	S, L, N,	[36]
Eupatorium lecheaefolium	Asteraceae	no	GA	A	yes	S, L, N	[36]
Eupatorium leucolepis	Asteraceae	no	GA	A	yes	S, L, N	[36]
Eupatorium pilosum	Asteraceae	no	GA	A	yes	S, L, N	[36]
Eupatorium rotundifolium	Asteraceae	no	GA	A	yes	S, L, N	[36]
Eupatorium sessilifolium	Asteraceae	no	GA	A	yes	S, L, N, G	[36]
Hieracium alpinum	Asteraceae	yes	GA	A	yes	S, L, N, G	[37,38]
Hieracium intybaceum	Asteraceae	yes	GA	A	yes	S	[15]
Hieracium pilosella s.l.	Asteraceae	yes	GA	A	yes	S, N	[2,39,40]
Parthenium argentatum	Asteraceae	no	GA	P	yes	S, L, N	[41,42]
Taraxacum Sect. *Alpestria*	Asteraceae	yes	GA	A	yes	S, L, N, G	[43]
Taraxacum Sect. *Erythrosperma*	Asteraceae	yes	GA	A	yes	S, L, N	[44]
Taraxacum Sect. *Ruderalia*	Asteraceae	yes	GA	A	yes	S, L, N	[7,44]
Townsendia condensata	Asteraceae	yes	GA	A	yes	S, L, N, G	[45]
Townsendia exscapa	Asteraceae	yes	GA	A	yes	S, L, N, G	[45]
Townsendia grandiflora	Asteraceae	yes	GA	A	yes	no	[45]
Townsendia hookeri	Asteraceae	yes	GA	A	yes	S, L, G	[45–47]
Townsendia incana	Asteraceae	yes	GA	A	yes	S, L, N	[45]
Townsendia leptotes	Asteraceae	yes	GA	A	yes	S, L, N, G	[45]
Townsendia montana	Asteraceae	yes	GA	A	yes	S, L, N	[45]
Townsendia parryi	Asteraceae	yes	GA	A	yes	S, L, N, G	[45]
Townsendia rothrockii	Asteraceae	yes	GA	A	yes	S, G	[45]

Table 1. *Cont.*

Taxon	Family	Arctic/Alpine	Type of Apomixis	Type of Pollination *	Ploidy Difference of Apomicts/Sexuals	GP Pattern **	References
Townsendia scapigera	Asteraceae	yes	GA	A	unknown	no	[45]
Townsendia spathulata	Asteraceae	yes	GA	A	unknown	no	[45]
Townsendia strigosa	Asteraceae	yes	GA	A	unknown	S, N	[45]
Handroanthus ochraceus	Bignoniaceae	no	SA	P	yes	S	[48]
Boechera collinsii	Brassicaceae	no	GA	P	no	no	[49]
Boechera crandallii	Brassicaceae	yes	GA	P	no	S	[49]
Boechera divaricarpa	Brassicaceae	yes	GA	P	no	no	[49]
Boechera fendleri	Brassicaceae	yes	GA	P	no	no	[49]
Boechera holboellii	Brassicaceae	yes	GA	P	no	no	[50]
Boechera lemmonii	Brassicaceae	yes	GA	P	no	no	[49]
Boechera lyallii	Brassicaceae	yes	GA	P	no	no	[49]
Boechera microphylla	Brassicaceae	yes	GA	P	no	no	[49]
Boechera pallidifolia	Brassicaceae	yes	GA	P	no	no	[49]
Boechera pauciflora	Brassicaceae	yes	GA	P	no	no	[49]
Boechera pendulocarpa	Brassicaceae	yes	GA	P	no	no	[49]
Boechera perennans	Brassicaceae	no	GA	P	no	no	[49]
Boechera pinetorum	Brassicaceae	yes	GA	P	no	no	[49]
Boechera puberula	Brassicaceae	yes	GA	P	no	no	[49]
Boechera retrofracta	Brassicaceae	yes	GA	P	yes	S	[49]
Boechera sparsiflora	Brassicaceae	yes	GA	P	no	no	[49]
Boechera spatifolia	Brassicaceae	yes	GA	P	no	no	[16]
Boechera stricta	Brassicaceae	yes	GA	P	no	S	[49]
Boechera williamsii	Brassicaceae	yes	GA	P	no	no	[49]
Hypericum maculatum	Hypericaceae	yes	GA	A or P	yes	no	[51,52]
Hypericum perforatum	Hypericaceae	yes	GA	A or P	yes	S, L, N	[51,52]
Eriotheca gracilipes/pubescens	Malvaceae	no	SA	P	yes	S, L	[53]
Nigritella nigra complex	Orchidaceae	yes	SA	A	yes	S, L, N, G	[54,55]
Spiranthes magnicamporum	Orchidaceae	no	SA	A	unknown	S	[56]
Spiranthes ochroleuca	Orchidaceae	no	SA	A	unknown	S, N	[56]
Limonium algarvense	Plumbaginaceae	no	GA	A	yes	S	[57]
Limonium binervosum/ovalifolium	Plumbaginaceae	no	GA	A	yes	S, L, N, G	[57,58]
Bouteloua curtipendula	Poaceae	no	GA	P	yes	no	[2,59]
Calamagrostis stricta	Poaceae	yes	GA	A	yes	S	[60]
Paspalum intermedium	Poaceae	no	GA	P	yes	S, L, N	[61]
Paspalum simplex	Poaceae	no	GA	P	yes	S, L, N	[62]
Poa cusickii ssp. *cusickii*	Poaceae	yes	GA	A	yes	S	[63]
Poa pringlei	Poaceae	yes	GA	A	unknown	S, N	[63]
Ranunculus auricomus agg.	Ranunculaceae	yes	GA	P	yes	S, L, N, G	[64,65]
Ranunculus kuepferi	Ranunculaceae	yes	GA	P	yes	S, L, N, G	[66,67]
Ranunculus parnassifolius s.l.	Ranunculaceae	yes	GA	P	yes	S, L, N, G	[68]
Amelanchier bartramiana	Rosaceae	no	GA	P	yes	S, N	[69,70]
Amelanchier cusickii/alnifolia	Rosaceae	yes	GA	P	yes	S, N	[69,70]
Crataegus ser. Douglasiana	Rosaceae	no	GA	P	yes	S, L, N	[71,72]
Potentilla crantzii	Rosaceae	yes	GA	P	yes	S, L, N, G	[73]
Potentilla puberula	Rosaceae	yes	GA	P	yes	S, N	[74]
Rubus subg. *Rubus*	Rosaceae	no	GA	P	yes	S, L, N	[75,76]

* usually A and P are not completely exclusive, here the predominant mode is reported. ** meaning that a part of the distribution area is allopatric between sexuals and apomicts, but not completely exclusive. S = most general pattern, L, N, and G are more specific categories.

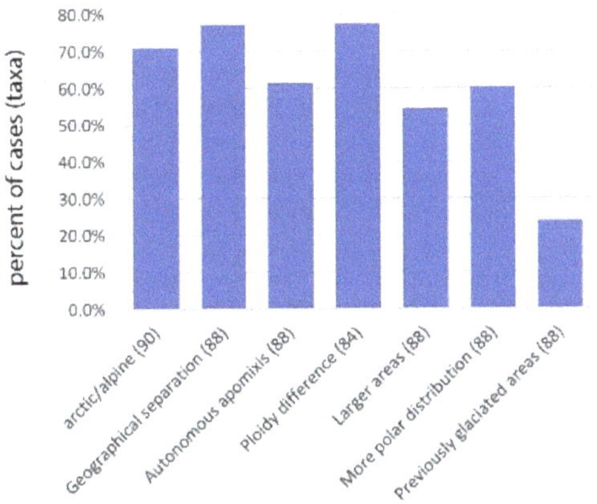

Figure 1. Percentages of taxa (no. of respective total cases in parentheses, based on presence/absence matrix of occurrences in Supplementary Table S1) with (at least partial) arctic/alpine distribution, geographical separation of apomicts and sexuals, autonomous apomixis, ploidy difference between sexuals and apomicts, larger areas of apomicts than sexuals, apomicts with a more polar (northern/southern) distribution than sexuals, and exclusive occurrence of apomicts in previously glaciated areas (see references in Table 1).

In most plant families apomictic seed formation is still dependent on pollination because sperm nuclei have to fertilize the polar nuclei for proper development of the endosperm, the nutritious tissue for the embryo (pseudogamy). Most plants are hermaphrodites and can use their own pollen for this purpose [71,77–82], and hence they are not only independent from another individual as pollen donor, but also are largely independent of a vector carrying pollen (wind or animals). This independence from pollen transfer and another mating partner gives a double advantage to colonization, as a single individual can found a population [5,18]. This advantage appears to overcome even cases of lower fertility of apomictic plants compared to sexuals, as observed in the alpine herb *Ranunculus kuepferi* with GP in the European Alps [66,67]. A simulation study considering modes of reproduction, fertility, ecology, and colonization history of this species revealed that the apomicts clearly showed a much more efficient postglacial recolonization of the previously glaciated areas of the European Alps despite a lower seed set [83].

A potential disadvantage of self-pollination and pseudogamy could be caused by herkogamy within flowers, ie. pollen would not come into contact with the stigmas, or dichogamy, ie. pollen maturation would be asynchronous with that of the stigma. In these cases, pollinators would be still required for successful pseudogamy, because insects moving around within the flower would disperse the pollen, and pollinator visits between flowers in different developmental stages within the same individual would overcome dichogamy. No detailed case studies addressed this question so far for pseudogamy (one case study on breeding systems of *Aechmea* revealed autonomous apomixis [84]). A second major constraint of pseudogamy is potential endosperm imbalance after fertilization of the polar nuclei, i.e., deviations from the optimal 2:1 ratio of maternal to paternal genome copies in the endosperm (e.g., [85]). Pseudogamous plants show a great variation in maternal:paternal genome contributions in the endosperm [66,85–88], and endosperm imbalance can result in a lower seed set [85]. Hence, although pseudogamy does allow for uniparental reproduction, it is constrained by potential pollinator limitation and/or by potentially reduced female fitness.

In Asteraceae, many species reproduce via autonomous apomixis, which means that pollination and fertilization of polar nuclei are not necessary for proper seed formation [4,89].

Asteraceae are strongly represented in the cases of GP (see Table 1) which supports the hypothesis that uniparental reproduction without any need for pollen is an important advantage to asexuality [4]. Endosperm imbalance is no more an issue as other tissues take over the nutritious function of the embryo. Some plant families like orchids do not form endosperm at all. Under these considerations, autonomous apomixis is expected to be more efficient for GP. Indeed, autonomous apomixis correlates significantly with larger distribution areas, and with more polar (northern/southern) distribution patterns (Supplementary Table S2). Pollen-independent apomixis appears to be an advantage to expand spatial distributions.

However, there was no significant correlation between autonomous apomixis to previously glaciated areas. There might be too few case studies to reveal an association. *Hieracium pilosella* is a widespread species with a GP pattern over the whole of Europe [37]. In a study comparing both sexual and apomictic cytotypes on glacier fore fields in the Swiss Alps, apomixis was more frequent in odd-number ploidy levels, and these occurred closer to the glacier snout than other cytotypes [90]. The authors concluded that apomixis provided reproductive assurance in odd-ploid cytotypes by avoiding meiosis disturbances and loss of fertility, and hence a colonization advantage in the most recent ice-free areas. Reproductive assurance without pollination was further confirmed in *Hieracium alpinum*, a widespread herb distributed in the European mountain system and in Northern Europe. Diploids are self-sterile and reproduce sexually, whereas triploids are autonomous (pollen-independent) obligate apomicts [38]; female fertility does not differ under natural conditions. Only triploids colonized previously glaciated areas in the Alps and in Scandinavia, whereas diploids occur in the Carpathians. The GP pattern was referred to as a better colonization ability and a more stable reproductive system of apomicts [38]. In sympatrically cultivated apomictic and sexual populations, apomicts had much higher fertility, which was explained by mate limitation acting for sexuals, but not for apomicts [91]. However, colonization history may influence GP patterns. In *Hieracium intybaceum*, diploids appeared to be sexual outcrossers, whereas tetraploids are autonomously apomictic. The diploids colonized most parts of the Alps whereas apomicts were confined to a small, disjunct area in the Western parts of the Alps, the Vosges and Schwarzwald mts. [15]. The authors explain this untypical pattern which only involves geographical separation of cytotypes (sensu [1]) with a very young evolutionary origin of apomicts [15].

Nevertheless, uniparental reproduction is also possible for sexual plants via selfing, and this taxonomically widespread trait reduces the advantage of asexuality over sexuality [5,92]. For instance, regular selfing of sexual plants may hinder the evolution of GP in the genus *Boechera* (Brassicaceae), which comprises both sexual regular selfers and apomictic species in North America [87]. But, only a few species show some geographical separation (Table 1). A further peculiarity of *Boechera* is the occurrence of apomixis on the diploid level [93], which means that positive side effects of polyploidy for apomicts would disappear [16]. The almost lack of the GP pattern in this genus might be due to a combinational effect as well as its presence in other genera.

Taken together, all studies so far confirmed the advantage of uniparental reproduction of asexuals for colonization. The effect is apparently stronger in autonomous than in pseudogamous apomicts, and probably stronger in the case of self-sterility of sexual relatives. However, uniparental reproduction alone is not the only causal factor as some apomictic species do not exhibit GP despite the respective modes of reproduction.

2.2. *Reproductive Interactions of Cytotypes*

Sexual reproduction is in angiosperms the ancestral state [94], and apomixis arises in sexual populations [11]. This means that at least initially sexual and apomictic individuals coexist and can interact with each other. In a mixed population of diploid sexuals and polyploid apomicts, apomictic polyploid individuals can overcome minority cytotype disadvantages [95] better than sexual polyploids, because they would not or only to a minor extent be fertilized by the pollen of the surrounding diploid sexuals. With apomixis, even a single plant can produce clonal offspring with the maternal ploidy level. This factor was also seen as a major advantage of GP for establishing polyploid apomictic populations in the alpine herb *Ranunculus kuepferi* [83].

On the other hand, pseudogamous apomicts usually maintain functional pollen, and hence can fertilize surrounding sexual individuals without being fertilized themselves by diploids. This unidirectional hybridization would theoretically result in the introgression of apomixis into sexuals and hence the extinction of sexuality [19,96]. However, experimental crossings in *Ranunculus auricomus* revealed that pollinations of diploid sexuals with pollen from tetraploid apomicts results in a very low seed set, probably because the different ploidy levels and also induced selfing of sexuals (Mentor effect) acted as efficient crossing barriers [97]. At higher ploidy levels, however, these crossing barriers between cytotypes appear to be less efficient. The crossability of polyploid cytotypes was studied in *Potentilla puberula*, a widely distributed species in the European Alps with tetraploid sexuals and higher polyploid apomictic cytotypes (5x, 6x, 7x, 8x, and 9x) [98,99]. In experimental heteroploid crosses, the seed set was reduced in sexuals, but not in apomicts, compared to homoploid crosses. This asymmetrical interference gives an advantage to apomicts [99]. The sexuals and apomicts in *Potentilla* show a sympatric distribution in the Alps, but with differentiation of reproduction modes along altitude and habitats [74]. Because of sympatry, interactions of cytotypes influence distribution patterns in natural populations, but ecological differentiation plays a role as well [74,100].

Pollen fertility is usually much reduced in autonomous apomicts [77], but nevertheless, remnant viable pollen can still cause asymmetric interference in mixed populations [101]. Simulations of local reproductive interference in sympatric sexual and apomictic *Crepis* confirmed that sexuals occur mostly in spatial isolation, whereas apomictic taxa can co-occur [102]. Hence, the presence of apomicts can lead to local exclusion of sexuals and drive the spatial separation of reproductive types in the initial stages. However, most extant cases already do have spatially separated sexual/apomictic populations, so the effects of interactions are probably only initially important (Table 1).

3. Polyploidy and Hybridization

Polyploidy was traditionally seen as a major factor fostering GP patterns [2,5,24]. However, its role as the only causal factor was questioned by the lack of general patterns of geographical polyploidy in the many sexual plants ([5]). In the updated review (Table 1), a significant correlation appeared only for Vandel's [1] basic pattern that diploid sexuals and polyploid apomicts have separated distributions (Supplementary Table S2). However, there was no significant association of polyploid apomicts with any of the other GP features (larger distributions, more polar distributions, or previously glaciated areas; Supplementary Table S2). The association with autonomous apomixis was also not significant. However, combinational effects were postulated [5], and actually confirmed in a simulation study on *R. kuepferi* on postglacial colonization of the Alps [83].

The indirect role of polyploidy for establishing separate distributions of apomicts and sexuals could be manifold. First, during initial sympatry, diploid sexuals might disappear in mixed populations because of a unilateral minority cytotype disadvantage (see Section 3). Second, polyploidy could be connected with a shift to self-fertility, which is an advantage for pseudogamous apomicts (see Section 2). Due to the bypass of meiosis, apomicts avoid loss of heterozygosity and inbreeding depression in the offspring, which otherwise reduces the fitness of selfing diploid populations. Inbreeding in small, diploid self-incompatible populations results in lower fitness and is disadvantageous to diploid sexual progenitors [76,103–105]. Third, polyploids, specifically allopolyploids, have higher levels of heterozygosity, both on the population level [106], but also as intragenomic heterozygosity ([65]). However, GP is also found in case studies with the same heterozygosity levels of sexuals and apomicts (e.g., in the autopolyploids *Antennaria friesiana* [29], and *R. kuepferi* [66,80,107], but also in allopolyploids of *Crataegus* [72,82]). Heterozygosity could help polyploids to buffer deleterious mutations and tolerate more extreme climatic conditions ([65]), or enable them to adapt to novel niches (see Section 5). Specifically, a recent global study on distributions of polyploid plants (sexual and apomictic) found a significant correlation between polyploids to colder climates [108].

Previous authors emphasized also the influence of hybridity on GP patterns, mainly based on examples from animals [3]. However, for plants, this correlation is less straightforward, as not only allopolyploids, but also many autopolyploids are known to exhibit GP (e.g., *Antennaria friesiana, Ranunculus kuepferi, Hieracium alpinum, Potentilla pusilla*). Moreover, most *Boechera* apomicts are diploid hybrids [49] but do not show typical GP patterns (Table 1). Altogether the information on the evolutionary origin of apomictic polyploids is still too incomplete for a more detailed evaluation of the influence of hybridity on GP patterns.

Both polyploidy and hybridization are regarded as the major evolutionary pathways for the origins of apomixis in natural populations [11]. However, successful natural origins are probably rare, as various genetic and epigenetic changes of regulatory mechanisms have to be combined for the establishment of functional apomictic seed formation, and there is no universal pathway to apomixis [109–113]. The rarity of the evolutionary origins of apomictic lineages may also constrain the frequency and expression of GP patterns.

4. Ecology

4.1. FNV and GPG: Does Genetic Diversity Really Matter?

Two classical models, frozen niche variation (FNV) and general-purpose genotypes (GPG), both rely on the assumption of clonality. In the first case, broad arrays of clones would occupy a larger resource space, while in the second scenario, single widely distributed clones would cover the larger resource space. Already previous literature surveys [7,64,106] demonstrated that the assumption of clonality is often not met because facultative sexuality and high genotypic diversity are widespread in apomictic plants. A "little bit of sex" is for plants probably under positive selection as even low frequencies of recombination can efficiently counteract the accumulation of deleterious mutations, even in polyploid genomes [114]. More recent population genetic surveys confirmed that GP is usually not associated with clonality of apomicts [16,51,65,80,100,115]. In such cases, the FNV and GPG models are simply not applicable. In *Boechera* species, range size is not correlated to allelic diversity and heterozygosity within populations [116]. Widespread generalist clones, however, were found in North American *Crataegus* [82]. In the alpine orchid genus *Nigritella*, high clonal diversity was found, but only with populations in the vicinity sharing the same genotypes [54]. In general, the influence of genotypic diversity or clonality on GP patterns appears to be less relevant for plants than predicted by classical models.

4.2. Niche Shifts of Cytotypes

In the last decade niche modeling based on climatic data from WordCLIM or other databases became a novel and widely used method in evolutionary and ecological research (e.g., [108,117]). Several studies on GP applied these methods and revealed a general tendency to climatic niche shifts of apomictic polyploids. That is, no single clones, but rather the sexual vs. apomictic cytotypes as a whole exhibit niche divergence. A tendency to colder climatic conditions was observed in alpine *R. kuepferi* [67,118], in the widespread common dandelion *Taraxacum officinale* [44], in *Crataegus suksdorfii* widespread in North America [82], in boreal *Boechera stricta* in North America [49], in the coastal *Limonium ovalifolium/binervosum* complex in Europe [57], and in south American *Paspalum intermedium* [61]. This trait would fit the general pattern that polyploids globally occur more frequently in colder climates [108], and to the frequent trend of asexuals towards more polar distributions (see above). However, this trait to colder climates is not universal. In *Potentilla puberula*, asexuals tended to wetter and to more anthropogenic habitats, and niche divergence contributed significantly to the small-scale differentiation patterns of sexuals and asexuals [74,100]. In the first survey of apomicts in the Himalayan flora, apomicts occurred in habitats with higher soil moisture [119]. In *Ranunculus auricomus*, the distribution of apomicts tended to regions with lower light intensity [65]. Niche expansion of asexuals was observed in *Taraxacum* Sect. *Erythrosperma* [44], and in *Crataegus douglasii* [82]. Alien sexual and apomictic plants do not differ in their niche spectra [120].

In general, the data support the hypothesis that ecological niche shifts of apomictic cytotypes are associated with GP patterns (whereby the data from the older literature are too incomplete to perform statistical tests for all cases). However, there is not a uniform trend to a certain niche, and also different patterns appear in different taxa of the same genus (e.g., in *Ranunculus* and *Taraxacum*). Little is known about the physiological background of these niche divergence patterns (see Section 5) Understanding ecological niche divergence needs more case-wise studies on the most relevant abiotic factors.

4.3. Do Apomicts Tend to Higher Elevations?

The classical reviews proposed a correlation of GP not only to higher latitudes but also to higher elevations [2,5], which would be also expected from the respective temperature gradients in mountains. However, the older literature did not provide detailed information on elevations for statistical tests and also the classical survey by Bierzychudek [2] discussed only one genus (*Townsendia*) as an example. In Central Europe, a survey of vegetation data suggested a correlation of apomicts to higher altitudes as the only ecological trait [6]. However, this study was based on a taxonomic assignment of apomixis and not on a site-per-site assessment of apomixis, and hence did not regard the infraspecific differentiation patterns of GP. Detailed examination of mode of reproduction per site (usually via flow cytometric seed screening; [121]) rather gives an equivocal picture: a study of all Asteraceae species of the European Alps (229 taxa) did not reveal a correlation of apomixis to higher elevations [122]. Likewise, in subnivale plants of the European Alps, representing the angiosperm flora at the highest elevations, only one of 12 species expressed apomixis [123]. On the other hand, *R. kuepferi* and *H. alpinum* showed clear correlations of apomixis to higher elevations in the alpine zone [38,67], but only for populations within temperate European mountains. A study on *Potentilla puberula* in the Alps, however, did not reveal differences in elevation between sexuals and apomicts [74]. The first detailed investigation of apomixis in the Himalayas in the Ladakh region revealed that apomixis is not correlated to elevation [119].

A problem with elevation is that comparisons only make sense within a mountain system and within the same latitudinal zone (e.g., Alps, Himalayas), because, with higher latitudes, the respective alpine vegetational belts move down to low elevations [124]. Therefore, for taxa with a large latitudinal range expansion, differences in elevation in the South disappear towards the North. For instance, apomicts of *Hieracium alpinum* occur in mountains of Central Europe between 1080–2570 m a.s.l. (mean 2020 m), the diploid sexual cytotypes between 1380–2120 m (mean 1777 m). The apomicts in Northern Europe grow between 200–1280 m (mean 738 m) (data from [38]). For complexes with a large north-south distribution, this effect might eliminate all elevational differences. Likewise, in *Amelanchier* widespread in North America, no correlation of apomixis to elevation was found, only to latitude [69]. In the *Ranunculus auricomus* complex, ranging from the Mediterranean to the Arctic of Europe, there was even a reverse pattern that sexuals occurred in higher elevations in Mediterranean mountains [65]. In altitudinal transects of *Taraxacum* sect. *Ruderalia* in Switzerland, sexuals were found in higher elevations [125–127]. Also in the South American cerrados, sexuals of the genus *Eriotheca* occur at higher elevations [53]. Such cases also fit global surveys of polyploid plants that did not reveal a clear correlation between polyploidy and elevation [108].

Within mountain systems, the actual ecological conditions depend not only on elevation but also on the microhabitat, e.g., north/south exposition, soil conditions, duration of snow layer, human influence, etc. Other effects of high elevation, e.g., high UV solar irradiation, or lower atmosphere pressure [124], are largely unexplored for apomictic plants. So far, detailed investigations within mountain systems are needed to get a full understanding of the effects of elevation and the associated ecological trends.

4.4. Do Apomicts Tend More to Previously Glaciated Areas Than Sexuals?

Previous reviews also postulated that apomicts would colonize previously devastated areas such as glaciated or arid areas [2,3,5]. For plants, this means mostly re-colonization of areas that were covered by ice shields during cold periods of the Pleistocene. However, evaluation of pairs with apomicts only in glaciated areas and sexuals not, only about 24% of cases had such exclusive patterns, and the association to GP is statistically not significant (Supplementary Table S2). This is simply due to the observation that sexual relatives often occur in previously glaciated areas as well [2]. Only a few detailed phylogeographic studies are available but suggest that biogeographical history depends also on mountain topography. In *Ranunculus kuepferi* in the European Alps, only apomicts managed to overcome a general topographical barrier in the southwestern Alps, mostly by a niche shift to colder conditions, but not by faster spread [83]. The study on *Potentilla puberula* in European also found a stronger influence of niche shifts than of biogeographical history [100]. An interesting case is *Townsendia hookeri* in western North America, where sexuals were found in unglaciated refugia north and south of the previously glaciated areas, where only apomicts occur [47]. Glaciations probably caused vicariance of sexuals, whereas postglacial recolonization of the Central parts of the western North American mountain chains was conducted only by apomicts [47]. This study also underlines that the topography of mountain systems and locations of ice-free Pleistocene refugia influence the realized GP patterns.

4.5. Biotic Interactions: No Support for the Red Queen Model

The so-called Red Queen hypothesis postulates that sexuals have an advantage in areas/situations with high frequencies of biotic interactions because they produce genetically variable offspring that could respond better to co-evolving parasites and pathogen infections in a co-evolutionary arms race [128–130]. Accordingly, asexuals should have an advantage in areas with low biotic interactions, as it is regarded as typical for higher latitudes and altitudes. Few studies on plants with GP are available on this aspect and they would not support the theory [5]. An experimental study on diploid and polyploid *Parthenium argentatum* with respect to their tolerance to the pathogenic fungus *Verticillium dahliae* did not reveal differences between cytotypes [131]. Hartman et al. [17] analyzed the GP pattern of *Hieracium alpinum* in natural populations concerning the intensity and frequency of seed predators and found no significant differences between sexuals and apomicts, and no latitudinal gradients, despite a higher genotypic diversity of the sexuals. Seed predation per population was even positively associated with intra-population diversity. In a study on common dandelion populations (*Taraxacum* sect. *Ruderalia*) over a large latitudinal gradient, northern populations showed reduced infections of a seed-eating weevil species compared to southern populations, but no difference appeared in pathogenic rust fungi and pathogenic soil microbes [132]. But it remained unclear whether the patterns can be attributed to a less efficient response of asexuals in plants to their predators or pathogens.

The problem with the Red Queen is that in plants many basic assumptions of the model are not met. First, clonality is not a general feature of apomictic plants due to facultative sexuality and some diversity of genotypes (see Section 4.1). Second, the assumption of a weaker defense of asexual plants to herbivory or other biotic attacks in higher latitudes is simply not supported by empirical data [133,134]. Large-scale surveys of chemical compounds rather suggest a greater resistance to herbivory of high-latitude plant species than of tropical species [134]. There is no difference in physical defenses [133]. Third, in biotic interactions, polyploidy of organisms could infer even better resistance or tolerance to biotic stressors, and this has been shown for plants and animals [135]. For plants, polyploidy alters profiles of chemical compounds and hence defense against biotic attacks [136]. So far, polyploidy appears to be an important indirect player in biotic interaction scenarios. Therefore, the Red Queen model is probably too simplistic or even inapplicable to explain GP in plants.

5. Physiology: Phenotypic Plasticity and Epigenetic Response

5.1. Cold Stress

The frequent occurrence of apomixis towards polar regions, and the general tendency of polyploids to colder climates [108] could have different physiological backgrounds: first, apomictic plants are more cold-tolerant because of polyploidy, and second, cold temperatures could directly induce apomictic reproduction [137]. These hypotheses are not mutually exclusive. Cold adaptation is in plants strongly under epigenetic control (see Section 5.3). The induction of apomixis by temperature extremes is most directly related to apomictic development. Low-temperature shocks are known to trigger unreduced gamete formation in plants by disturbing spindle formation at meiosis [138]. Although these mechanisms are better studied for male development, female development would be similarly affected resulting in unreduced megaspores (the diplosporous pathway) or failure of meiosis and somatic development (apospory). In experimental cold treatments in climate growth chambers, diploid alpine *R. kuepferi* produced few but significantly more apomictic seeds than in the warm-treated control, whereas the tetraploid apomicts remained unaffected [137]. Seed screening of diploid populations from field samples revealed a female triploid bridge with unreduced egg cell formation, which could be regarded as the first step for the establishment of apomixis in natural populations [139].

However, the reproductive development of apomictic plants remains to be studied. Many alpine plants perform flower preformation in the previous or even pre-previous summer season, whereby the timing of the critical phase of meiosis and gametogenesis can vary from species to species [124]. Flower preformation during the warm season could avoid the cold shocks during meiosis that would cause unreduced gamete formation. Flower preformation, but also higher freezing tolerance, would mean that plants occurring in the extreme subnival zones are so well adapted to low-temperature conditions that developmental pathways in flower buds would not be affected [123]. This adaptation could explain why apomixis is rare in plants of the highest elevations of the Alps. Under these assumptions, rather plants adapted to lower altitudes and moving upwards in re-colonization processes would experience cold shocks and express cold-induced apomixis. This hypothesis still needs to be tested.

5.2. Light Stress

Light stress is specifically of interest for mountain plants because solar radiation (annual means) increases with altitude, and also the fraction of UV light increases with high altitude [124]. Moreover, alpine plants above the tree line and also above the dwarf shrub zone grow often in completely unshaded conditions. Many typical alpine plants are well adapted to solar radiation and even to high UV light exposure. The influence of light stress on the mode of reproduction is not well known. It has been shown experimentally in facultative apomictic lowland plants of the *Ranunculus auricomus* complex that light stress during the flowering bud stage can increase frequencies of sexual ovule development [88,140–142]. Photosynthetic stress in the reproductive tissues can result in oxidative stress within ovules which would initiate meiosis and sexual development [143,144]. Also, other kinds of oxidative stress (drought, starvation, hydrogen peroxide treatments) can trigger sexual reproduction in facultative apomictic plants [145,146]. In the widespread *Ranunculus auricomus* complex, the GP pattern showed a significant correlation of sexual seed formation to higher light intensities, which is due to the occurrence of sexual species in southern mountain systems at higher elevations than most apomictic taxa [65]. Experimental light treatments showed that polyploidy reduces the effect of light stress on sexual ovule formation [141], which might be due to a better quenching of excess light [141,144]. In *Hieracium alpinum*, lower foliar nitrogen and carbon content was found in triploid apomicts of Scandinavia compared to diploids in the Carpathians, which was related to lower solar radiation and lower precipitation in the Northern areas [147].

These observations could also explain why the association of apomixis to elevation is equivocal (see Section 4.3): higher solar radiation and stronger UV light exposure of high alpine plants would rather cause oxidative stress in reproductive tissues and stimulate the maintenance of obligate sexuality [123].

5.3. Epigenetic Control Mechanisms

In plants, response to biotic and abiotic stresses is strongly modulated by epigenetic control mechanisms that influence gene expression patterns [148–150]. The best-known epigenetic mechanism is DNA methylation which can be involved in direct phenotypic response to abiotic and biotic stresses as well as in heritable changes of regulatory mechanisms [25,149–153]. The stress memory kept via methylation changes allows plants to rapidly acclimate and adapt to changed environments [150,153,154]. DNA-methylation changes are important for phenotypic plasticity and adaptation of evolutionarily young, apomictic lineages, despite low genetic divergence and variability [25]. Moreover, asexual reproduction would bypass the epigenetic resetting mechanisms during meiosis and gametogenesis, and hence could enhance the transgenerational transmission of epialleles [25,154]. DNA methylations can further silence transposable elements (TEs), which make up large portions of plants' genomes; hence, asexual genome evolution might be strongly affected by TE dynamics [155]. In dandelions, heritable methylation changes in apomictic plant lineages could explain the high divergence of methylations and accession-specific TE profiles [156]. TE profiles have a strong influence on heritable gene expression patterns, some of which could reflect the functional divergence of lineages [157].

Epigenetic research has always focused on model organisms and crops, and few studies are available on wild apomictic plants in the context of ecological adaptations. The most comprehensive studies related to GP are available on alpine *R. kuepferi* by using methylation-sensitive AFLPs (MSAPs). Wild populations differed in their MSAP profiles between diploid sexual and tetraploid apomictic cytotypes, but also between obligate and facultative apomictic tetraploids [158]. Controlled temperature experiments in climate growth chambers revealed a significant loss of methylations after temperature changes in both cytotypes and also influenced growth [159,160]; gene expression analysis further confirmed a stress response after cold treatments, which was stronger in diploids than in tetraploids [161]. All results confirmed a better cold acclimation of tetraploids compared to diploids as a major factor for the GP pattern in this species [161].

Another model system is the common dandelion (*Taraxacum* sect. *Ruderalia*). Experimental abiotic stress treatments of apomictic clones resulted in partly heritable methylation changes as measured with MSAP profiles [162]. A study of methylations over a large geographical scale from Central Europe to Scandinavia revealed a strong dependence of epigenetic variation on the genetic background, but also a small fraction of sequence-independent epigenetic variation that may contribute to phenotypic variation and local adaptation [163]. A more detailed study on the heritability of stress-induced methylation changes revealed that transgenerational modification of methylation patterns is genotype and context-dependent, and mostly undirected [164]. In wild populations of dandelions, differences in flowering time are associated with methylation patterns but not with geographical distance [165].

A limitation of MSAP-based studies was the low number of markers and the lack of sequence-based information for the anonymous MSAP fragments which strongly limits functional information of methylation patterns. The resources needed for bisulfite-based sequencing methods to analyze complete methylomes were so far not feasible for non-model organisms, and for the number of samples that are required for biogeographical studies. However, detailed experimental work on apomictic plants testing acclimation/adaptation to key environmental factors compared to epigenetic profiling will be promising to understand the ecophysiological background of GP.

6. Summary and Outlook

Research of the last decade has confirmed that GP is a widespread phenomenon for sexual-apomictic plant complexes and has shed light on general aspects of its causality.

Most patterns show a geographical separation and range expansion of apomicts, while associations of GP to elevation and previous glaciations are rather equivocal, probably due to the topographical and climatic complexity of mountain systems. While the advantages of uniparental reproduction for apomicts are largely confirmed, the role of polyploidy seems to be an indirect one and is probably connected to the frequently observed climatic niche shifts of asexual polyploid cytotypes compared to their sexual progenitors. The background of these niche shifts might be rather a phenotypic plasticity and short-term adaptation of asexual lineages to altered abiotic conditions, probably driven by epigenetic variability rather than by genetic changes.

Still, only a few model systems are comprehensively explored for these aspects, and further information on reproductive data, ecological, experimental, and molecular work is needed on many genera to confirm the here proposed traits. A major challenge is still to perform large-scale screenings of a representative sampling from large geographical areas, often involving complex mountain systems, and also high arctic regions that are not easily accessible and therefore not well explored. More detailed studies on alpine plants in different mountain systems of the world are wanted to understand the uncertain relations of apomixis to higher elevations and the specific physiological adaptations to extreme conditions at high altitudes. A broad application of the flow cytometric seed screening method is most promising for this task. However, for understanding the causality of the pattern, ecological and physiological studies, experimental work, and genetic/epigenetic studies are wanted.

Funding: Research for this review was funded by Deutsche Forschungsgemeinschaft DFG, grant number Ho4395/10-2 within the priority program "Taxon-Omics: New Approaches for Discovering and Naming Biodiversity" (SPP 1991).

Supplementary Materials: The following supporting information can be downloaded at: https://www.mdpi.com/article/10.3390/plants12040844/s1, Supplementary Table S1: Presence/absence Table of data reported in Table 1 and candidates of putative GP patterns with uncertain data; Supplementary Table S2: Associations between the main putative causal factors. References [2,12,28,122,123] are cited in the Supplementary Materials.

Data Availability Statement: Not applicable.

Conflicts of Interest: The author declares no conflict of interest.

References

1. Vandel, A. La parthenogenese geographique: Contribution a l/'etude biologique et cytologique dc la parthenogenese naturelle. *Bull. Biol. Fr. Belg.* **1928**, *62*, 164–281.
2. Bierzychudek, P. Patterns in plant parthenogenesis. *Experientia* **1985**, *41*, 1255–1264. [CrossRef]
3. Kearney, M. Hybridization, glaciation and geographical parthenogenesis. *Trends Ecol. Evol.* **2005**, *20*, 495–502. [CrossRef] [PubMed]
4. Hörandl, E.; Cosendai, A.-C.; Temsch, E.M. Understanding the geographic distributions of apomictic plants: A case for a pluralistic approach. *Plant Ecol. Divers.* **2008**, *1*, 309–320. [CrossRef] [PubMed]
5. Hörandl, E. The complex causality of geographical parthenogenesis. *New Phytol.* **2006**, *171*, 525–538. [CrossRef] [PubMed]
6. Gregor, T. Apomicts in the vegetation of Central Europe. *Tuexenia* **2013**, *33*, 233–257.
7. Van Dijk, P.J. Ecological and evolutionary opportunities of apomixis: Insights from Taraxacum and Chondrilla. *Philos. Trans. R. Soc. Lond. Ser. B-Biol. Sci.* **2003**, *358*, 1113–1120. [CrossRef] [PubMed]
8. Hewitt, G.M. Some genetic consequences of ice ages, and their role in divergence and speciation. *Biol. J. Linn. Soc.* **1996**, *58*, 247–276. [CrossRef]
9. Hewitt, G.M. Genetic consequences of climatic oscillations in the Quaternary. *Philos. Trans. R. Soc. Lond. B Biol. Sci.* **2004**, *359*, 183–195. [CrossRef] [PubMed]
10. Hewitt, G.M. Quaternary phylogeography: The roots of hybrid zones. *Genetica* **2011**, *139*, 617–638. [CrossRef] [PubMed]
11. Hojsgaard, D.; Hörandl, E. The Rise of Apomixis in Natural Plant Populations. *Front. Plant Sci.* **2019**, *10*, 358. [CrossRef] [PubMed]
12. Hörandl, E. Evolution and biogeography of alpine apomictic plants. *Taxon* **2011**, *60*, 390–402. [CrossRef]
13. Johnson, M.T.J.; Smith, S.D.; Rausher, M.D. Effects of plant sex on range distributions and allocation to reproduction. *N. Phytol.* **2010**, *186*, 769–779. [CrossRef] [PubMed]

14. Duchoslav, M.; Jandová, M.; Kobrlová, L.; Šafářová, L.; Brus, J.; Vojtěchová, K. Intricate distribution patterns of six cytotypes of *Allium oleraceum* at a continental scale: Niche expansion and innovation followed by niche contraction with increasing ploidy level. *Front. Plant Sci.* **2020**, *11*, 591137. [CrossRef]
15. Zahradnicek, J.; Chrtek, J. Cytotype distribution and phylogeography of *Hieracium intybaceum* (Asteraceae). *Bot. J. Linn. Soc.* **2015**, *179*, 487–498. [CrossRef]
16. Lovell, J.T.; Grogan, K.; Sharbel, T.F.; McKay, J.K. Mating system and environmental variation drive patterns of adaptation in *Boechera spatifolia* (Brassicaceae). *Mol. Ecol.* **2014**, *23*, 4486–4497. [CrossRef] [PubMed]
17. Hartmann, M.; Stefanek, M.; Zdvorak, P.; Herman, P.; Chrtek, J.; Mraz, P. The Red Queen hypothesis and geographical parthenogenesis in the alpine hawkweed *Hieracium alpinum* (Asteraceae). *Biol. J. Linn. Soc.* **2017**, *122*, 681–696. [CrossRef]
18. Baker, H.G. Support for Baker's Law-as a rule. *Evolution* **1967**, *21*, 853–856. [CrossRef] [PubMed]
19. Mogie, M. *The Evolution of Asexual Reproduction in Plants*; Chapman and Hall: London, UK, 1992; p. 292.
20. Mogie, M. Premeiotic endomitosis and the costs and benefits of asexual reproduction. *Biol. J. Linn. Soc.* **2013**, *109*, 487–495. [CrossRef]
21. Lynch, M. Destabilizing hybridization, general-purpose genotypes and geographic parthenogenesis. *Q. Rev. Biol.* **1984**, *59*, 257–290. [CrossRef]
22. Vrijenhoek, R.C.; Parker, J.E.D. Geographical Parthenogenesis: General Purpose Genotypes and Frozen Niche Variation. In *Lost Sex*; Schön, I., Martens, K., Van Dijk, P., Eds.; Springer: Berlin/Heidelberg, Germany, 2009; pp. 99–131.
23. Vrijenhoek, R.C. Factors affecting clonal diversity and coexistence. *Am. Zool.* **1979**, *19*, 787–797. [CrossRef]
24. Lundmark, M. Polyploidization, hybridization and geographical parthenogenesis. *Trends Ecol. Evol.* **2006**, *21*, 9. [CrossRef]
25. Verhoeven, K.J.F.; Preite, V. Epigenetic variation in asexually reproducing organisms. *Evol. Int. J. Org. Evol.* **2014**, *68*, 644–655. [CrossRef]
26. Zhang, Y.Y.; Fischer, M.; Colot, V.; Bossdorf, O. Epigenetic variation creates potential for evolution of plant phenotypic plasticity. *N. Phytol.* **2013**, *197*, 314–322. [CrossRef]
27. Leon-Martinez, G.; Vielle-Calzada, J.P. Apomixis in Flowering Plants: Developmental and Evolutionary Considerations. In *Plant Development and Evolution*; Grossniklaus, U., Ed.; Elsevier: Amsterdam, The Netherlands, 2019; Volume 131, pp. 565–604. [CrossRef]
28. Hammer, O. Past 4—The Past of the Future. (Paleontological Statistics). Available online: https://www.nhm.uio.no/english/research/resources/past/ (accessed on 30 November 2022).
29. Bayer, R.J. Allozymic and morphological variation in Antennaria (Asteraceae, Inuleae) from the low arctic of northwestern North-America. *Syst. Bot.* **1991**, *16*, 492–506. [CrossRef]
30. Bayer, R.J.; Stebbins, G.L. Distribution of sexual and apomictic populations of Antennaria-parlinii. *Evolution* **1983**, *37*, 555–561. [CrossRef] [PubMed]
31. Bayer, R.J. Patterns of clonal diversity in geographically marginal populations of Antennaria-rosea (Asteraceae, Inuleae) from sub-arctic Alaska and Yukon territory. *Bot. Gaz.* **1991**, *152*, 486–493. [CrossRef]
32. Bayer, R.J.; Chandler, G.T. Evolution of polyploid agamic complexes: A case study using the *Catipes* group of *Antennaria*, including the *A. rosea* complex (Asteraceae: Gnaphalieae). In *Apomixis: Evolution, Mechanisms and Perspectives*; Hörandl, E., Grossniklaus, U., Van Dijk, P.J., Sharbel, T., Eds.; International Association for Plant Taxonomy: Ruggel, Liechtenstein, 2007; pp. 317–336.
33. Wolf, S.J. Cytogeographical studies in the genus Arnica (Compositae: Senecioneae). I. *Am. J. Bot.* **1980**, *67*, 300–308. [CrossRef]
34. Sears, C.J.; Whitton, J. A reexamination of the North American Crepis agamic complex and comparison with the findings of Babcock and Stebbins' classic biosystematic monograph. *Am. J. Bot.* **2016**, *103*, 1289–1299. [CrossRef]
35. Noyes, R.D. Apomixis in Erigeron sect. Phalacroloma. In *Apomixis: Evolution, Mechanisms and Perspectives*; Hörandl, E., Grossniklaus, U., Van Dijk, P., Sharbel, T., Eds.; A.R.G. Gantner Verlag: Rugell, Liechtenstein, 2007; pp. 337–358.
36. Sullivan, V.I. Diploidy, polyploidy, and agamospermy among species of Eupatorium (Compositae). *Can. J. Bot. Rev. Can. Bot.* **1976**, *54*, 2907–2917. [CrossRef]
37. Mraz, P.; Chrtek, J.; Singliarova, B. Geographical parthenogenesis, genome size variation and pollen production in the arctic-alpine species Hieracium alpinum. *Bot. Helv.* **2009**, *119*, 41–51. [CrossRef]
38. Mraz, P.; Zdvorak, P.; Hartmann, M.; Stefanek, M.; Chrtek, J. Can obligate apomixis and more stable reproductive assurance explain the distributional successes of asexual triploids in *Hieracium alpinum* (Asteraceae)? *Plant Biol.* **2019**, *21*, 227–236. [CrossRef]
39. Fehrer, J.; Gemeinholzer, B.; Chrtek, J., Jr.; Braeutigam, S. Incongruent plastid and nuclear DNA phylogenies reveal ancient intergeneric hybridization in Pilosella hawkweeds (Hieracium, Cichorieae, Asteraceae). *Mol. Phylogenetics Evol.* **2007**, *42*, 347–361. [CrossRef] [PubMed]
40. Mráz, P.; Šingliarová, B.; Urfus, T.; Krahulec, F. Cytogeography of *Pilosella officinarum* (Compositae): Altitudinal and longitudinal differences in ploidy level distribution in the Czech Republic and Slovakia and the general pattern in Europe. *Ann. Bot.* **2008**, *101*, 59–71. [CrossRef] [PubMed]
41. Rollins, R.C. Sources of genetic variation in *Parthenium argentatum* Gray (Compositae). *Evolution* **1949**, *3*, 358–368. [CrossRef]
42. Cruz, V.V.; Lynch, A.L.; Ray, D.T.; Niaura, W.S.; Purdy, P.H.; Dierig, D.A. Analysis of mode of reproduction of guayule (*Parthenium argentatum* A. Gray) using flow cytometry and identification of polyhaploids for breeding. *Ind. Crops Prod.* **2017**, *107*, 618–623. [CrossRef]

43. Stepanek, J.; Kirschner, J.; Jarolimova, V.; Kirschnerova, L. *Taraxacum nigricans, T. alpestre* and allies in the Taraxacum sect. Alpestria: Taxonomy, geography and conservation status. *Preslia* **2011**, *83*, 537–564.
44. Meirmans, P.G. Niche divergence contributes to geographical parthenogenesis in two dandelion taxa. *J. Evol. Biol.* **2021**, *34*, 1071–1086. [CrossRef]
45. Beaman, J.H. The systematics and evolution of Townsendia (Compositae). *Contrib. Gray Herb. Harv. Univ.* **1957**, *183*, 1–159. [CrossRef]
46. Thompson, S.L.; Choe, G.; Ritland, K.; Whitton, J. Cryptic sex within male-sterile polyploid populations of the Easter daisy, Townsendia hookeri. *Int. J. Plant Sci.* **2008**, *169*, 183–193. [CrossRef]
47. Thompson, S.L.; Whitton, J. Patterns of recurrent evolution and geographic parthenogenesis within apomictic polyploid Easter daises (Townsendia hookeri). *Mol. Ecol.* **2006**, *15*, 3389–3400. [CrossRef] [PubMed]
48. Mendes, M.G.; de Oliveira, A.P.; Oliveira, P.E.; Bonetti, A.M.; Sampaio, D.S. Sexual, apomictic and mixed populations in *Handroanthus ochraceus* (Bignoniaceae) polyploid complex. *Plant Syst. Evol.* **2018**, *304*, 817–829. [CrossRef]
49. Mau, M.; Lovell, J.T.; Corral, J.M.; Kiefer, C.; Koch, M.A.; Aliyu, O.M.; Sharbel, T.F. Hybrid apomicts trapped in the ecological niches of their sexual ancestors. *Proc. Natl. Acad. Sci. USA* **2015**, *112*, E2357–E2365. [CrossRef]
50. Sharbel, T.F.; Mitchell-Olds, T.; Dobes, C.; Kantama, L.; de Jong, H. Biogeographic distribution of polyploidy and B chromosomes in the apomictic *Boechera holboellii* complex. *Cytogenet. Genome Res.* **2005**, *109*, 283–292. [CrossRef] [PubMed]
51. Molins, M.P.; Corral, J.M.; Aliyu, O.M.; Koch, M.A.; Betzin, A.; Maron, J.L.; Sharbel, T.F. Biogeographic variation in genetic variability, apomixis expression and ploidy of St. Johns wort (*Hypericum perforatum*) across its native and introduced range. *Ann. Bot.* **2014**, *113*, 417–427. [CrossRef] [PubMed]
52. Koch, M.A.; Scheriau, C.; Betzin, A.; Hohmann, N.; Sharbel, T.F. Evolution of cryptic gene pools in *Hypericum perforatum*: The influence of reproductive system and gene flow. *Ann. Bot.* **2013**, *111*, 1083–1094. [CrossRef]
53. Mendes-Rodrigues, C.; Marinho, R.C.; Balao, F.; Arista, M.; Ortiz, P.L.; Carmo-Oliveira, R.; Oliveira, P.E. Reproductive diversity, polyploidy, and geographical parthenogenesis in two Eriotheca (Malvaceae) species from Brazilian Cerrado. *Perspect. Plant Ecol. Evol. Syst.* **2019**, *36*, 1–12. [CrossRef]
54. Hedren, M.; Lorenz, R.; Teppner, H.; Dolinar, B.; Giotta, C.; Griebl, N.; Hansson, S.; Heidtke, U.; Klein, E.; Perazza, G.; et al. Evolution and systematics of polyploid Nigritella (Orchidaceae). *Nord. J. Bot.* **2018**, *36*, njb-01539. [CrossRef]
55. Teppner, H. Adventitious embryony in Nigritella (Orchidaceae). *Folia Geobot. Phytotaxon.* **1996**, *31*, 323–331. [CrossRef]
56. Catling, P.M. Breeding systems of northeastern North-American Spiranthes (Orchidaceae). *Can. J. Bot. Rev. Can. Bot.* **1982**, *60*, 3017–3039. [CrossRef]
57. Caperta, A.D.; Castro, S.; Loureiro, J.; Rois, A.S.; Conceicao, S.; Costa, J.; Rhazi, L.; Santo, D.E.; Arsénio, P. Biogeographical, ecological and ploidy variation in related asexual and sexual Limonium taxa (Plumbaginaceae). *Bot. J. Linn. Soc.* **2017**, *183*, 75–93.
58. Róis, A.S.; Sádio, F.; Paulo, O.S.; Teixeira, G.; Paes, A.P.; Espírito-Santo, D.; Sharbel, T.F.; Caperta, A.D. Phylogeography and modes of reproduction in diploid and tetraploid halophytes of Limonium species (Plumbaginaceae): Evidence for a pattern of geographical parthenogenesis. *Ann. Bot.* **2016**, *117*, 37–50. [CrossRef]
59. Grant, V. *Plant Speciation*, 2nd ed.; Columbia University Press: New York, NY, USA, 1981.
60. Greene, C.W. Sexual and apomictic reproduction in Calamagrostis (Gramineae) from eastern North-America. *Am. J. Bot.* **1984**, *71*, 285–293. [CrossRef]
61. Karunarathne, P.; Schedler, M.; Martinez, E.J.; Honfi, A.I.; Novichkova, A.; Hojsgaard, D. Intraspecific ecological niche divergence and reproductive shifts foster cytotype displacement and provide ecological opportunity to polyploids. *Ann. Bot.* **2018**, *121*, 1183–1196. [CrossRef] [PubMed]
62. Urbani, M.H.; Quarin, C.L.; Espinoza, F.; Penteado, M.I.O.; Rodrigues, I.F. Cytogeography and reproduction of the *Paspalum simplex* polyploid complex. *Plant Syst. Evol.* **2002**, *236*, 99–105. [CrossRef]
63. Soreng, R.J. Systematics of the Epiles group of Poa (Poaceae). *Syst. Bot.* **1991**, *16*, 507–528. [CrossRef]
64. Hörandl, E. Geographical parthenogenesis: Opportunities for asexuality. In *Lost Sex*; Schön, I., Martens, K., Van Dijk, P.J., Eds.; Springer: Berlin/Heidelberg, Germany, 2009; pp. 161–186.
65. Karbstein, K.; Tomasello, S.; Hodac, L.; Lorberg, E.; Daubert, M.; Hörandl, E. Moving beyond assumptions: Polyploidy and environmental effects explain a geographical parthenogenesis scenario in European plants. *Mol. Ecol.* **2021**, *30*, 2659–2675. [CrossRef]
66. Cosendai, A.-C.; Hörandl, E. Cytotype stability, facultative apomixis and geographical parthenogenesis in *Ranunculus kuepferi* (Ranunculaceae). *Ann. Bot.* **2010**, *105*, 457–470. [CrossRef] [PubMed]
67. Schinkel, C.C.F.; Kirchheimer, B.; Dellinger, A.S.; Klatt, S.; Winkler, M.; Dullinger, S.; Hörandl, E. Correlations of polyploidy and apomixis with elevation and associated environmental gradients in an alpine plant. *Aob Plants* **2016**, *8*, plw064. [CrossRef] [PubMed]
68. Vuille, C.; Küpfer, P. Aposporie chez le *Ranunculus parnassifolius* L. I. Etude cytoembryologique. *Bull. Soc. Neuchâteloise Sci. Nat.* **1985**, *108*, 123–134.
69. Burgess, M.B.; Cushman, K.R.; Doucette, E.T.; Talent, N.; Frye, C.T.; Campbell, C.S. Effects of apomixis and polyploidy on diversification and geographic distribution in Amelanchier (Rosaceae). *Am. J. Bot.* **2014**, *101*, 1375–1387. [CrossRef]
70. Cushman, K.R.; Burgess, M.B.; Doucette, E.T.; Nelson, G.A.; Campbell, C.S. Species Delimitation in Tetraploid, Apomictic Amelanchier (Rosaceae). *Syst. Bot.* **2017**, *42*, 234–256. [CrossRef]

71. Lo, E.Y.Y.; Stefanovic, S.; Dickinson, T.A. Geographical parthenogenesis in Pacific Northwest hawthorns (Crataegus; Rosaceae). *Botany* **2013**, *91*, 107–116. [CrossRef]
72. Coughlan, J.M.; Stefanovic, S.; Dickinson, T.A. Relative resource allocation to dispersal and competition demonstrates the putative role of hybridity in geographical parthenogenesis. *J. Biogeogr.* **2014**, *41*, 1603–1613. [CrossRef]
73. Smith, G.L. Studies in Potentilla, L. II. Cytological aspects of apo- mixis in P. crantzii (Cr.) Beck ex Fritsch. *New Phytol.* **1963**, *62*, 283–300. [CrossRef]
74. Alonso-Marcos, H.; Nardi, F.D.; Scheffknecht, S.; Tribsch, A.; Hulber, K.; Dobes, C. Difference in reproductive mode rather than ploidy explains niche differentiation in sympatric sexual and apomictic populations of Potentilla puberula. *Ecol. Evol.* **2019**, *9*, 3588–3598. [CrossRef]
75. Weber, H.E. Former and modern taxonomic treatment of the apomictic Rubus complex. *Folia Geobot. Phytotaxon.* **1996**, *31*, 373–380. [CrossRef]
76. Sochor, M.; Sarhanova, P.; Pfanzelt, S.; Travnicek, B. Is evolution of apomicts driven by the phylogeography of the sexual ancestor? Insights from European and Caucasian brambles (*Rubus*, Rosaceae). *J. Biogeogr.* **2017**, *44*, 2717–2728. [CrossRef]
77. Asker, S.; Jerling, L. *Apomixis in Plants*; CRC Press: Boca Raton, FL, USA, 1992.
78. Richards, J.A. *Plant Breeding Systems*, 2nd ed.; Chapman and Hall: London, UK, 1997; p. 529.
79. Hörandl, E. The evolution of self-fertility in apomictic plants. *Sex. Plant Reprod.* **2010**, *23*, 73–86. [CrossRef] [PubMed]
80. Cosendai, A.C.; Wagner, J.; Ladinig, U.; Rosche, C.; Hörandl, E. Geographical parthenogenesis and population genetic structure in the alpine species *Ranunculus kuepferi* (Ranunculaceae). *Heredity* **2013**, *110*, 560–569. Available online: http://www.nature.com/hdy/journal/vaop/ncurrent/suppinfo/hdy20131s1.html (accessed on 7 February 2013). [CrossRef] [PubMed]
81. Hörandl, E. Evolutionary implications of self-compatibility and reproductive fitness in the apomictic *Ranunculus auricomus* polyploid complex (Ranunculaceae). *Int. J. Plant Sci.* **2008**, *169*, 1219–1228. [CrossRef]
82. Coughlan, J.M.; Han, S.; Stefanovic, S.; Dickinson, T.A. Widespread generalist clones are associated with range and niche expansion in allopolyploids of Pacific Northwest Hawthorns (Crataegus L.). *Mol. Ecol.* **2017**, *26*, 5484–5499. [CrossRef]
83. Kirchheimer, B.; Wessely, J.; Gattringer, G.; Hülber, K.; Moser, D.; Schinkel, C.C.F.; Appelhans, M.; Klatt, S.; Caccianiga, M.; Dellinger, A.; et al. Reconstructing geographical parthenogenesis: Effects of niche differentiation and reproductive mode on Holocene range expansion of an alpine plant. *Ecol. Lett.* **2018**, *21*, 392–401. [CrossRef]
84. Pool-Chale, M.; Ramirez-Morillo, I.; Fernandez-Concha, G.C.; Hornung-Leoni, C.T. Reproductive biology of *Aechmea bracteata* (Sw.) Griseb. (Bromelioideae: Bromeliaceae). *Plant Biol.* **2018**, *20*, 113–120. [CrossRef]
85. Spielman, M.; Vinkenoog, R.; Scott, R.J. Genetic mechanisms of apomixis. *Philos. Trans. R. Soc. London. Ser. B Biol. Sci.* **2003**, *358*, 1095–1103. [CrossRef]
86. Talent, N.; Dickinson, T.A. Endosperm formation in aposporous Crataegus (Rosaceae, Spiraeoideae, tribe Pyreae): Parallels to Ranunculaceae and Poaceae. *N. Phytol.* **2007**, *173*, 231–249. [CrossRef]
87. Aliyu, O.M.; Schranz, M.E.; Sharbel, T.F. Quantitative variation for apomictic reproduction in the genus *Boechera* (Brassicaceae). *Am. J. Bot.* **2010**, *97*, 1719–1731. [CrossRef]
88. Klatt, S.; Hadacek, F.; Hodač, L.; Brinkmann, G.; Eilerts, M.; Hojsgaard, D.; Hörandl, E. Photoperiod extension enhances sexual megaspore formation and triggers metabolic reprogramming in facultative apomictic *Ranunculus auricomus* . *Front. Plant Sci.* **2016**, *7*, 278. [CrossRef]
89. Noyes, R.D. Apomixis in the Asteraceae: Diamonds in the Rough. *Funct. Plant Sci. Biotechnol.* **2007**, *1*, 207–222.
90. Sailer, C.; Stöcklin, J.; Grossniklaus, U. Dynamics of apomictic and sexual reproduction during primary succession on a glacier forefield in the Swiss Alps. *Sci. Rep.* **2020**, *10*, 8269. [CrossRef]
91. Mraz, P.; Mrazova, V. Greater reproductive assurance of asexual plant compared with sexual relative in a low-density sympatric population: Experimental evidence for pollen limitation. *J. Evol. Biol.* **2021**, *34*, 1503–1509. [CrossRef]
92. Reutemann, A.V.; Martinez, E.J.; Schedler, M.; Davina, J.R.; Hojsgaard, D.H.; Honfi, A.I. Uniparentality: Advantages for range expansion in diploid and diploid-autopolyploid species. *Bot. J. Linn. Soc.* **2022**, *200*, 563–585. [CrossRef]
93. Kantama, L.; Sharbel, T.F.; Schranz, M.E.; Mitchell-Olds, T.; de Vries, S.; de Jong, H. Diploid apomicts of the Boechera holboellii complex display large-scale chromosome substitutions and aberrant chromosomes. *Proc. Natl. Acad. Sci. USA* **2007**, *104*, 14026–14031. [CrossRef]
94. Hojsgaard, D.; Klatt, S.; Baier, R.; Carman, J.G.; Hörandl, E. Taxonomy and biogeography of apomixis in angiosperms and associated biodiversity characteristics. *Crit. Rev. Plant Sci.* **2014**, *33*, 414–427. [CrossRef] [PubMed]
95. Levin, D.A. Minority cytotype exclusion in local plant populations. *Taxon* **1975**, *24*, 35–43. [CrossRef]
96. Mogie, M.; Britton, N.F.; Stewart-Cox, A.J. Asexuality, Polyploidy and the Male Function. In *Apomixis: Evolution, Mechanisms and Perspectives*; Hörandl, E., Grossniklaus, U., Van Dijk, P., Sharbel, T.F., Eds.; Gantner: Rugell, Liechtenstein, 2007; pp. 169–194.
97. Hörandl, E.; Temsch, E. Introgression of apomixis into sexual species is in the *Ranunculus auricomus* complex inhibited by mentor effects and ploidy barriers. *Ann. Bot.* **2009**, *104*, 81–89. [CrossRef] [PubMed]
98. Dobes, C.; Milosevic, A.; Prohaska, D.; Scheffknecht, S.; Sharbel, T.F.; Hulber, K. Reproductive differentiation into sexual and apomictic polyploid cytotypes in *Potentilla puberula* (Potentilleae, Rosaceae). *Ann. Bot.* **2013**, *112*, 1159–1168. [CrossRef]
99. Dobeš, C.; Scheffknecht, S.; Fenko, Y.; Prohaska, D.; Sykora, C.; Hülber, K. Asymmetric reproductive interference: The consequences of cross-pollination on reproductive success in sexual-apomictic populations of *Potentilla puberula* (Rosaceae). *Ecol. Evol.* **2018**, *8*, 365–381. [CrossRef]

100. Nardi, F.D.; Hulber, K.; Moser, D.; Alonso-Marcos, H.; Tribsch, A.; Dobes, C. Occurrence of apomictic conspecifics and ecological preferences rather than colonization history govern the geographic distribution of sexual *Potentilla puberula* . *Ecol. Evol.* **2020**, *10*, 7306–7319. [CrossRef]
101. Hersh, E.; Grimm, J.; Whitton, J. Attack of the clones: Reproductive interference between sexuals and asexuals in the Crepis agamic complex. *Ecol. Evol.* **2016**, *6*, 6473–6483. [CrossRef]
102. Whitton, J.; Sears, C.J.; Maddison, W.P. Co-occurrence of related asexual, but not sexual, lineages suggests that reproductive interference limits coexistence. *Proc. R. Soc. B* **2017**, *284*, 20171579. [CrossRef]
103. Pinc, J.; Chrtek, J.; Latzel, V.; Mraz, P. Negative effect of inbreeding on fitness of an arctic-alpine *Hieracium alpinum* (Asteraceae), a species with a geographical parthenogenesis distribution pattern. *Plant Syst. Evol.* **2020**, *306*, 62. [CrossRef]
104. Karbstein, K.; Rahmsdorf, E.; Tomasello, S.; Hodač, L.; Hörandl, E. Breeding system of diploid sexuals within the *Ranunculus auricomus* complex and its role in a geographical parthenogenesis scenario. *Ecol. Evol.* **2020**, *10*, 14435–14450. [CrossRef]
105. Haag, C.R.; Ebert, D. A new hypothesis to explain geographic parthenogenesis. *Ann. Zool. Fenn.* **2004**, *41*, 539–544.
106. Hörandl, E.; Paun, O. Patterns and sources of genetic diversity in apomictic plants: Implications for evolutionary potentials. In *Apomixis: Evolution, Mechanisms and Perspectives*; Hörandl, E., Grossniklaus, U., Van Dijk, P.J., Sharbel, T., Eds.; International Association for Plant Taxonomy: Ruggel, Liechtenstein, 2007; pp. 169–194.
107. Cosendai, A.C.; Rodewald, J.; Hörandl, E. Origin and distribution of autopolyploids via apomixis in the alpine species *Ranunculus kuepferi* (Ranunculaceae). *Taxon* **2011**, *60*, 355–364. [CrossRef]
108. Rice, A.; Smarda, P.; Novosolov, M.; Drori, M.; Glick, L.; Sabath, N.; Meiri, S.; Belmaker, J.; Mayrose, I. The global biogeography of polyploid plants. *Nat. Ecol. Evol.* **2019**, *3*, 265–273. [CrossRef]
109. Schmidt, A. Controlling Apomixis: Shared Features and Distinct Characteristics of Gene Regulation. *Genes* **2020**, *11*, 329. [CrossRef]
110. Grimanelli, D. Epigenetic regulation of reproductive development and the emergence of apomixis in angiosperms. *Curr. Opin. Plant Biol.* **2012**, *15*, 57–62. [CrossRef]
111. Paetzold, C.; Barke, B.H.; Horandl, E. Evolution of Transcriptomes in Early-Generation Hybrids of the Apomictic Ranunculus auricomus Complex (Ranunculaceae). *Int. J. Mol. Sci.* **2022**, *23*, 13881. [CrossRef]
112. Barcaccia, G.; Palumbo, F.; Sgorbati, S.; Albertini, E.; Pupilli, F. A Reappraisal of the Evolutionary and Developmental Pathway of Apomixis and Its Genetic Control in Angiosperms. *Genes* **2020**, *11*, 859. [CrossRef]
113. Hand, M.L.; Koltunow, A.M.G. The genetic control of apomixis: Asexual seed formation. *Genetics* **2014**, *197*, 441–450. [CrossRef]
114. Hodač, L.; Klatt, S.; Hojsgaard, D.; Sharbel, T.; Hörandl, E. A little bit of sex prevents mutation accumulation even in apomictic polyploid plants. *BMC Evol. Biol.* **2019**, *19*, 170. [CrossRef]
115. Nardi, F.D.; Dobes, C.; Muller, D.; Grasegger, T.; Myllynen, T.; Alonso-Marcos, H.; Tribsch, A. Sexual intraspecific recombination but not de novo origin governs the genesis of new apomictic genotypes in *Potentilla puberula* (Rosaceae). *Taxon* **2018**, *67*, 1108–1131. [CrossRef]
116. Lovell, J.T.; McKay, J.K. Ecological genetics of range size variation in *Boechera* spp. (Brassicaceae). *Ecol. Evol.* **2015**, *5*, 4962–4975. [CrossRef] [PubMed]
117. Glennon, K.L.; Ritchie, M.E.; Segraves, K.A. Evidence for shared broad-scale climatic niches of diploid and polyploid plants. *Ecol. Lett.* **2014**, *17*, 574–582. [CrossRef]
118. Kirchheimer, B.; Schinkel, C.C.F.; Dellinger, A.S.; Klatt, S.; Moser, D.; Winkler, M.; Lenoir, J.; Caccianiga, M.; Guisan, A.; Nieto-Lugilde, D.; et al. A matter of scale: Apparent niche differentiation of diploid and tetraploid plants may depend on extent and grain of analysis. *J. Biogeogr.* **2016**, *43*, 716–726. [CrossRef]
119. Brozova, V.; Koutecky, P.; Dolezal, J. Plant apomixis is rare in Himalayan high-alpine flora. *Sci. Rep.* **2019**, *9*, 14386. [CrossRef] [PubMed]
120. Dellinger, A.S.; Essl, F.; Hojsgaard, D.; Kirchheimer, B.; Klatt, S.; Dawson, W.; Pergl, J.; Pyšek, P.; van Kleunen, M.; Weber, E.; et al. Niche dynamics of alien species do not differ among sexual and apomictic flowering plants. *New Phytol.* **2016**, *209*, 1313–1323. [CrossRef] [PubMed]
121. Matzk, F.; Meister, A.; Schubert, I. An efficient screen for reproductive pathways using mature seeds of monocots and dicots. *Plant J.* **2000**, *21*, 97–108. [CrossRef]
122. Pegoraro, L.; Baker, E.C.; Aeschimann, D.; Balant, M.; Douzet, R.; Garnatje, T.; Guignard, M.; Leitch, I.; Leitch, A.; Palazzesi, L.; et al. The correlation of phylogenetics, elevation and ploidy on the incidence of apomixis in Asteraceae in the European Alps. *Bot. J. Linn. Soc.* **2021**, *194*, 410–422, Erratum in *Bot. J. Linnean Soc.* **2021**, *196*, 143. [CrossRef]
123. Hörandl, E.; Dobes, C.; Suda, J.; Vit, P.; Urfus, T.; Temsch, E.M.; Cosendai, A.-C.; Wagner, J.; Ladinig, U. Apomixis is not prevalent in subnival to nival plants of the European Alps. *Ann. Bot.* **2011**, *108*, 381–390. [CrossRef] [PubMed]
124. Körner, C. *Alpine Plant Life. Functional Plant Ecology of High Mountain Systems*, 3rd ed.; Springer: Berlin/Heidelberg, Germany, 2021; p. 500.
125. Meirmans, P.G.; Calame, F.G.; Bretagnolle, F.; Felber, F.; den Nijs, J.C.M. Anthropogenic disturbance and habitat differentiation between sexual diploid and apomictic triploid Taraxacum sect. Ruderalia. *Folia Geobot.* **1999**, *34*, 451–469. [CrossRef]
126. Meirmans, P.G.; Vlot, E.C.; Den Nijs, J.C.M.; Menken, S.B.J. Spatial ecological and genetic structure of a mixed population of sexual diploid and apomictic triploid dandelions. *J. Evol. Biol.* **2003**, *16*, 343–352. [CrossRef]

127. Calame, F.G.; Felber, F. Distribution of diploid sexual and triploid apomictic dandelions (Taraxacum sect. Ruderalia) along two altitudinal gradients in Switzerland. *Bot. Helv.* **2000**, *110*, 109–114.
128. Glesener, R.R.; Tilman, D. Sexuality and components of environmental uncertainty—Clues from geographic parthenogenesis in terrestrial animals. *Am. Nat.* **1978**, *112*, 659–673. [CrossRef]
129. Bell, G. *The Masterpiece of Nature: The Evolution and Genetics of Sexuality*; California Press: Berkely, CA, USA, 1982.
130. Lively, C.M. A Review of Red Queen Models for the Persistence of Obligate Sexual Reproduction. *J. Hered.* **2010**, *101*, S13–S20. [CrossRef]
131. Ray, D.T.; Orum, T.V.; Bigelow, D.M.; Alcorn, S.M. Selection of diploid and screening of polyploid guayule lines for Verticillium tolerance. *Ind. Crops Prod.* **1995**, *4*, 303–310. [CrossRef]
132. Verhoeven, K.J.F.; Biere, A. Geographic parthenogenesis and plant-enemy interactions in the common dandelion. *BMC Evol. Biol.* **2013**, *13*, 23. [CrossRef]
133. Moles, A.T.; Bonser, S.P.; Poore, A.G.B.; Wallis, I.R.; Foley, W.J. Assessing the evidence for latitudinal gradients in plant defence and herbivory. *Funct. Ecol.* **2011**, *25*, 380–388. [CrossRef]
134. Moles, A.T.; Wallis, I.R.; Foley, W.J.; Warton, D.I.; Stegen, J.C.; Bisigato, A.J.; Cella-Pizarro, L.; Clark, C.J.; Cohen, P.S.; Cornwell, W.K.; et al. Putting plant resistance traits on the map: A test of the idea that plants are better defended at lower latitudes. *N. Phytol.* **2011**, *191*, 777–788. [CrossRef]
135. Van de Peer, Y.; Ashman, T.L.; Soltis, P.S.; Soltis, D.E. Polyploidy: An evolutionary and ecological force in stressful times. *Plant Cell* **2021**, *33*, 11–26. [CrossRef] [PubMed]
136. Drummond, C.P.; Renner, T. Genomic insights into the evolution of plant chemical defense. *Curr. Opin. Plant Biol.* **2022**, *68*, 102254. [CrossRef]
137. Klatt, S.; Schinkel, C.C.; Kirchheimer, B.; Dullinger, S.; Hörandl, E. Effects of cold treatments on fitness and mode of reproduction in the diploid and polyploid alpine plant *Ranunculus kuepferi* (Ranunculaceae). *Ann. Bot.* **2018**, *121*, 1287–1298. [CrossRef]
138. De Storme, N.; Geelen, D. The impact of environmental stress on male reproductive development in plants: Biological processes and molecular mechanisms. *Plant Cell Environ.* **2014**, *37*, 1–18. [CrossRef]
139. Schinkel, C.C.F.; Kirchheimer, B.; Dullinger, S.; Geelen, D.; De Storme, N.; Hörandl, E. Pathways to polyploidy: Indications of a female triploid bridge in the alpine species *Ranunculus kuepferi* (Ranunculaceae). *Plant Syst. Evol.* **2017**, *303*, 1093–1108. [CrossRef]
140. Quarin, C.L. Seasonal changes in the incidence of apomixis of diploid, triploid, and tetraploid plants of *Paspalum cromyorrhizon*. *Euphytica* **1986**, *35*, 515–522. [CrossRef]
141. Ulum, F.B.; Costa Castro, C.; Hörandl, E. Ploidy-dependent effects of light stress on the mode of reproduction in the *Ranunculus auricomus* complex (Ranunculaceae). *Front. Plant Sci.* **2020**, *11*, 104. [CrossRef] [PubMed]
142. Saran, S.; de Wet, J.M.J. Environmental control of reproduction in Dichanthium intermedium. *J. Cytol. Genet.* **1976**, *11*, 22–28.
143. Hörandl, E.; Hadacek, F. The oxidative damage initiation hypothesis for meiosis. *Plant Reprod.* **2013**, *26*, 351–367. [CrossRef] [PubMed]
144. Ulum, F.B.; Hadacek, F.; Hörandl, E. Polyploidy improves photosynthesis regulation within the *Ranunculus auricomus* complex (Ranunculaceae). *Biology* **2021**, *10*, 811. [CrossRef]
145. Mateo de Arias, M.; Gao, L.; Sherwood, D.A.; Dwivedi, K.; Price, B.J.; Jamison, M.; Kowallis, B.M.; Carman, J.G. Whether gametophytes are reduced or unreduced in angiosperms might be determined metabolically. *Genes* **2020**, *11*, 1449. [CrossRef]
146. Shah, J.N.; Kirioukhova, O.; Pawar, P.; Tayyab, M.; Mateo, J.L.; Johnston, A.J. Depletion of Key Meiotic Genes and Transcriptome-Wide Abiotic Stress Reprogramming Mark Early Preparatory Events Ahead of Apomeiotic Transition. *Front. Plant Sci.* **2016**, *7*, 1539. [CrossRef] [PubMed]
147. Hartmann, M.; Jandova, K.; Chrtek, J.; Stefanek, M.; Mraz, P. Effects of latitudinal and elevational gradients exceed the effects of between-cytotype differences in eco-physiological leaf traits in diploid and triploid Hieracium alpinum. *Alp. Bot.* **2018**, *128*, 133–147. [CrossRef]
148. Dowen, R.H.; Pelizzola, M.; Schmitz, R.J.; Lister, R.; Dowen, J.M.; Nery, J.R.; Dixon, J.E.; Ecker, J.R. Widespread dynamic DNA methylation in response to biotic stress. *Proc. Natl. Acad. Sci. USA* **2012**, *109*, E2183–E2191. [CrossRef] [PubMed]
149. Sun, M.H.; Yang, Z.; Liu, L.; Duan, L. DNA Methylation in Plant Responses and Adaption to Abiotic Stresses. *Int. J. Mol. Sci.* **2022**, *23*, 6910. [CrossRef]
150. Arora, H.; Singh, R.K.; Sharma, S.; Sharma, N.; Panchal, A.; Das, T.; Prasad, A.; Prasad, M. DNA methylation dynamics in response to abiotic and pathogen stress in plants. *Plant Cell Rep.* **2022**, *41*, 1931–1944. [CrossRef] [PubMed]
151. Law, J.A.; Jacobsen, S.E. Establishing, maintaining and modifying DNA methylation patterns in plants and animals. *Nat. Rev. Genet.* **2010**, *11*, 204–220. [CrossRef]
152. Bossdorf, O.; Richards, C.L.; Pigliucci, M. Epigenetics for ecologists. *Ecol. Lett.* **2008**, *11*, 106–115. [CrossRef]
153. Akhter, Z.; Bi, Z.Z.; Ali, K.; Sun, C.; Fiaz, S.; Haider, F.U.; Bai, J.P. In Response to Abiotic Stress, DNA Methylation Confers EpiGenetic Changes in Plants. *Plants* **2021**, *10*, 1096. [CrossRef]
154. Paszkowski, J.; Grossniklaus, U. Selected aspects of transgenerational epigenetic inheritance and resetting in plants. *Curr. Opin. Plant Biol.* **2011**, *14*, 195–203. [CrossRef]
155. Hörandl, E.; Bast, J.; Brandt, A.; Scheu, S.; Bleidorn, C.; Cordellier, M.; Nowrousian, M.; Begerow, D.; Sturm, A.; Verhoeven, K.; et al. Genome Evolution of Asexual Organisms and the Paradox of Sex in Eukaryotes. In *Evolutionary Biology: A transdisciplinary Approach*; Pontarotti, P., Ed.; Springer: Basel, Switzerland, 2020. [CrossRef]

156. De Carvalho, J.F.; de Jager, V.; van Gurp, T.P.; Wagemaker, N.; Verhoeven, K.J.F. Recent and dynamic transposable elements contribute to genomic divergence under asexuality. *BMC Genom.* **2016**, *17*, 884. [CrossRef]
157. De Carvalho, J.F.; Oplaat, C.; Pappas, N.; Derks, M.; de Ridder, D.; Verhoeven, K.J.F. Heritable gene expression differences between apomictic clone members in Taraxacum officinale: Insights into early stages of evolutionary divergence in asexual plants. *BMC Genom.* **2016**, *17*, 203. [CrossRef] [PubMed]
158. Schinkel, C.C.F.; Syngelaki, E.; Kirchheimer, B.; Dullinger, S.; Klatt, S.; Hörandl, E. Epigenetic patterns and geographical parthenogenesis in the alpine plant species *Ranunculus kuepferi* (Ranunculaceae). *Int. J. Mol. Sci.* **2020**, *21*, 3318. [CrossRef] [PubMed]
159. Syngelaki, E.; Daubert, M.; Klatt, S.; Hörandl, E. Phenotypic responses, reproduction mode and epigenetic patterns under temperature treatments in the alpine plant species *Ranunculus kuepferi* (Ranunculaceae). *Biology* **2020**, *9*, 315. [CrossRef]
160. Syngelaki, E.; Schinkel, C.C.F.; Klatt, S.; Hörandl, E. Effects of temperature treatments on cytosine-methylation profiles of diploid and autotetraploid plants of the alpine species *Ranunculus kuepferi* (Ranunculaceae) *Front . Plant Sci.* **2020**, *11*, 435. [CrossRef]
161. Syngelaki, E.; Paetzold, C.; Horandl, E. Gene Expression Profiles Suggest a Better Cold Acclimation of Polyploids in the Alpine Species *Ranunculus kuepferi* (Ranunculaceae). *Genes* **2021**, *12*, 1818. [CrossRef] [PubMed]
162. Verhoeven, K.J.F.; Jansen, J.J.; van Dijk, P.J.; Biere, A. Stress-induced DNA methylation changes and their heritability in asexual dandelions. *New Phytol.* **2010**, *185*, 1108–1118. [CrossRef]
163. Preite, V.; Snoek, L.B.; Oplaat, C.; Biere, A.; van der Putten, W.H.; Verhoeven, K.J.F. The epigenetic footprint of poleward range-expanding plants in apomictic dandelions. *Mol. Ecol.* **2015**, *24*, 4406–4418. [CrossRef]
164. Preite, V.; Oplaat, C.; Biere, A.; Kirschner, J.; van der Putten, W.H.; Verhoeven, K.J.F. Increased transgenerational epigenetic variation, but not predictable epigenetic variants, after environmental exposure in two apomictic dandelion lineages. *Ecol. Evol.* **2018**, *8*, 3047–3059. [CrossRef] [PubMed]
165. Wilschut, R.A.; Oplaat, C.; Snoek, L.B.; Kirschner, J.; Verhoeven, K.J.F. Natural epigenetic variation contributes to heritable flowering divergence in a widespread asexual dandelion lineage. *Mol. Ecol.* **2016**, *25*, 1759–1768. [CrossRef] [PubMed]

plants

MDPI

Article

Insights into the Taxonomically Challenging Hexaploid Alpine Shrub Willows of *Salix* Sections *Phylicifoliae* and *Nigricantes* (Salicaceae)

Natascha D. Wagner *, Pia Marinček, Loïc Pittet and Elvira Hörandl

Department of Systematics, Biodiversity and Evolution of Plants (with Herbarium), University of Goettingen, Untere Karspüle 2, D-37073 Göttingen, Germany
* Correspondence: natascha.wagner@uni-goettingen.de

Abstract: The complex genomic composition of allopolyploid plants leads to morphologically diverse species. The traditional taxonomical treatment of the medium-sized, hexaploid shrub willows distributed in the Alps is difficult based on their variable morphological characters. In this study, RAD sequencing data, infrared-spectroscopy, and morphometric data are used to analyze the phylogenetic relationships of the hexaploid species of the sections *Nigricantes* and *Phylicifoliae* in a phylogenetic framework of 45 Eurasian *Salix* species. Both sections comprise local endemics as well as widespread species. Based on the molecular data, the described morphological species appeared as monophyletic lineages (except for *S. phylicifolia* s.str. and *S. bicolor*, which are intermingled). Both sections *Phylicifoliae* and *Nigricantes* are polyphyletic. Infrared-spectroscopy mostly confirmed the differentiation of hexaploid alpine species. The morphometric data confirmed the molecular results and supported the inclusion of *S. bicolor* into *S. phylicifolia* s.l., whereas the alpine endemic *S. hegetschweileri* is distinct and closely related to species of the section *Nigricantes*. The genomic structure and co-ancestry analyses of the hexaploid species revealed a geographical pattern for widespread *S. myrsinifolia*, separating the Scandinavian from the alpine populations. The newly described *S. kaptarae* is tetraploid and is grouped within *S. cinerea*. Our data reveal that both sections *Phylicifoliae* and *Nigricantes* need to be redefined.

Keywords: biogeography; genomics; morphometrics; polyploidy; taxonomy

Citation: Wagner, N.D.; Marinček, P.; Pittet, L.; Hörandl, E. Insights into the Taxonomically Challenging Hexaploid Alpine Shrub Willows of *Salix* Sections *Phylicifoliae* and *Nigricantes* (Salicaceae). *Plants* **2023**, *12*, 1144. https://doi.org/10.3390/plants12051144

Academic Editor: Ulrike Lohwasser

Received: 8 February 2023
Revised: 22 February 2023
Accepted: 23 February 2023
Published: 2 March 2023

1. Introduction

Polyploidy occurs across the tree of life [1–3]. The presence of multiple gene copies in polyploids allows for gene neo- and subfunctionalizations, epigenetic changes, and consequently a differential expression of homeologous genes [4,5]. Additionally, polyploidy provides larger physiological and phenotypic flexibility to respond to different environmental conditions [3,6–8], which facilitates the colonization of various ecosystems [9–11].

Especially allopolyploidization (formed by hybridization between different species/lineages followed by genome duplication) is considered particularly likely to create biotypes with novel genomic features [2,12,13]. Polyploidy creates an immediate reproductive barrier against the diploid progenitors [12,14] and sexual polyploids can form distinct evolutionary lineages and be recognized as species [15]. However, the gene flow between polyploid lineages is possible especially on the same ploidy level (e.g., [16]). Thus, the evolutionary origins of polyploid clades can be complex and involve multiple origins, more than two progenitors, but also hybridization between progenitors before polyploidization (e.g., [17]). For genera with polyploid clades, the subgeneric classification has been notoriously difficult, because of the reticulate, non-hierarchical relationships of groups (e.g., [18,19]).

The genus *Salix* L. (Salicaceae) comprises about 450 species of dioecious trees and shrubs mainly occurring in the Northern Hemisphere [18,20]. About 40% of species are polyploids [21]. Willows are important elements of various kinds of natural wetlands, riparian vegetation, and arctic–alpine tundras, and are involved in many biotic interactions (e.g., [22–24]). The *Chamaetia*/*Vetrix* clade comprises about three quarters of the described species diversity in the genus *Salix*, containing more than 350 species classified in about 40 sections. Some species of this clade are adapted to cold, hostile environments of the arctic–alpine zone and show a decumbent or creeping growth ("dwarf shrubs"), while others are medium-sized shrubs and small trees. The European Alps harbor 33 willow species, of which 28 belong to the *Chamaetia*/*Vetrix* clade (so-called "shrub willows" [25]), showing that willows are an important part of alpine plant diversity. However, the determination of alpine shrub willows is difficult because of dioecy, simple, reduced flowers, common natural formation of hybrids, and large intraspecific phenotypic variation [20,24,26]. Especially the latter point led to the description of many species, and the overall taxonomy of *Salix* is still far from being resolved [27]. The presence of polyploidy further increases these difficulties. Especially the described hexaploid species tend to be morphologically diverse [24,25], although taxonomically accepted [20,24,28]. However, the phylogenetic relationships of the medium-sized, hexaploid shrub willows of the sections *Nigricantes* and *Phylicifoliae* have never been tested using a comprehensive sampling.

Recent molecular studies based on RAD sequencing data were able to produce robust phylogenetic trees, resolving the relationships of diploid and tetraploid Eurasian shrub willows [29,30]. The included alpine species did form monophyletic lineages in these molecular studies, which show that SNP-based methods based on RAD loci were suitable to analyze polyploid *Salix* species [30]. In this paper, we apply this technique to conduct molecular analyses on the two sections containing hexaploids. However, there are limitations to such methods, especially regarding the simplifications of SNP calling in a consensus sequence that might blur some of the results for the higher polyploids. Therefore, we combine the data with non-molecular approaches. Infrared spectroscopy (sometimes also called "NIRS" (near-infrared spectroscopy)) of leaves is emerging as a promising tool to analyze plant phylogenetic diversity as well as certain chemical compounds [31–33]. This method uses the structural and chemical traits of leaves assessed by specific reflectance patterns to differentiate groups at the species or population level with a high accuracy. Leaf spectroscopy is non-destructive, fast, cheap, and reliable [34], and thus can be easily applied to dried leaf material (herbarium material). Additionally, it can be applied to any ploidy level.

Previous taxonomic treatments of species in willows have been exclusively based on descriptive morphology. Morphological characters, especially of vegetative parts, are highly variable in willows due to high phenotypic plasticity and gradual changes during development in the course of one season [20,35]. Therefore, even leaves of the same individual exhibit vast variation of diagnostic characters. Morphological studies resulted in several different taxonomic circumscriptions, specifically in the species complex of the European *Phylicifoliae* group sensu Rechinger [28] (including *S. phylicifolia* L., *S. bicolor* Ehrh. ex Willd., and *S. hegetschweileri* Heer) by different authors (Table 1). The three taxa were diagnosed only by a few leaf characters (shape, margin, stipules, and indumentum). To shed light on questionable species circumscriptions, comparative morphometric studies can help to quantify objectively the similarity/dissimilarity of taxa and estimate the taxonomic value of diagnostic traits [36]. Therefore, we apply a morphological analysis on the target species in addition to molecular data and infrared spectroscopy.

In this paper, we use a comprehensive dataset to test the taxonomic classification and to analyze the phylogenetic relationships of several taxonomically challenging polyploid species of the section *Nigricantes* sensu Skvortsov [20] and section *Phylicifoliae* sensu Rechinger [28]. The first comprises *S. myrsinifolia* (6x), widespread in Eurasia, *S. mielichhoferi* (6x), endemic to the Eastern Alps, *S. apennina* (6x) from the Apennines, *S. cantabrica* (4x) from Spain, and the recently described *S. kaptarae* endemic to Crete (Figure 1). For section *Phylicifoliae* (in Eurasia subsect. *Bicolores* in [20] and *S. phylicifolia* group sensu Rechinger [28]; see Table 1), we consider widespread *S. phylicifolia* L. (6x) and *S. bicolor* Ehrh. ex Willd. (3x, 6x) with a scattered distribution in Europe as well as the tetra- to hexaploid *S. hegetschweileri* Heer em. Buser, an alpine endemic species (Figure 1). The two sections are morphologically similar, and previous authors hypothesized close relationships between the groups [18,20,28]. We include additional species of both sections and combine the molecular data with the already published RAD sequencing data for European *Salix* species to analyze the position of the species in a phylogenetic framework. Subsequently, we apply genetic structure and co-ancestry analyses to obtain insights into the genomic composition of the hexaploids. In addition, we test whether leaf reflectance spectra bear phylogenetic signals within the hexaploid species and provide further information on taxonomic classification. Finally, we test species delimitation within the *Phylicifoliae* group with a morphometric analysis. By including specimens from the type locations of *S. bicolor* and *S. hegetschweileri* into both molecular, infrared spectroscopic and morphometric analyses (see Materials and Methods, Supplementary Table S1), we can fix the nomenclature for final taxonomic conclusions.

Table 1. Taxonomic treatments of polyploid willows of sect. *Nigricantes* and sect. *Phylicifoliae* by different authors and our results. See the Materials and Methods Section for more details.

Skvortsov (1999) [20]		**Rechinger (1964) [28]; Aeschimann et al. (2004) [37]**		**Results of This Study**	**Distribution**
Subsect. Bicolores		***S. phylicifolia* Group**		***Sect. Phylicifoliae***	
S. phylicifolia	*subsp. phylicifolia*	*S. phylicifolia*	N. Europe, Russia	*S. phylicifolia*	N. Eurasia
S. phylicifolia	*subsp. rhaetica (incl. S. bicolor)*	*S. hegetschweileri*	Alps	*(incl. S. bicolor, excl. S. hegetschweileri)*	and Central European Mts.
S. basaltica (? incl. S. cantabrica)		*S. bicolor* (incl. *S. basaltica*)	European mts.	not analyzed	Not analyzed Pyrenees, Massif Central
Sect. Nigricantes		***S. myrsinifolia group***		***Sect. Nigricantes***	
				S. hegetschweileri	Alps (endemic)
S. myrsinifolia	*subsp. myrsinifolia*	*S. myrsinifolia*	Eurasia	*S. myrsinifolia s.l.*	Eurasia
S. myrsinifolia	*subsp. borealis*	*S. borealis*	N. Europe	*S. myrsinifolia s.l.*	Eurasia
S. mielichhoferi		*S. mielichhoferi*	Alps	*S. mielichhoferi*	Alps (endemic)
S. apennina		*S. apennina*	Apennines, S. Alps	*S. apennina*	Apennines, S. Alps
				Sect. Vetrix	
-		*S. cantabrica*	Iberian Peninsula	*S. cantabrica*	Iberian Peninsula (endemic)
-				*S. kaptarae* (*syn. of S. cinerea?*)	Crete

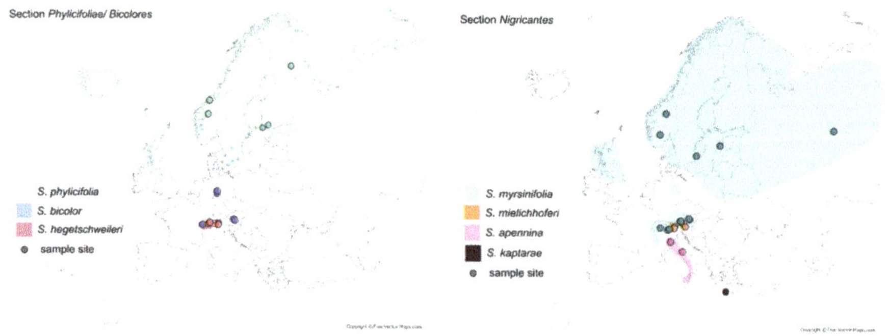

Figure 1. Distribution maps of the included hexaploid species of the sections *Nigricantes* and *Phylicifoliae*. Sample sites of the included accessions indicated as circles. Color coding according to legend within each map.

2. Results

2.1. Phylogenetic Analyses and Genetic Structure Based on RAD Sequencing Data

RAD sequencing yielded an average of 7.26 Mio reads per sample. We used up to five accessions per species for the analyses of the backbone topology to reveal the position of members of the sections of interest in the phylogeny without obscuring the results too much by mixing several ploidy levels. The ipyrad pipeline of 145 samples representing 45 species revealed 46,568 RAD sequencing loci containing 630,764 SNPs. The concatenated sequence alignment comprised 66.95% of the missing data. The results of the RAxML analyses are presented in Figure 2, and the same dataset including QS scores is shown in Supplementary Figure S1. The species *S. triandra* was used as the outgroup to root the tree. Three major clades could be observed, while *S. reticulata* was in a sister position to all remaining samples. The hexaploid species *S. myrsinifolia, S. mielichhoferi, S. apennina*, and *S. hegetschweileri* form a clade that showed slightly discordant and skewed QS values and medium-supported BS support. It occurred in a sister position to a clade comprising the species *S. caprea, S. aurita, S. appendiculata, S. cinerea, S. atrocinerea, S. laggeri*, and *S. salviifolia*. *Salix cantabrica*, taxonomically assigned to the section *Phylicifoliae*, appeared in a sister position to the just mentioned clade. *Salix kaptarae*, a recently described species from Crete and assigned to the section *Nigricantes*, occurred within the widespread lowland species *S. cinerea*.

The Fennoscandian hexaploid species *S. phylicifolia* and triploid/hexaploid *S. bicolor* did not form clades according to their species identity but appeared intermingled within a paraphyletic group. Each subclade showed only low QS and BS support. However, accessions of both species were situated in a sister position to a clade containing two subclades, one including *S. lapponum, S. helvetica, S. foetida*, and *S. waldsteiniana* and a subclade containing the species *S. viminalis, S. schwerinii*, and *S. rehderiana* (Figures 2 and S1). The remaining topology is in accordance with previously published data on the *Salix* subg. *Chamaetia/Vetrix* clade [25,30].

Subsequently, a RAxML analysis, including all collected samples of the analyzed sections as well as a reduced backbone sampling, was conducted to analyze the tree topology of both sections in more detail. The results are shown in Supplementary Figure S2. The samples of *S. myrsinifolia, S. hegetschweileri, S. apennina*, and *S. mielichhoferi* form one clade. Within this clade, *S. myrsinifolia* was separated into two subclades: One containing samples from Scandinavia as well as Russian *S.* cf. *jenisseensis*. The other clade contained samples from the Alps. Samples of *S. mielichhoferi* formed a monophyletic group with a strong support. The accessions of *S. apennina* were grouped in a sister position to *S. mielichhoferi*. *Salix hegetschweileri* is not monophyletic but forms two clades as a paraphyletic group in a basal position to the remaining species. The accessions of *S. bicolor* also form two clades, one clade including samples from Switzerland and the other samples from Germany and Austria in a sister position to samples of *S. phylicifolia*.

The results of the co-ancestry analysis of the species belonging to the section *Nigricantes* and *Phylicifoliae* are shown in Figure 3. The analysis was based on 43 samples and 49,754 RAD loci. *Salix phylicifolia* s.l., including *S. phylicifolia* and *S. bicolor*, was clearly separated from the remaining species. *Salix hegetschweileri* shared co-ancestry with species of the section *Nigricantes*, however, the analysis also revealed some co-ancestry with the *Phylicifoliae* group indicated by the orange color. *Salix myrsinifolia* was divided into two groups. One includes all samples from the European Alps. The other group comprises samples originating in Fennoscandia and Estonia as well as *S.* cf. *jenisseensis* from Russia. The latter clade showed some shared ancestry with Nordic *S. phylicifolia*.

To analyze the two sections containing the alpine hexaploid species in more detail, we conducted a separate analysis for the two observed major clades of each section *Nigricantes* plus *S. hegetschweileri*, and *Phylicifoliae* (Supplementary Figure S3). For the section *Nigricantes*, 29 samples of *S. mielichhoferi, S. myrsinifolia, S. apennina*, and *S. hegetschweileri* were included. The ipyrad pipeline obtained 8,164 RAD loci showing 90,756 SNPs and 15.7% missing data on a restrictive filtering. For twelve samples of *S. phylicifolia* and *S. bicolor*

(*Phylicifoliae*), 27,010 RAD loci (149,614 SNPs, 6.51% missing data) were observed with the ipyrad pipeline. The results of the co-ancestry analysis with fineRADstructure for the sections *Nigricantes* and *Phylicifoliae* are illustrated in Supplementary Figure S3. The co-ancestry analysis based on 8149 shared loci revealed separate clusters for each species included. Overall, however, the red–orange colors showed moderate levels of co-ancestry within all hexaploids included in this paper. Shared loci were observed for *S. mielichhoferi* and *S. myrsinifolia*. Additionally, shared loci were present for *S. hegetschweileri* with *S. myrsinifolia*. For the subclade of the section *Phylicifoliae*, the samples from Harz Mountain were separated and showed very low levels of co-ancestry (yellow color). Triploid samples from Austria showed only low levels of co-ancestry with the remaining samples, while hexaploid samples of *S. phylicifolia* and *S. bicolor* from Switzerland showed a high number of shared loci indicated by orange and red colors.

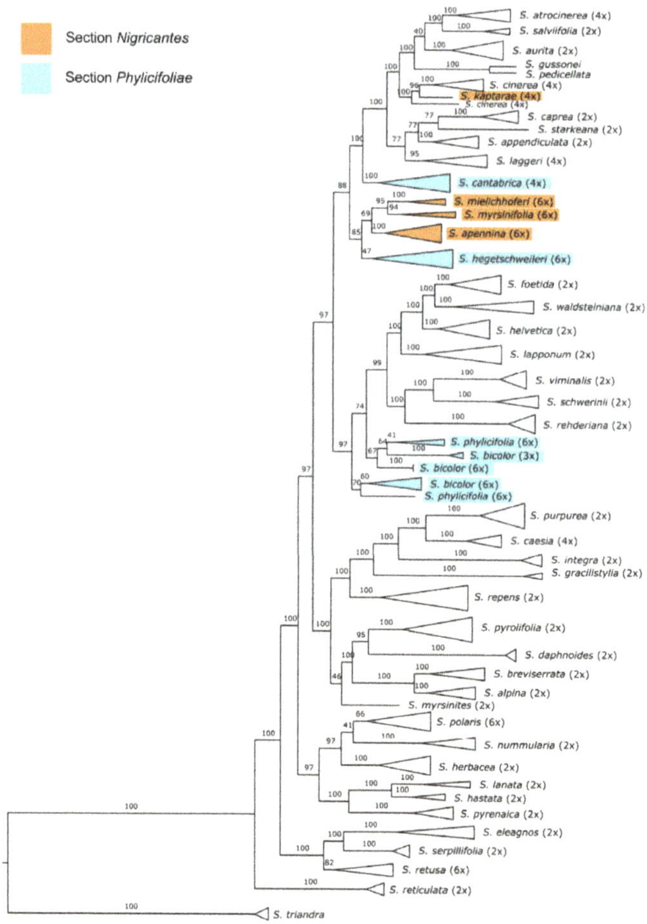

Figure 2. Backbone topology of the *Chamaetia*/*Vetrix* clade including 145 samples representing 45 Eurasian species plus *S. triandra* as outgroup based on a RAxML analyses on 46,568 RAD loci (630,764 SNPs) in a concatenated alignment of 5,341,180 bp and 66.95% missing sites. Bootstrap support values above branches. Ploidy level of species behind the species names. Species of the sections *Nigricantes* and *Phylicifoliae* are displayed in colors according to the legend. (Additional quartet sampling scores are displayed in Supplementary Figure S1).

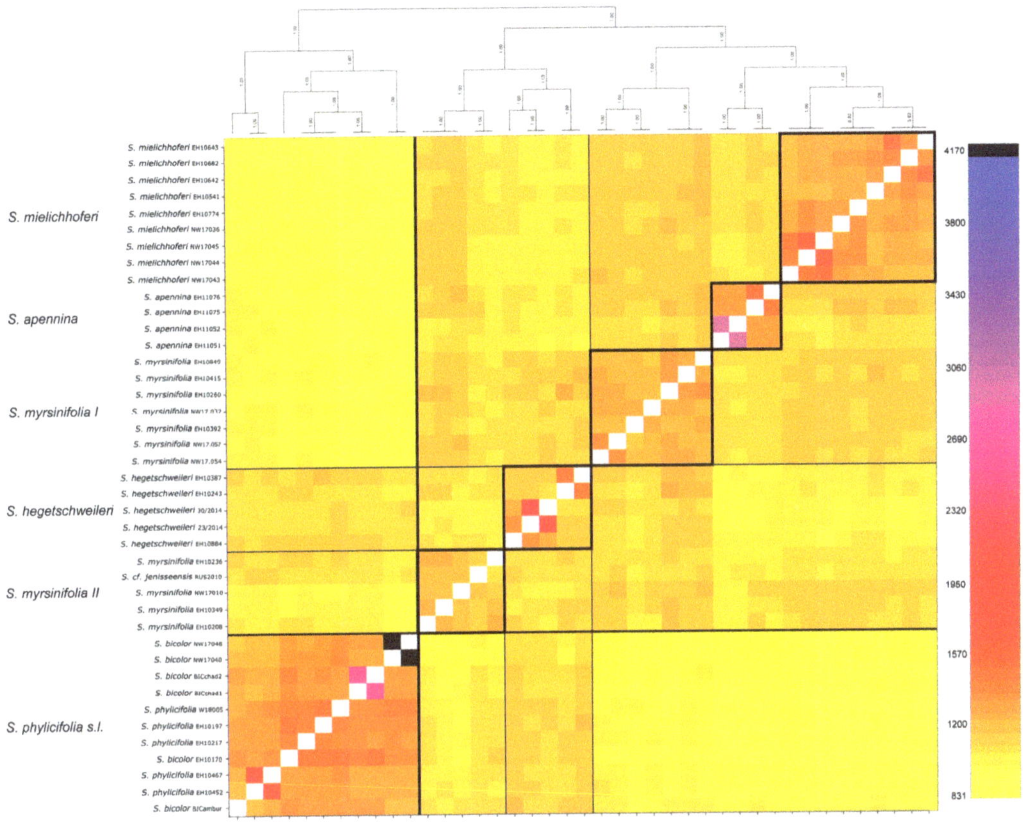

Figure 3. Co-ancestry analysis with fineRADstructure of 43 samples belonging to the sections *Nigricantes* and *Phylicifoliae* s.l. based on 49,754 RAD loci shared by at least 10 individuals. The amount of co-ancestry is illustrated by the color shade (legend left side). The sections and species are clearly separated. *Salix myrsinifolia* appears in two groups, one including samples from the European Alps (*S. myrsinifolia I*) and the other group includes samples from Fennoscandia and *S.* cf. *jenisseensis* from Russia (*S. myrsinifolia II*). *Salix bicolor* is included in *S. phylicifolia*. Detailed clade-wise results are presented in Supplementary Figure S3.

The results of the genetic structure analyses with sNMF for all samples and both clades observed in the Maximum Likelihood tree are displayed in Supplementary Figure S4.

The overall analysis revealed some admixture in *S. hegetschweileri* and *S. jenisseensis c.f.* for k = 2, while the other individuals share genetic partitions according to their sectional classification (see Supplementary Figure S4a). The overall as well as the clade-wise results indicate that the individuals of *S. bicolor* and *S. phylicifolia* do not form two distinct groups according to their taxonomic species identity. Instead, two samples of *S. bicolor* from Harz Mountain share the same structure (light blue), while all remaining accessions show a different genomic composition (dark blue) independent of species determination (k = 2). A population-wise structure is present for k = 6 (Supplementary Figure S4b).

For the *Nigricantes* clade, the sNMF results for k = 6 are presented in Supplementary Figure S4c. *Salix mielichhoferi* and the majority of samples of *S. myrsinifolia* share a genetic cluster (yellow), while *S. myrsinifolia* shows a certain amount of admixture. Two samples of *S. mielichhoferi* show a distinct cluster (light blue). Samples of *S. myrsinifolia* originating from Fennoscandia as well as *S.* cf. *jenisseensis* form a distinct cluster (orange). *Salix apennina* and *S. hegetschweileri* share a genetic cluster (dark blue).

2.2. Infrared Spectroscopy Data

The results of the infrared spectroscopy data of all hexaploid species are shown in Figure 4. *Salix phylicifolia* and *S. bicolor* are distinct from *S. myrsinifolia*, which showed a variable spectrum, and *S. mielichhoferi*. The two included samples of *S. hegetschweileri* (after filtering) were bridging *S. myrsinifolia* and *S. phylicifolia*. The prediction table showed high prediction values for *S. mielichhoferi* and for *S. bicolor* and *S. phylicifolia* when combining the latter two into *S. phylicifolia* s.l.. Most *S. myrsinifolia* samples were assigned to *S. mielichhoferi*. The two measurements of *S. hegetschweileri* were assigned to *S. phylicifolia* s.l. The more detailed results of the infrared spectroscopy of leaf material for the section *Nigricantes* (*S. mielichhoferi*, *S. myrsinifolia*, and *S.* cf. *jenessiensis*) as well as for the section *Phylicifoliae* (*S. bicolor*, *S. hegetschweileri*, and *S. phylicifolia*) are illustrated in Supplementary Figure S5. The results shown as PCA and density plots show discrimination between the species *S. mielichhoferi* and *S. myrsinifolia* based on their filtered spectra. However, the measurement of *S.* cf. *jenessiensis* falls into *S. myrsinifolia*. For the section *Phylicifoliae*, *S. bicolor* and *S. phylicifolia* are in close proximity based on the PCA. The first coordinate explains 54.1% of the variation in the dataset. When including information on ploidy (3n and 6n), the results show a similar, but slightly more distinct pattern.

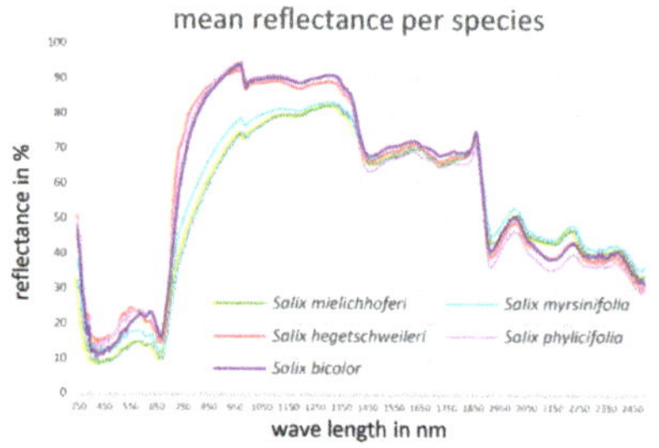

(a)

prediction table (PLS-DA)

Salix species	hegetsch.	mielich.	myrsinif.	phylicif. s.l.
hegetschweileri	0	0	0	1.00
mielichhoferi	0	0.90	0.10	0
myrsinifolia	0	0.67	0.17	0.17
phylicifolia s.l.	0	0.05	0	0.95

(b)

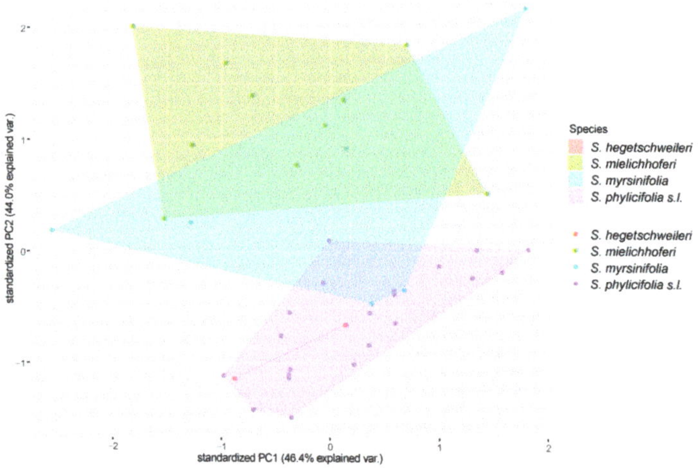

(c)

Figure 4. Infrared spectroscopy data. (**a**) Mean reflectance per species; species color-coded according to legend. (**b**) PLS-DA prediction table on spectral data showing relative numbers (percentage) of prediction. (**c**) PCA of spectral data analyses (convex hulls) of hexaploid species. The individual samples are color-coded according to the legend on the right. The displayed outcomes combine *S. bicolor* and *S. phylicifolia*.

2.3. Ploidy Level Determinations for S. bicolor and S. kaptarae

The FC results (Supplementary Table S2) show a hexaploid genome size for *S. bicolor* samples from Switzerland as well as for the two samples from the Harz Mountains, Germany, which represent the only remaining population from the locus classicus. *Salix kaptarae* showed a tetraploid genome size.

2.4. Morphometric Data of S. phylicifolia, S. bicolor and S. hegetschweileri

The characters of leaf length, leaf width, angle of the blade base, and length of stipules showed significant differences ($p > 0.05$) between *S. hegetschweileri* and both *S. bicolor*/*S. phylicifolia*, whereas the two latter species did not differ from each other (Figure 5). *Salix hegetschweileri* has longer and wider leaves, a broader angle of the blade base, and more pronounced stipules compared to the other taxa. Length/width ratios of leaves, however, did not show any differences between taxa (Figure 5c). Length of teeth differed in all three taxa among each other, with *S. bicolor* having entire margins or very short teeth, followed by tiny teeth in *S. phylicifolia* and pronounced teeth in *S. hegetschweileri* (Figure 5f).

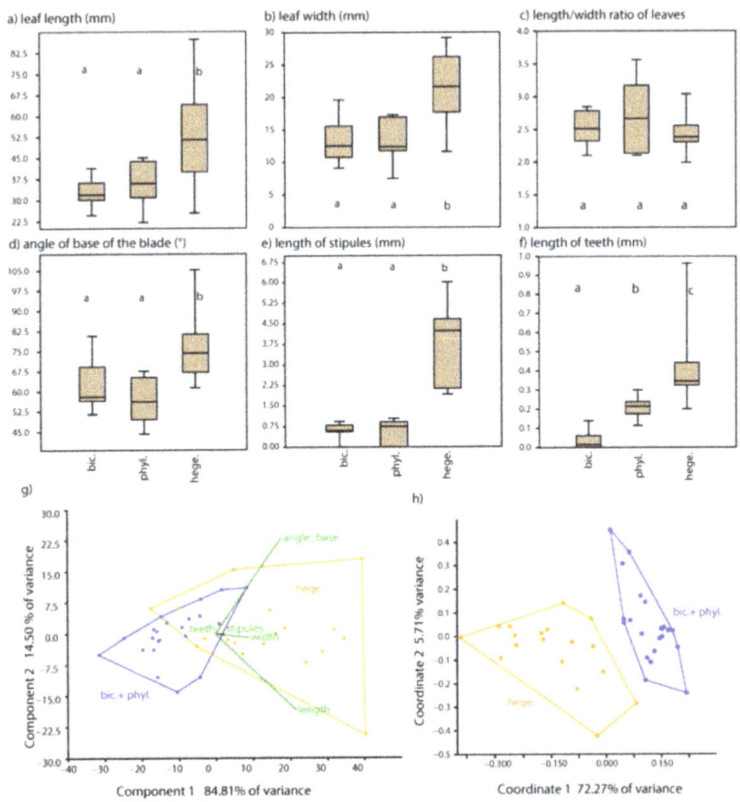

Figure 5. Morphometric analyses. (**a–f**) Boxplots of six characters plotted for *S. bicolor* (n = 15, bic.), *S. phylicifolia* (n = 7, phyl.), and *S. hegetschweileri* (n = 18, hege.). Significant pairwise differences according to Mann–Whitney tests are shown with different letters ((**a**,**b**); see Supplementary Table S3 for all *p*-values). Boxes show the median and the quartile length, and whiskers the range. (**g**,**h**) Ordinations using the characters above except for (**c**); symbols and convex hulls according to the best grouping in discriminant analyses (97.5% correctly classified with two groups: *S. bicolor* and *S. phylicifolia* in blue, and *S. hegetschweileri* in orange). (**g**) PCA and (**h**) PCoA of five morphological traits.

PCA using five characters and plotted with convex hulls for the three taxa suggested strong overlaps of *S. bicolor* and S. *phylicifolia* (Supplementary Figure S6), whereas PCA with convex hulls for only two taxa according to genetic data (*bicolor* + *phylicifolia* and *hegetschweileri*) revealed only a slight overlap of the two clusters (Figure 5g). Leaf length and angle of the base have the strongest correlations (highest loading values) with the first two axes. PCoA with convex hulls for two taxa (*bicolor* + *phylicifolia and hegetschweileri*) revealed two clearly distinct clusters in the scatter plot (Figure 5h), whereas *S. phylicifolia* and *S. bicolor* visualized separately did not show a dissimilarity gap. The discriminant analysis with three predefined taxa revealed a confusion matrix with 92.5% of samples correctly classified, whereas with two predefined groups (*S. bicolor* + *phylicifolia* and *S. hegetschweileri* as in Figure 5g,h) revealed 97.5% of samples correctly classified.

3. Discussion

The complexity of the allopolyploid genome, especially for higher polyploids as studied in this paper, is a source of variable phenotypic and physiological traits [3,7]. However, although hexaploid *Salix* species are morphologically variable and thus taxonomically challenging, the reduced representation data of genomes analyzed with bioinformatic tools were able to resolve the relationships of the hexaploid species analyzed in this paper. Our study showed that spectral data can be a useful supplement to analyze phylogenetic relationships, which is in line with other studies [31–33]. In combination with molecular and morphometric data, we were able to answer questions on the taxonomic treatment of alpine hexaploid *Salix* species. Despite the mosaic-like genomic composition, our results show clear species delimitation for all species, except *S. phylicifolia/bicolor*. This might be due to successful sexual reproduction and the establishment of the polyploid lineages as observed in tetraploid willows [30]. Nevertheless, our data indicate introgression and geneflow between the hexaploid genomes (e.g., Figure 3). In the following, we discuss our methodical approach, the evolutionary history, and the impact of our results on the taxonomic treatment of the two sections *Nigricantes* and *Phylicifoliae*.

3.1. Comparison of Datasets

In this study, we used RAD loci and SNPs to analyze the phylogenomic relationships in hexaploid species. RAD sequencing is a reduced representation method and, because we used a de novo assembly, we generated short anonymous loci. However, the advantage of this technique is the considerable number of generated SNPs that could be successfully used for species delimitation in hexaploids. The reduction in the complexity of up to six alleles into one single (ambiguous) consensus sequence allowed us to combine samples with different ploidy levels. To find the best balance between missing data and number of informative SNPs, we tested different parameter sets. The high amount of missing data (~70%) in the final dataset used for the backbone topology did not affect the phylogenetic reconstruction. This is in accordance with [38]. However, to display (allo)polyploid species in a bifurcating tree is not the best way to address a network-like evolutionary pattern [2]. Thus, we implemented genetic structure analyses with sNMF and co-ancestry analyses with fineRADstructure. Both tools can handle different ploidy levels [39,40]. The close relationships of species within subclades allowed for more conserved settings with less locus dropout and, thus, less missing data in the final datasets. Our study is one of the few available studies applying these two tools to hexaploid non-model plant species. We showed that these population genomic tools are suitable for (polyploid) species complexes and/or closely related species to reveal the genomic structure and co-ancestry, which is in accordance with other studies [41–43]. However, we realized that genetically (almost) identical individuals (=clones) from the same populations blurred the results of both tools and tended to form separate clusters (see, e.g., *S. bicolor* from Harz Mountain or *S. mielichhoferi* from Austria). In these cases, the signal is stronger than the co-ancestry with ally species. Similar patterns were observed in a study on an alpine hybrid zone of two willow species [44]. We, therefore, recommend excluding clonal or highly similar individuals from this sort of analyses.

Subsequent to the molecular dataset, we applied infrared spectroscopy to the target species, which is not sensitive to ploidy. This technique (also called "near-infrared spectroscopy (NIRS)") became a popular tool in forestry in the 2000s, where leaf spectra were used for tree species discrimination and for (the prediction of) foliage chemical compounds in the field (reviewed in [34]). The main reasons for the frequent use of spectroscopy are its low cost as well as the non-destructive, fast, and reliable measurements. The utility of spectral data for taxonomic purposes has been demonstrated in several plant species [45,46]. Stasinski et al. [31] used spectral data in combination with molecular data (genotyping-by-sequencing, GBS) for a fine-scale diversity analysis in two hybridizing arctic shrub species of genus *Dryas*. The authors demonstrated that reflectance spectroscopy captured genetic information that can be used to accurately classify leaves of species, hybrids, and populations in a taxonomically challenged group. Our data showed good differentiation between the two analyzed species groups (Figure 4) but only low differentiation within species. However, our dataset comprised much fewer measurements per species than in the studies mentioned above and we experienced a bias in the prediction results towards the species with a higher number of measurements (e.g., *S. myrsinifolia* samples were predicted as *S. mielichhoferi*, and *S. hegetschweileri* as *S. phylicifolia*, while we measured in both cases more samples for the latter one). Additional samples might stabilize the prediction in the PLS-DA. The combination of spectral data and morphological data in the genus *Myrcia* (Myrtaceae) by Gaem et al. [47] demonstrated that the multidimensional natures of the entire spectra were very efficient in assigning individuals to species categories. Our results support the ability of this technique to discriminate different *Salix* species within one section. Even better, we are able to discriminate different species groups and, thus, our results are in line with the above-mentioned studies. Additionally, we observed a considerable within-species variability in *S. myrsinifolia* and *S. bicolor*. However, we did not consider collection time (leaf age) and collection site (ecological factors), which might affect the chemical composition of the leaves of these highly variable plant species. The leaves of *S. myrsinifolia* and *S. mielichhoferi* turn black when dried and this might further contribute to the observed diversity of measurements (see Figure 4a). Pigments absorb light in the visible region (400–700 nm), whereas light in the near infrared region (700–1100 nm) is scattered by leaf anatomical, tissue, water, and surface features, and light in the short-wave infrared region (1400–2500 nm) is scattered and absorbed by anatomical features and biochemicals, such as cellulose, phenolics, and water [31]. Thus, leaf spectra are complex datasets influenced by both environmental and genetic factors. Overall, considering the fast and easy use of this technique, it is a valuable supplement to molecular and morphological datasets.

Finally, we addressed the delimitation of species of the previous section *Phylicifoliae* with a morphological analysis, which confirmed *S. hegetschweileri* being distinct from *S. phylicifolia*/*S. bicolor*. At the species and individual levels, we can confirm a high variability of characters that is typical for willows [20,48]. A certain overlap of phenotypic variation of genetically clearly distinct species was also observed in other molecular–morphometric studies on willows [49,50]. In addition to gradual changes during seasonal development, phenotypic plasticity depending on habitat conditions causes large individual variation. For instance, leaf size usually decreases on dry soils, which applies to our dataset, specifically to *S. hegetschweileri* individuals found on the rather dry glacier moraines of the Gletschboden population. Only these small-leaved individuals caused the overlap of variation with the *S. bicolor*/*S. phylicifolia* cluster in the displayed PCA (Figure 5g). However, morphometric data in combination with molecular data are powerful to separate closely related taxa, as it has been demonstrated in many other polyploid complexes [51–53].

3.2. Evolutionary History and Biogeography

Previous studies on polyploid willows suggested allopolyploid origins for most polyploid species [30]. Skvortsov [20] already emphasized that willow species are characterized by distinct eco-geographical distributions. In the *Nigricantes* clade, it is plausible that all polyploids speciated via geographical and ecological isolation. *Salix apennina*, *S. hegetschweileri*, and *S. mielichhoferi* have allopatric distributions. Only in the Alps, *S. myrsinifolia* is sympatric with *S. hegetschweileri* in the western Central Alps and with *S. mielichhoferi* in the Eastern Central Alps, but in both areas, the two latter species occur in subalpine shrubberies, whereas *S. myrsinifolia* occurs mostly in the montane forest zone in wet meadows and along rivulets [37,54]. The alpine endemic species *S. mielichhoferi* seems to have evolved as a separate evolutionary lineage out of the Central European populations of widespread *S. myrsinifolia* via ecological speciation. It could have originated after the last glacial maximum, since the current distribution area of *S. mielichhoferi* was mostly covered by ice at the last glacial maximum (LGM), except for the southernmost parts in Northern Italy [37,55]. Alternatively, the species predated the LGM and survived in the southern peripheral calcicolous refugia between the Lake Como and the Dolomites [55]. Both species share a certain amount of their genomes (Figure 3 and Figure S4) but are clearly separated into two lineages (Figure 2 and Figure S3). However, the shared genomic compartments might also be explained by recent gene flows, since they overlap in their elevational distribution in the Alps in the subalpine zone, where intermediate forms between both species were observed in the contact zone [54]. The hybrid origin of such individuals, however, needs to be confirmed by further studies.

The evolutionary origin of *S. hegetschweileri* has long been under dispute, also due to unclear delimitation against *S. bicolor* and *S. phylicifolia* and the notorious confusion of taxa (Table 1; reviewed in [54]). Based on our data, we considered *S. hegetschweileri* a distinct lineage and alpine endemic species (see Section 3.4). The survival of glacial maxima could have been possible in southwestern peripheral refugia, where many siliceous bedrock conditions are available [55]. This silicolous species is very successful as a colonizer of glacier forefields and could have been one of the first postglacial pioneers of the alpine flora after glacier retreat.

The evolution of the *S. phylicifolia–bicolor* alliance can be best explained by geographical isolation. During the LGM (or during previous cold periods), *S. phylicifolia* s.l. probably colonized the whole ice-free tundra zone between the big ice shields covering the Alps and Northern Europe. During the postglacial recolonization, *S. phylicifolia* migrated mainly northwards and left some isolated relic populations in the Jura and adjacent western Alps, in the Harz Mountains, the Riesengebirge (not included in this paper), and in the Eastern Alps in a refugial area on siliceous bedrock (see [55]). The disjunct clades of the Swiss and Austrian populations of "*S. bicolor*" might be due to geographic isolation during LGM. Postglacial reforestation hindered range expansion and gene flow between the isolated relic Central European willow populations. The population structure observed in the co-ancestry analyses could be explained by the disjunct distribution of these populations classified as "*S. bicolor*".

3.3. *Taxonomy of the Section Nigricantes*

Most taxonomically described species within this section can be confirmed by our results. The infrared spectroscopy data were able to distinguish *S. mielichhoferi* and *S. myrsinifolia*, and in the molecular datasets, species of this section were clearly discriminated. The widespread and morphologically diverse *S. myrsinifolia* is divided into two clades, separating the Scandinavian accessions from the alpine accessions (Figures 2 and 3). According to [20], the widespread *S. myrsinifolia* was divided into three subspecies. For Fennoscandia, *S. myrsinifolia* ssp. *borealis* was described. We did not infer subspecific treatment in this paper since our main interest was on the alpine hexaploids. However, the separate clade for Fennoscandian *S. myrsinifolia* could be due to the geographical separation of these samples. The observed low degree of co-ancestry (Figure 3) with Scandinavian *S. phylicifolia* could be due to recent introgression. Whether this group merits a taxonomic treatment as subspecies should be tested with an expanded sampling. *S. myrsinifolia* and *S. mielichhoferi* were accepted by all authors as distinct species. Morphologically, they are differentiated by the characteristics of leaves and twigs [24] and the separation was supported by our results. *Salix apennina* was nested within the Nigricantes clade, and all accessions form a monophyletic group. *Salix hegetschweileri*, which was previously regarded as part of Sect. *Phylicifolia* (see discussion below), is sister to the *S. apennina–myrsinifolia–mielichhoferi* clade.

Salix kaptarae, a recently described species from Crete [56], was described as closely related to *S. apennina* based on morphological characters and taxonomically assigned to section *Nigricantes*. However, our data did neither confirm the proximity to *S. apennina* as suggested by the authors, nor the assignment to the section *Nigricantes*. Instead, our data indicate the inclusion of this species into *S. cinerea* (Figure 2, and Supplementary Figures S1 and S2) within the section *Vetrix* sensu [20]. The morphological characters of *S. kaptarae* such as pubescent branches, leaf morphology, and pronounced striae on decorticated wood (pers. observation of N. Wagner), fit the assignment to this section. Additionally, the sample included is tetraploid based on our flow cytometry data (Supplementary Table S2), as it is also documented for *S. cinerea* ($2n = 76$; [57–59]). Although the flora of Crete has been well studied for decades, in the vascular plant checklist of Crete only *S. alba* was mentioned [60]. The population of *S. kaptarae* was mentioned in the year 2000 by Jahn in [61] and later by [62]. That leads to the assumption that the individuals of *S. kaptarae* (*S. cinerea*) were introduced to Crete either via long-distance dispersal or, more likely, by humans. However, we included only one sample from the type location in our molecular dataset. More studies will be needed to analyze the spatio-temporal evolution and taxonomic treatment of *S. kaptarae* in more detail.

The previously circumscribed Sect. *Nigricantes* appears to be polyphyletic in phylogenomic analyses and should be refined to the clade including *S. apennina*, *S. myrsinifolia*, *S. mielichhoferi*, and *S. hegetschweileri*. A shared character of this section that differs from the *Phylicifoliae* is the formation of pronounced stipules [20]. The position of putative *S. jenisseensis*, which was classified in Section *Glabrella* by [20], needs to be clarified with more samples.

3.4. Taxonomy of the Section Phylicifoliae

Traditionally, this section included *S. phylicifolia*, *S. bicolor*, *S. cantabrica*, and *S. hegetschweileri* along with other species not included in this paper [20]. Based on our results, this section is polyphyletic. The Iberian *S. cantabrica* was not the focus of our study on alpine hexaploids of both sections, so we included only two samples. However, the results suggest that this species is closely related to members of section *Vetrix* (Figure 2).

The morphological analyses of the section *Phylicifoliae* clearly separate *S. hegetschweileri* from *S. phylicifolia* and *S. bicolor* (Figure 5). Less clear are the results of the spectral data that show high similarity between *S. hegetschweileri* and *S. phylicifolia* (Figure 4 and Figure S5). However, only two samples were included, which is not enough to cover the species diversity. Skvortsov [20] regarded *S. hegetschweileri* a geographically separated, Central European subspecies of *S. phylicifolia* (as ssp. *rhaetica*). Lautenschlager [63] proposed that *S. hegetschweileri* is a recent hybrid between "*S. bicolor*" and *S. myrsinifolia*. He crossed *S. hegetschweileri* individuals from the locus classicus (Urserental, Switzerland), which he misidentified as *S. bicolor*, with *S. myrsinifolia* and obtained *S. hegetschweileri*-like offspring, which he then misinterpreted as "*bicolor*" x *myrsinifolia*-hybrids. Our molecular analyses clearly indicate that populations from Urserental, Gletschboden, and Tyrol have nothing to do with "*S. bicolor*" but are *S. hegetschweileri*. Our data do not support the hypotheses of a homoploid hybrid origin out of two hexaploid extant parents. Our phylogenetic data (Figure 2, Figure S2, and Figure 5) instead separate *S. hegetschweileri* from other members of this section. The alpine endemic is a distinct lineage and shows close relationships to the *Nigricantes* clade in phylogenetic analyses. Based on the molecular and morphological data, we assumed the post-origin local introgression of *S. hegetschweileri* with *S. myrsinifolia* in subalpine contact zones, which also sometimes blurs the morphological differences between these two distinct species (Supplementary Table S4). However, *S. hegetschweileri* does show a genetic contribution from *S. bicolor*/*phylicifolia* (Figure 3 and Figure S4). Thus, we cannot rule out ancient contributions from the ancestors of the *phylicifolia* lineage nor recent gene flows. Taken together, we support the taxonomical treatment following [28] and [24], treating *S. hegetschweileri* as a distinct species and alpine endemic.

The taxonomic treatment of *S. bicolor* has long been under discussion (Table 1). Our data suggest that "*S. bicolor*" represents disjunct relic populations of former widespread *S. phylicifolia*. The flow cytometry data showed a hexaploid genome size for the populations of the locus classicus from the Harz Mountains as well as from the Jura and adjacent Swiss Alps (Supplementary Table S2). The triploid population from Austria seems to be a single isolated clone of female plants [54] that either evolved from a hexaploid parent via haploid parthenogenesis (meiosis complete, but embryogenesis without fertilization of reduced egg cells), which can happen occasionally in many angiosperms (e.g., [64]). Alternatively, the triploids represent a population that originated from a diploid parent, underwent polyploidization, and became stuck in a triploid bridge stage. A "triploid bridge" is usually formed in the process of polyploidization via unreduced gametes, but in dioecious plants, this process could severely disturb the balance of proportions of sex chromosomes, and hence proportions of male and female individuals [65]. We regard the first scenario as more likely as all other *S. bicolor* samples from the western Alps, Jura, and Harz Mountains were hexaploid. In any case, the triploids represent a special population. Our morphological analysis also includes *S. bicolor* into *S. phylicifolia* (Figure 5). The discrimination of both species in the spectral data analyses as well as the molecular data is rather weak (Figures 2 and S5).

Salix phylicifolia s.l. (incl. *S. bicolor*) is, in our present phylogeny, a paraphyletic group (Figure 2). Paraphyly is a frequent phenomenon in polyploid species and can be referred to progenitor-derivative relationships within a dataset [66,67]. Paraphyletic groups are based on shared ancestry and, hence, can be used for species classification [66,67]. The section *Phylicifolia*, however, is polyphyletic, and needs a new circumscription, which requires further studies, including other Eurasian and North American species [18] that we did not sample here.

4. Materials and Methods

4.1. Studied Material

We sampled species of the sections *Nicricantes* A. Kern and *Phylicifoliae* (Fries) Andersson with specific emphasis on the alpine species of both sections. *Salix myrsinifolia* Salisb., which is hexaploid, widely distributed in Europe, and morphologically highly diverse, has a long tradition of taxonomic discussion. The leaves of this shrub or small tree turn black when they are dried. *S. myrsinifolia* is probably the most variable species of all the European willows [20,24,54]. This species hybridizes with other willows, e.g., *S. glabra*, *S. mielichhoferi*, and *S. cinerea*; however, most plants that have been treated as hybrids are, according [20], "variants within the species variability range". Several species names were synonymized to *S. myrsinifolia* (e.g., *S. nigricans* Sm. is a younger synonym). Two subspecies are currently accepted (by the authors of Table 1): *S. myrsinifolia* ssp. *myrsinifolia*, and ssp. *borealis* (Flod.) A. Skvortsov. In this study, we sampled *S. myrsinifolia* originating from the Alps and from Fennoscandia. One of our samples matches *S. jenisseensis* (F.Schmidt) Flod. from Russia, a species related to *S. myrsinifolia*. Due to the lack of samples, the identification is provisional.

S. mielichhoferi Saut. is a hexaploid species [57,59] that occurs in Austria and Northern Italy and is endemic to the Alps, and also belongs to the section *Nigricantes*. Morphologically, it differs from *S. myrsinifolia* by glabrous, more oblong leaves without a pruinose layer on the lower surface. Both species share the blackening of leaves when dried. However, hybridization with *S. myrsinifolia* is possible [54], leading to intermediate forms.

Additionally, the accessions of *S. apennina* A. K. Skvortsov, a hexaploid species from the Apennines, and one sample from the locus classicus of *S. kaptarae* Cambr.Brullo&Brullo were included. The latter species was recently described from Crete [56] and assigned to the section *Nigricantes*. Based on the morphological characters, it was assumed as being related to *S. apennina*.

The species *S. phylicifolia* L., *S. bicolor* Ehrh. ex Willd. as well as tetra- to hexaploid *S. hegetschweileri* Heer em. Buser belong to the *S. phylicifolia* group sensu Rechinger [28], which can be assigned to the section *Phylicifoliae*. All three species are medium-sized shrubs. *S. bicolor* shows a disjunct distribution, while *S. hegetschweileri* is endemic to the central Alps. One stand of *S. bicolor* from the Alps is triploid [59]. For the other locations, neither chromosome counts nor genome size estimation exist to date. *Salix bicolor* has been the subject of controversial taxonomic treatments (see Table 1). Skvortsov [20] separated the Iberian species *S. basaltica* Coste from *S. bicolor*, and considered that *S. cantabrica* Rech.f., a tetraploid species from the Cordillera Cantabrica [57], should be included there. For *S. hegetschweileri*, chromosome counts of 4x, 5x, and 6x were reported, sometimes within one stand [58,59]. *Salix hegetschweileri* was separated by [28,54,68] and described as a variable, but clearly distinct species of the central Alps. However, during identification, confusion with glabrous forms of *S. myrsinifolia* is possible (Supplementary Table S4). Hybridization between both species can happen in overlapping distribution areas and was reported several times [54]. More problematic is the discrimination of *S. bicolor* from *S. phylicifolia*, which was subject to controversial discussions in recent decades. While both species were accepted by several authors [28,69,70], Skvortsov [20] treated the plants from the mountains of Central Europe not as distinct species but as subsp. of *S. phylicifolia* (*ssp. rhaetica*). Hexaploid *S. phylicifolia* L. *s.str.* is distributed in Northern Europe and Western Siberia.

4.2. Molecular Analyses

The leaves of a total of 64 newly sampled individuals, including material from type locations for *S. bicolor*, *S. hegetschweileri*, and *S. kaptarae*, were dried in silica gel and herbarium voucher specimens were deposited in the herbarium of the University of Goettingen (GOET). The sampling was supplemented by already published data to a final dataset of 145 samples (45 species). Detailed information for the sampling is summarized in Supplementary Table S1.

The DNA of all samples was extracted using the Qiagen DNeasy Plant Mini Kit following the manufacturer's instructions (Valencia, CA, USA). After the quality check, the DNA was sent to Floragenex, Inc. (Portland, OR, USA) where the sequencing library preparation was conducted after [71] using the restriction enzyme *Pst*I (see [29] for details). Polyploids require an increased depth of coverage based on the larger genome size and the higher number of alleles [72,73]. Thus, we sequenced polyploid taxa on a separate plate to avoid loss of coverage. The quality of the resulting single-end 100bp long sequence reads was checked using FastQC v.0.10.1 [74]. After de-multiplexing, the reads were used to run ipyrad v.0.9.52 [75] with a clustering threshold of 85% and a minimum depth of eight reads for base calling. The clustering was performed with VSEARCH as implemented in ipyrad v.0.9.52. The maximum number of SNPs per locus was set to 20, and the maximum number of indels to 8. We set a threshold of a maximum of four alleles per site in the final cluster filtering. Filtering settings were optimized as described in [29]. Eventually, the m15 dataset (loci shared by at least 15 individuals) was used for phylogenetic analyses of the backbone topology. Additionally, we performed ipyrad for all samples of the sections *Nigricantes* and *Phylicifoliae* with a reduced backbone sampling to test for within-section relationships. Finally, it was conducted for each "hexaploid clade" separately applying the same settings.

We inferred phylogenetic relationships on concatenated alignments of the complete dataset as well as the reduced backbone dataset by using the GTR+ Γ model of nucleotide substitution implemented in RAXML v.8.2.4 [76]. We conducted, for each ML analysis, a rapid bootstrapping (BS) analysis with 100 replicates. In addition to BS, we applied quartet sampling (QS; [77]) with default settings to test the statistical support of a given topology. We ran 300 replicates using the –L option (minimum likelihood differential). QS is able to distinguish between conflicting signals and poor phylogenetic information. For each phylogeny shown in this paper, the observed QS values (QC/QD/QI) were visualized along with the BS values above and below branches.

4.3. Co-Ancestry Analysis and Genetic Structure

To explore the genetic structure of the hexaploid species, we employed RADpainter included in the package fineRADstructure [39], which infers population structure from RAD sequencing data. The program creates a co-ancestry similarity matrix based on haplotypes. The analysis compares nearest neighbor haplotypes by finding the closest relative for each allele for a given sample set using SNP data. For more details, see [39]. We performed the analyses using default settings based on the 'alleles.loci' file resulting from clade-specific ipyrad runs. First, we prepared the input file using the Python script fnRADstructure_input.py included in 'finRADstructure-tools' (https://github.com/edgardomortiz/fineRADstructure-tools (accessed on 8 January 2023)). With the input file, we calculated the co-ancestry matrix employing RADpainter. We then used 'finestructure' for clustering and tree assembly using 100,000 MCMC replicates and a burn-in of 100,000 applying the clustering approach '-m T'. The results were visualized with the R script 'fineRADstructurePlot'.

Moreover, we performed genetic structure analyses applying SNMF within the R package LEA v.3.0.0 [78], which handles mixed-ploidy datasets, does not rely on assumptions of Hardy–Weinberg equilibrium, and is thus particularly suitable for analyzing polyploid apomicts [40,78,79]. We used .ustr files (one SNP per locus) generated by ipyrad and set the number of genetic clusters (K) from 1 to 10, maximal ploidy to 6, and repetitions to 8. To choose the number of ancestral K values, we used the implemented cross-entropy criterion. We retained and plotted multiple K values as the cross-entropy criterion only increased with increasing K values. The best run for each K value, i.e., the run with the lowest cross-entropy criterion, was selected to create bar plots in MS Excel.

4.4. Leaf Spectroscopy and Spectral Analyses

To measure the leaf spectra, we exclusively used well-dried herbarium material. Water might strongly affect the measurements and, by using herbarium samples, we could avoid any natural bias caused by different water contents. The sampling included 21 samples representing the six target species that were also used for molecular treatment supplemented by 17 additional samples to increase the sample size per species (Supplementary Table S1), totaling 38 samples from GOET. The reflectance measurements were conducted with a PSR+ portable spectroradiometer (Spectral Evolution, Haverhill, MA, USA) with bifurcated fiber optic cable and leaf probe clip. An "empty" white reflectance measurement was performed for calibration before each round of measurements. The reflectance spectra scanned from 350 to 2500 nm wavelength. We measured three times upper (adaxial) and three times lower (abaxial) surface to capture within-sample variability and statistical bias for each sample resulting in six scans per specimen. The measurement plots (spectra) were collected in the software DARWin SP Application Software v1.3 (Spectral Evolution).

The reflectance data were processed with a customized R script following [33]. In the first step, the data were filtered to remove spectra that resulted from erroneous scans with measured reflectance over 100%. Next, an average of all scans per sample was generated. The mean spectra per species were visualized as graphs. The results were forwarded to partial least squares discriminant analyses (PLS-DA), a multivariate analysis that classifies observation from PLS regression on indicator variables and proved to work well with high-dimensional multicollinear datasets as spectral data [31,80]. A prediction table was calculated based on the PLS scores. Then, we investigated the dataset with a principal component analysis (PCA). For this analysis, we defined different groups: "species level" (assignment of measurements to species determination) and "individual level" (assignment of spectra for each individual). The results were visualized with scatter plots with convex hulls or 95% confidence ellipses and with density plots.

4.5. Flow Cytometry

To analyze the ploidy levels of *S. bicolor* and *S. kaptarae*, flow cytometry (FC) was conducted using silica-gel-dried leaf material. Samples (~1 cm^2) were reduced to small pieces using Tissue Lyzer II (Qiagen; 30 Hz, time 3 s). A total of 200 µL of 1% PVP Otto I extraction buffer were added, and samples were gently inverted for 1 min. Samples were filtered through CellTrics ® filter (30 µm mesh, Sysmex Partec GmbH, Görlitz, Germany) into a flow cytometry sample tube. A total of 800 µL of DAPI-containing Otto II buffer [81] were then added to strain the DNA. FC analyses were carried out on a CyFlow Ploidy Analyzer (Sysmex, Norderstedt, Germany) and the Software CUBE16 v.1.6 (Sysmex, Norderstedt, Germany) was used to analyze the results and compute DNA content. Gain was set at 530 nm for an optimal differentiation between DNA content. Three different diploid *S. caprea* samples were used as external standard.

4.6. Morphometric Analyses

We conducted morphometric analyses on herbarium material to test for differences of *S. phylicifolia*, *S. bicolor*, and *S. hegetschweileri*. We examined the voucher specimens of all individuals used for molecular analysis, plus additional specimens from the same population, and additional populations for distinctive morphological characters as given in the literature (Table 1). Altogether 40 specimens were included. The original *S. bicolor* type population at Brocken, Harz Mountain, is well documented in the herbarium of the University of Göttingen (GOET) with several male and female individuals collected in the 18th and 19th century, including collections from Ehrhard. One isotype specimen of Ehrhard from the herbarium of the University of Vienna (WU) can also be included. The population has been almost destroyed due to constructions of buildings at the top of Brocken in the 20th century, but the remaining two individuals outside and cultivated in the Brocken garden that we used also for molecular and flow cytometric analyses derived from the original population [82]. The locus classicus of *S. hegetschweileri* in Urserental in the Alps of Switzerland is still occupied by a large healthy population, and both recent and old specimens could be used. *S. phylicifolia* L. was described from northern Sweden without indication of locality; we included specimens from this region. For all localities, see Supplementary Table S1.

From each specimen being at the stage of fruit maturity, we selected 5–8 fully developed leaves from the middle parts of branches to cover intra-individual variation. We measured leaf (blade) length, leaf (blade) width, angle at the blade's base, length of longest teeth at leaf margin, and length of stipules. Length/width ratios of leaves were calculated. To estimate the density of indumentum, a character regarded as diagnostic for *S. bicolor* by [83], the hairs at lower leaf surface, were counted in a square of 5 × 5 mm under 10× magnification. However, in *S. bicolor* and in *S. phylicifolia*, indumentum density ceases continuously during the season, whereas *S. hegetschweileri* is glabrous. Therefore, the presence of hairs turned out to be extremely variable even between the leaves of one individual, and 87% of *S. bicolor* individuals also had glabrous leaves. Mean values were calculated for all measures for each individual, and these data were subjected to further statistical analyses using PAST vs. 4.03 [84]. Boxplots were created for the three species, and each variable was tested for normality using Shapiro–Wilk tests; since five of the six characters (except angle of blade base) were non-normally distributed, we calculated one-way ANOVAs, with Kruskal–Wallis tests and pairwise Mann–Whitney post hoc tests for statistical differences (for the angle of blade base, we additionally calculated Tukey tests that provided congruent results). Pairwise differences at the 0.05 significance level are indicated on boxplots with different letters; for *p*-values, see Supplement Table S3. Data for density of hairs showed strong deviations from normality and failure of one-way ANOVAs because of a lack of variance. Because of its liability, the indumentum was excluded from further statistical analyses. Since length–width ratios revealed no significant differences between the three groups, this character was also excluded from multivariate statistics. With the remaining five characters, principal component analysis (PCA) and principal coordinate analysis (PCoA) using Gower's distances were calculated, and scatter plots were produced with the respective first two axes. We visualized groups with different colors and convex hulls for three species and two species (*S. bicolor* + *S. phylicifolia* versus

S. hegetschweileri). Discriminant analyses (DA) were calculated for the respective three and two predefined groups, and the percentages of correctly classified samples were calculated with confusion matrices.

Supplementary Materials: The following supporting information can be downloaded at: https://www.mdpi.com/article/10.3390/plants12051144/s1. Table S1. Detailed sampling information for molecular, morphological, and infrared-spectroscopy analyses. Table S2. Results of the flow cytometry measurements. Table S3. Morphometric statistics. Table S4. Morphological differences between *S. hegetschweileri* and *S. myrsinifolia*. Figure S1. Backbone topology of the *Chamaetia/Vetrix* clade including 145 samples representing 45 Eurasian species plus *S. triandra* as outgroup based on a RAxML analyses on 46,568 RAD loci (630,764 SNPs) in a concatenated alignment of 5,341,180 bp and 66.95% missing sites. Quartet sampling values above branches. Species of the sections *Nigricantes* and *Phylicifoliae* are highlighted in colors according to the legend. Figure S2. RAxML analyses of all samples of the analyzed species of the sections *Nigricantes* (green) and *Phylicifoliae* (blue) with a reduced backbone sampling of taxa representing the *Chamaetia/Vetrix* clade. The dataset comprised 73 samples and was based on 29,395 RAD loci, resulting in a concatenated alignment of 3,466,391 bp including 179,212 SNPs and 72.08% missing sites. Bootstrap values >50 above branches; sectional assignment is color-coded. Figure S3. Co-ancestry matrix of the sections *Nigricantes* (*a*) and *Phylicifolia* (*b*) based on the clade-wise analyses. Figure S4. Genetic structure analysis with sNMF including all hexaploid species (a) and sNMF results of clade-wise analyses of the sections *Phylicifolia* (b) and *Nigricantes* (c). Displayed are bar plots for different k-values. Each bar represents one sample and the respective ID is indicated below each bar. Figure S5. PCA and DAPC density plots of the infrared spectroscopy measurements of different sampling subgroups (a–d). Figure S6. PCA plot of morphological data with convex hulls for three taxa.

Author Contributions: Conceptualization, N.D.W. and E.H.; methodology, N.D.W. and E.H.; software, N.D.W., P.M. and L.P., data analyses, N.D.W., P.M., L.P. and E.H.; writing—original draft preparation, N.D.W.; writing—review and editing, all authors; visualization, N.D.W.; project administration, N.D.W. and E.H.; funding acquisition, N.D.W. and E.H. All authors have read and agreed to the published version of the manuscript.

Funding: This research was funded by DFG (Ho 5462/7-1 to E.H.) and by the priority program Taxon-Omics SPP1991 of the DFG (Wa3684/2-1 to N.W.).

Data Availability Statement: The raw reads of molecular analysis can be found at NCBI Genbank SRA; Bioproject Accession Number PRJNA433286.

Acknowledgments: We would like to thank Tommi Nyman, Sophie Karrenberg, Li He, Maria Tomoshevich, Christophe Bornand, and Susanne Gramlich for collecting plant material. We also would like to thank Mareike Daubert and Dirk Albach for support with the infrared spectroscopy data assessment. We acknowledge the financial support by the Open Access Publication Funds of Göttingen University. We thank the anonymous reviewers for their useful comments.

Conflicts of Interest: The authors declare no conflict of interest. The funders had no role in the design of the study; in the collection, analyses, or interpretation of data; in the writing of the manuscript; or in the decision to publish the results.

References

1. Otto, S.P.; Whitton, J. Polyploid incidence and evolution. *Annu. Rev. Genet.* **2000**, *34*, 401–437. [CrossRef] [PubMed]
2. Rothfels, C.J. Polyploid phylogenetics. *New Phytol.* **2021**, *230*, 66–72. [CrossRef] [PubMed]
3. Van de Peer, Y.; Mizrachi, E.; Marchal, K. The evolutionary significance of polyploidy. *Nat. Rev. Genet.* **2017**, *18*, 411–424. [CrossRef] [PubMed]
4. Comai, L. The advantages and disadvantages of being polyploid. *Nat. Rev. Genet.* **2005**, *6*, 836–846. [CrossRef]
5. Blischak, P.D.; Mabry, M.E.; Conant, G.C.; Pires, J.C. Integrating networks, phylogenomics, and population genomics for the study of polyploidy. *Annu. Rev. Ecol. Evol. Syst.* **2018**, *49*, 253–278. [CrossRef]
6. Hörandl, E. The complex causality of geographical parthenogenesis. *New Phytol.* **2006**, *171*, 525–538. [CrossRef]
7. Blaine Marchant, D.; Soltis, D.E.; Soltis, P.S. Patterns of abiotic niche shifts in allopolyploids relative to their progenitors. *N. Phytol.* **2016**, *212*, 708–718. [CrossRef]

8. Karbstein, K.; Tomasello, S.; Hodač, L.; Lorberg, E.; Daubert, M.; Hörandl, E. Moving beyond assumptions: Polyploidy and environmental effects explain a geographical parthenogenesis scenario in European plants. *Mol. Ecol.* **2021**, *30*, 2659–2675. [CrossRef]
9. Te Beest, M.; Le Roux, J.J.; Richardson, D.M.; Brysting, A.K.; Suda, J.; Kubesova, M.; Pysek, P. The more the better? The role of polyploidy in facilitating plant invasions. *Ann. Bot.* **2012**, *109*, 19–45. [CrossRef]
10. Rice, A.; Šmarda, P.; Novosolov, M.; Drori, M.; Glick, L.; Sabath, N.; Meiri, S.; Belmaker, J.; Mayrose, I. The global biogeography of polyploid plants. *Nat. Ecol. Evol.* **2019**, *3*, 265–273. [CrossRef]
11. Meudt, H.M.; Albach, D.C.; Tanentzap, A.J.; Igea, J.; Newmarch, S.C.; Brandt, A.J.; Lee, W.G.; Tate, J.A. Polyploidy on islands: Its emergence and importance for diversification. *Front. Plant Sci.* **2021**, *12*, 637214. [CrossRef] [PubMed]
12. Abbott, R.; Albach, D.; Ansell, S.; Arntzen, J.W.; Baird, S.J.E.; Bierne, N.; Boughman, J.; Brelsford, A.; Buerkle, C.A.; Buggs, R.; et al. Hybridization and speciation. *J. Evol. Biol.* **2013**, *26*, 229–246. [CrossRef] [PubMed]
13. Van de Peer, Y.; Ashman, T.-L.; Soltis, P.S.; Soltis, D.E. Polyploidy: An evolutionary and ecological force in stressful times. *Plant Cell* **2021**, *33*, 11–26. [CrossRef] [PubMed]
14. Ramsey, J.; Schemske, D.W. Pathways, mechanisms, and rates of polyploid formation in flowering plants. *Annu. Rev. Ecol. Syst.* **1998**, *29*, 467–501. [CrossRef]
15. Hörandl, E. Novel approaches for species concepts and delimitation in polyploids and hybrids. *Plants* **2022**, *11*, 204. [CrossRef]
16. Melichárková, A.; Šlenker, M.; Zozomová-Lihová, J.; Skokanová, K.; Šingliarová, B.; Kačmárová, T.; Caboňová, M.; Kempa, M.; Šrámková, G.; Mandáková, T.; et al. So closely related and yet so different: Strong contrasts between the evolutionary histories of species of the *Cardamine pratensis* polyploid complex in Central Europe. *Front. Plant Sci.* **2020**, *11*, 1988. [CrossRef]
17. Kelly, L.J.; Leitch, A.R.; Fay, M.F.; Renny-Byfield, S.; Pellicer, J.; Macas, J.; Leitch, I.J. Why size really matters when sequencing plant genomes. *Plant Ecol. Divers.* **2012**, *5*, 415–425. [CrossRef]
18. Argus, G.W. Infrageneric classification of *Salix* (Salicaceae) in the New World. *Syst. Bot. Monogr.* **1997**, *52*, 1–121. [CrossRef]
19. Chase, M.W.; Knapp, S.; Cox, A.V.; Clarkson, J.J.; Butsko, Y.; Joseph, J.; Savolainen, V.; Parokonny, A.S. Molecular Systematics, GISH and the origin of hybrid taxa in *Nicotiana* (Solanaceae). *Ann. Bot.* **2003**, *92*, 107–127. [CrossRef]
20. Skvortsov, A.K. *Willows of Russia and Adjacent Countries: Taxonomical and Geographical Revision*; Zinovjev, A.G., Argus, G.W., Tahvanainen, J., Roininen, H., Eds.; University of Joensuu: Joensuu, Finland, 1999; Volume 39, ISBN 9517087667.
21. Suda, Y.; Argus, G.W. Chromosome numbers of some North American *Salix*. *Brittonia* **1968**, *20*, 191–197. [CrossRef]
22. Sommerville, A.H.C. Willows in the environment. *Proc. R. Soc. Edinb.* **1992**, *98B*, 215–224. [CrossRef]
23. Pasteels, J.M.; Rowell-Rahier, M. The chemical ecology of herbivory on willows. *Proc. R. Soc. Edinburgh. Sect. B. Biol. Sci.* **1992**, *98*, 63–73. [CrossRef]
24. Hörandl, E.; Florineth, F.; Hadacek, F. *Weiden in Österreich und Angrenzenden Gebieten (Willows in Austria and Adjacent Regions)*, 2nd ed.; University of Agriculture, Vienna: Vienna, Austria, 2012.
25. Wagner, N.D.; He, L.; Hörandl, E. The evolutionary history, diversity, and ecology of willows (*Salix* l.) in the European Alps. *Diversity* **2021**, *13*, 146. [CrossRef]
26. Cronk, Q.; Ruzzier, E.; Belyaeva, I.; Percy, D. *Salix* transect of Europe: Latitudinal patterns in willow diversity from Greece to arctic Norway. *Biodivers. Data J.* **2015**, *3*, e6258. [CrossRef]
27. Dickmann, D.; Kuzovkina, Y.A. Poplars and willows in the World. In *Poplars and Willows: Trees for Society and the Environment*; Isebrands, J.G., Richardson, J.E., Eds.; FAO UN and CABI: Rome, Italy, 2014; ISBN 978 92 5 107185 4.
28. Rechinger, K.H. fil. Salix. In *Flora Europaea I*; Tutin, H.G., Heywood, V.H., Burges, N.A., Valentine, D.H., Walters, S.M., Webb, D.A., Eds.; University Press: Cambridge, UK, 1964; pp. 43–54.
29. Wagner, N.D.; Gramlich, S.; Hörandl, E. RAD sequencing resolved phylogenetic relationships in European shrub willows (*Salix* L. subg. *Chamaetia* and subg. *Vetrix*) and revealed multiple evolution of dwarf shrubs. *Ecol. Evol.* **2018**, *8*, 8243–8255. [CrossRef]
30. Wagner, N.D.; He, L.; Hörandl, E. Phylogenomic relationships and evolution of polyploid *Salix* species revealed by RAD sequencing data. *Front. Plant Sci.* **2020**, *11*, 1077. [CrossRef]
31. Stasinski, L.; White, D.M.; Nelson, P.R.; Ree, R.H.; Meireles, J.E. Reading light: Leaf spectra capture fine-scale diversity of closely related, hybridizing arctic shrubs. *N. Phytol.* **2021**, *232*, 2283–2294. [CrossRef]
32. Paiva, D.N.A.; Perdiz, R.d.O.; Almeida, T.E. Using Near-Infrared Spectroscopy to discriminate closely related species: A case study of neotropical ferns. *J. Plant Res.* **2021**, *134*, 509–520. [CrossRef]
33. Buono, D.; Albach, D.C. Infrared spectroscopy for ploidy estimation—An example in two species of *Veronica* (Plantaginaceae) using fresh and herbarium specimens. *Appl. Plant Sci.* 2023, accepted.
34. Wang, Y.; Xiang, J.; Tang, Y.; Chen, W.; Xu, Y. A Review of the application of Near-Infrared Spectroscopy (NIRS) in forestry. *Appl. Spectrosc. Rev.* **2022**, *57*, 300–317. [CrossRef]
35. Neumann, A. *Die Mitteleuropäischen Salix-Arten*; Mitteilung; Oesterreichischer Agrarverlag: Wien, Austria, 1981.
36. Jensen, R.J.; Ciofani, K.M.; Miramontes, L.C. Lines, outlines, and landmarks: Morphometric analyses of leaves of *Acer rubrum*, *Acer saccharinum* (Aceraceae) and their hybrid. *Taxon* **2002**, *51*, 475–492. [CrossRef]
37. Aeschimann, D.; Lauber, K. *Flora Alpina: Ein Atlas Sämtlicher 4500 Gefäßpflanzen der Alpen*; Haupt Verlag: Bern, Switzerland, 2004.
38. Eaton, D.A.R.; Spriggs, E.L.; Park, B.; Donoghue, M.J. Misconceptions on missing data in RAD-Seq phylogenetics with a deep-scale example from flowering plants. *Syst. Biol.* **2017**, *66*, 399–412. [CrossRef] [PubMed]

39. Malinsky, M.; Trucchi, E.; Lawson, D.J.; Falush, D. RADpainter and fineRADstructure: Population inference from RADseq data. *Mol. Biol. Evol.* **2018**, *35*, 1284–1290. [CrossRef] [PubMed]
40. Frichot, E.; Mathieu, F.; Trouillon, T.; Bouchard, G.; François, O. Fast and efficient estimation of individual ancestry coefficients. *Genetics* **2014**, *196*, 973–983. [CrossRef] [PubMed]
41. Karbstein, K.; Tomasello, S.; Hodač, L.; Wagner, N.; Marinček, P.; Barke, B.H.; Paetzold, C.; Hörandl, E. Untying gordian knots: Unraveling reticulate polyploid plant evolution by genomic data using the large *Ranunculus auricomus* species complex. *N. Phytol.* **2022**, *235*, 2081–2098. [CrossRef]
42. Spriggs, E.L.; Fertakos, M.E. Evolution of *Castanea* in North America: Restriction-site-associated DNA sequencing and ecological modeling reveal a history of radiation, range shifts, and disease. *Am. J. Bot.* **2021**, *108*, 1692–1704. [CrossRef]
43. Stobie, C.S.; Oosthuizen, C.J.; Cunningham, M.J.; Bloomer, P. Exploring the phylogeography of a hexaploid freshwater fish by RAD Sequencing. *Ecol. Evol.* **2018**, *8*, 2326–2342. [CrossRef]
44. Marinček, P.; Pittet, L.; Wagner, N.D.; Hörandl, E. Evolution of a hybrid zone of two willow species (*Salix* L.) in the European Alps analyzed by RAD-seq and morphometrics. *Ecol. Evol.* **2023**, *13*, e9700. [CrossRef]
45. Wang, Y.Z.; Dong, W.Y.; Kouba, A.J. Fast Discrimination of Bamboo Species Using VIS/NIR Spectroscopy. *J. Appl. Spectrosc.* **2016**, *83*, 826–831. [CrossRef]
46. Lang, C.; Costa, F.R.C.; Camargo, J.L.C.; Durgante, F.M.; Vicentini, A. Near infrared spectroscopy facilitates rapid identification of both young and mature Amazonian tree species. *PLoS ONE* **2015**, *10*, e0134521. [CrossRef]
47. Gaem, P.H.; Andrade, A.; Mazine, F.F.; Vicentini, A. tree species delimitation in tropical forest inventories: Perspectives from a taxonomically challenging case study. *For. Ecol. Manage.* **2022**, *505*, 119900. [CrossRef]
48. Neumann, A. Die Mitteleuropaeischen Salix-Arten. *Mitt. Der Forstl. Bundes-Vers.* **1981**, *134*, 1–151.
49. Hardig, T.M.; Brunsfeld, S.J.; Fritz, R.S.; Morgan, M.; Orians, C.M. Morphological and molecular evidence for hybridization and introgression in a willow (*Salix*) hybrid zone. *Mol. Ecol.* **2000**, *9*, 9–24. [CrossRef] [PubMed]
50. Triest, L. Hybridization in staminate and pistillate *Salix alba* and *S. fragilis* (Salicaceae): Morphology versus RAPDs. *Plant Syst. Evol.* **2001**, *226*, 143–154. [CrossRef]
51. Lihová, J.; Kudoh, H.; Marhold, K. Morphometric Studies of polyploid *Cardamine* species (Brassicaceae) from Japan: Solving a long-standing taxonomic and nomenclatural controversy. *Aust. Syst. Bot.* **2010**, *23*, 94. [CrossRef]
52. Morales, M.; Giannoni, F.; Inza, M.V.; Soldati, M.C.; Bessega, C.F.; Poggio, L.; Zelener, N.; Fortunato, R.H. Genetic and morphological diversity and population structure of a polyploid complex of *Mimosa* (Leguminosae). *Syst. Biodivers.* **2020**, *18*, 237–254. [CrossRef]
53. Laport, R.G.; Ramsey, J. Morphometric analysis of the North American creosote bush (*Larrea tridentata*, Zygophyllaceae) and the microspatial distribution of its chromosome races. *Plant Syst. Evol.* **2015**, *301*, 1581–1599. [CrossRef]
54. Hörandl, E. Die Gattung *Salix* in Österreich (mit Berücksichtigung angrenzender Gebiete). *Abh. Zool. Bot. Ges. Österreich* **1992**, *27*, 1–170.
55. Schönswetter, P.; Stehlik, I.; Holderegger, R.; Tribsch, A. Molecular evidence for glacial refugia of mountain plants in the European Alps. *Mol. Ecol.* **2005**, *14*, 3547–3555. [CrossRef]
56. Cambria, S.; Brullo, C.; Brullo, S. *Salix kaptarae* sp. nov. (Salicaceae) from Crete. *Nord. J. Bot.* **2019**, *37*, 1–9. [CrossRef]
57. Neumann, A.; Polatschek, A. Cytotaxonomischer Beitrag zur Gattung *Salix*. *Ann. Naturhist. Mus. Wien* **1972**, *76*, 612–633.
58. Büchler, W. Neue Chromosomenzählungen in der Gattung *Salix*. II. *Bot. Helv.* **1986**, *96*, 135–143. [CrossRef]
59. Dobeš, C.; Vitek, E. *Documented Chromosome Number Checklist of Austrian Vascular Plants*; Museum of Natural History Vienna: Vienna, Austria, 2000.
60. Barclay, C. Crete: Checklist of the vascular plants. *Englera* **1986**, *6*, I-XIII+1-138. [CrossRef]
61. Greuter, W.; Raus, T. Med-Checklist Notulae, 19. *Willdenowia* **2000**, *30*, 229–243. [CrossRef]
62. Bergmeier, E.; Abrahamczyk, S. Current and historical diversity and new records of wetland plants in Crete, Greece. *Willdenowia* **2008**, *38*, 433–453. [CrossRef]
63. Lautenschlager-Fleury, D.; Lautenschlager-Fleury, E. Zur Abklärung der *Salix hegetschweileri* Heer. *Bauhinia* **1991**, *9*, 265–271.
64. Asker, S.; Jerling, L. *Apomixis in Plants*; CRC Press: Boca Raton, FL, USA, 1992.
65. He, L.; Hörandl, E. Does polyploidy inhibit sex chromosome evolution in Angiosperms? *Front. Plant Sci.* **2022**, *13*. [CrossRef]
66. Hörandl, E.; Stuessy, T.F. Paraphyletic groups as natural units of biological classification. *Taxon* **2010**, *59*, 1641–1653. [CrossRef]
67. Rieseberg, L.H.; Brouillet, L. Are many plant species paraphyletic? *Taxon* **1994**, *43*, 21–32. [CrossRef]
68. Buser, R. Kritische Beiträge zur Kenntnis der Schweizerischen Weiden. *Ber. Schweiz. Bot. Ges.* **1940**, *50*, 567–788.
69. Rechinger, K.H. fil. *Salix*. In *Illustrierte Flora von Mitteleuropa III/1*; Hegi, G., Ed.; Carl Hanser: Munich, Germany, 1957.
70. Floderus, B. Two Linnean species of *Salix* and their allies. *Ark. f. Bot.* **1939**, *29A*, 1–54.
71. Baird, N.A.; Etter, P.D.; Atwood, T.S.; Currey, M.C.; Shiver, A.L.; Lewis, Z.A.; Selker, E.U.; Cresko, W.A.; Johnson, E.A. Rapid SNP discovery and genetic mapping using sequenced RAD markers. *PLoS ONE* **2008**, *3*, e3376. [CrossRef] [PubMed]
72. Hirsch, C.N.; Buell, C.R. Tapping the promise of genomics in species with complex, nonmodel genomes. *Annu. Rev. Plant Biol.* **2013**, *64*, 89–110. [CrossRef] [PubMed]
73. Clevenger, J.P.; Korani, W.; Ozias-Akins, P.; Jackson, S. Haplotype-based genotyping in polyploids. *Front. Plant Sci.* **2018**, *9*, 564. [CrossRef]

74. Andrews, S. FastQC: A Quality Control Tool for High Throughput Sequence Data. 2010. Available online: http://www.bioinformatics.babraham.ac.uk/projects/fastqc/ (accessed on 8 January 2023).
75. Eaton, D.A.R.; Overcast, I. Ipyrad: Interactive assembly and analysis of RADseq datasets. *Bioinformatics* **2020**, *36*, 2592–2594. [CrossRef]
76. Stamatakis, A. RAxML version 8: A tool for phylogenetic analysis and post-analysis of large phylogenies. *Bioinformatics* **2014**, *30*, 1312–1313. [CrossRef]
77. Pease, J.B.; Brown, J.W.; Walker, J.F.; Hinchliff, C.E.; Smith, S.A. Quartet sampling distinguishes lack of support from conflicting support in the green plant tree of life. *Am. J. Bot.* **2018**, *105*, 385–403. [CrossRef]
78. Frichot, E.; François, O. LEA: An R package for landscape and ecological association studies. *Methods Ecol. Evol.* **2015**, *6*, 925–929. [CrossRef]
79. Weiss, M.; Weigand, H.; Weigand, A.M.; Leese, F. Genome-wide single-nucleotide polymorphism data reveal cryptic species within cryptic freshwater snail species-The case of the *Ancylus fluviatilis* species complex. *Ecol. Evol.* **2018**, *8*, 1063–1072. [CrossRef]
80. Cavender-Bares, J.; Meireles, J.; Couture, J.; Kaproth, M.; Kingdon, C.; Singh, A.; Serbin, S.; Center, A.; Zuniga, E.; Pilz, G.; et al. Associations of leaf spectra with genetic and phylogenetic variation in oaks: Prospects for remote detection of biodiversity. *Remote Sens.* **2016**, *8*, 221. [CrossRef]
81. Otto, F. DAPI staining of fixed cells for high-resolution flow cytometry of nuclear DNA. In *Method in Cell Biology;* Darzynkiewicz, Z., Crissman, H.A., Eds.; Academic Press: San Diego, CA, USA, 1990; Volume 33, pp. 105–110.
82. Zander, M. Untersuchungen zur Identifizierung ausgewählter Vertreter der Gattung *Salix* L. im NO-deutschen Tiefland, unter besonderer Berücksichtigung des *Salix-repens*-Komplexes. *Mitt. florist. Kart. Sachs. Anhalt.* **2000**, *5*, 3–137.
83. Willdenow, C.L. 2. Salix. In *Berl. Baumz. In Species Plantarum Bd. 4*; Nabu Press: Berlin, Germany, 1796; p. 691.
84. Hammer, O.; Harper, D.A.T.; Ryan, P.D. PAST: Paleontological statistics software package for education and data analysis. *Palaeontol. Electron.* **2001**, *4*, 9.

plants MDPI

Article

Genome Size in the *Arenaria ciliata* Species Complex (Caryophyllaceae), with Special Focus on *A. ciliata* subsp. *bernensis*, a Narrow Endemic of the Swiss Northern Alps

Gregor Kozlowski [1,2,3,*], Yann Fragnière [1], Benoît Clément [1] and Conor Meade [4]

1 Department of Biology and Botanical Garden, University of Fribourg, Chemin du Musée 10, CH-1700 Fribourg, Switzerland
2 Natural History Museum Fribourg, Chemin du Musée 6, CH-1700 Fribourg, Switzerland
3 Eastern China Conservation Centre for Wild Endangered Plant Resources, Shanghai Chenshan Botanical Garden, Shanghai 201602, China
4 Molecular Ecology and Biogeography Laboratory, Biology Department, Maynooth University, W23 F2H6 Maynooth, Ireland
* Correspondence: gregor.kozlowski@unifr.ch; Tel.: +41-26-300-8842

Abstract: The genus *Arenaria* (Caryophyllaceae) comprises approximately 300 species worldwide; however, to date, just six of these taxa have been investigated in terms of their genome size. The main subject of the present study is the *A. ciliata* species complex, with special focus on *A. ciliata* subsp. *bernensis*, an endemic plant occurring in the Swiss Northern Alps. Altogether, 16 populations and 77 individuals of the *A. ciliata* complex have been sampled and their genome sizes were estimated using flow cytometry, including *A. ciliata* subsp. *bernensis*, *A. ciliata* s.str., *A. multicaulis*, and *A. gothica*. The *Arenaria ciliata* subsp. *bernensis* shows the highest 2c-value of 6.91 pg of DNA, while *A. gothica* showed 2c = 3.69 pg, *A. ciliata* s.str. 2c = 1.71 pg, and *A. multicaulis* 2c = 1.57 pg. These results confirm the very high ploidy level of *A. ciliata* subsp. *bernensis* (2n = 20x = 200) compared to other taxa in the complex, as detected by our chromosome counting and previously documented by earlier work. The genome size and, thus, also the ploidy level, is stable across the whole distribution area of this taxon. The present study delivers additional support for the taxonomic distinctiveness of the high alpine endemic *A. ciliata* subsp. *bernensis*, which strongly aligns with other differences in morphology, phylogeny, phenology, ecology, and plant communities, described previously. In affirming these differences, further support now exists to re-consider the species status of this taxon. Upgrading to full species rank would significantly improve the conservation prospects for this taxon, as, because of its precise ecological adaptation to alpine summit habitats, the *A. ciliata* subsp. *bernensis* faces acute threats from accelerated climate warming.

Keywords: arctic-alpine plants; disjunctions; flow cytometry; narrow endemism; polyploidy

Citation: Kozlowski, G.; Fragnière, Y.; Clément, B.; Meade, C. Genome Size in the *Arenaria ciliata* Species Complex (Caryophyllaceae), with Special Focus on *A. ciliata* subsp. *bernensis*, a Narrow Endemic of the Swiss Northern Alps. *Plants* **2022**, *11*, 3489. https://doi.org/10.3390/plants11243489

Academic Editor: Panayiotis Dimitrakopoulos

Received: 28 November 2022
Accepted: 10 December 2022
Published: 13 December 2022

1. Introduction

Nuclear DNA amount and genome size (C-values) are important biodiversity features that have essential biological significance and many practical and predictive uses [1]. C-value information is widely used in various domains of biology [2], as the knowledge of genome size in a given taxon is of great importance when framing scientific questions or planning research [1]. Consequently, there is a large demand for C-value estimates for plant species.

Flow cytometry has become the method of choice for measuring DNA content, particularly because its sample preparation and analysis protocols are convenient and rapid [3,4]. The estimation of genome size and ploidy using flow-cytometry is a key data source for the investigation of evolutionary and biogeographical processes, as well as taxonomic issues, both of wild and cultivated plant species and varieties (e.g., [5–8]).

A comprehensive database of recorded genome sizes in plants worldwide is published regularly by the Kew Royal Botanic Gardens [9], with the most recent update (Release 7.1., April 2019) providing 2c values for 11,500 vascular plant species. This comprises just 3.7% of the approximatively 308,000 described plant species, globally [10]. Despite progress with other more localized database initiatives (e.g., [11]), as well as national inventories (e.g., [12] for the Czech Republic, and [13] for The Netherlands), the list of species and taxonomic groups awaiting investigation remains very long.

This is also the case for the *Arenaria ciliata* (Caryophyllaceae) species complex. This is surprising, as the taxon is widespread in many European countries and has already been the subject of several karyological and taxonomic studies [14–16]. Although the genus *Arenaria* comprises approximatively 300 species worldwide [17], only six taxa have been investigated in terms of genome size [9], none of which belong to the *A. ciliata* species complex.

The *A. ciliata* species complex comprises a group of poorly-differentiated arctic-alpine herbaceous taxa with overlapping morphological and ploidy identities [16,18–20]. In northern Europe and in the Arctic, there are traditionally two taxa belonging to this group: *A. norvegica* Gunn. and *A. ciliata* subsp. *pseudofrigida* Ostenf. and O. C. Dahl, the latter reaching Svalbard and Franz Joseph Land to the North [20–22]. In the European Alpine System (EAS), and thus also in the Swiss Alps and the Jura Mountains, this species complex is represented by four taxa: *A. ciliata* s.str. L., *A. ciliata* subsp. *bernensis* Favarger, *A. gothica* Fr., and *A. multicaulis* L. [19,23]. The results of chloroplast DNA analyses suggest that the *A. ciliata* species complex is a monophyletic group [18,19].

The main focus of the present study is the status of the *A. ciliata* subsp. *bernensis* (Figure 1), an endemic taxon occurring exclusively in the Swiss Northern Alps [19,24,25]. It was discovered in 1955 by Swiss botanist Claude Favarger, a professor at the University of Neuchâtel, and described in 1963 [14]. Originally, the taxon was known only from the summit area of Gantrisch and Leiterenpass (Canton of Bern). However, recent studies have shown that it occurs on nearly all summits between Stockhorn in the Canton Bern and Moléson in the Canton of Fribourg (Figure 2), forming an arc of sky-island populations ca. 50 km across [19]. The taxon grows exclusively on shady, cool and steep slopes with northern exposition in the alpine zone above 1900–2000 m a.s.l. (Figure 1). The majority of populations are small (less than 100 individuals), and, globally, the taxon counts no more than ca. 4000 individuals [19].

According to Parisod [26], the *A. ciliata* subsp. *bernensis* is a neoendemic taxon, probably of allopolyploid origin, having recently formed close to the so-called Penninic-Savoyic zone of secondary contact in the North-Western Alps. Similarly, Favarger and Contandriopoulos [27] classified the *A. ciliata* subsp. *bernensis* as an apoendemic taxon of allopolyploid origin. Such endemics are often the result of a relatively rapid mixture between different floristic elements, for example, due to rapid migration events. Due to the geographic position of the *A. ciliata* subsp. *bernensis*, in relation to *A. multicaulis* and *A. ciliata* s.str. [23], a post glacial hybrid origin involving these two species is possible. Interestingly, Berthouzoz et al. [19], using chloroplast DNA markers, demonstrated that the *A. ciliata* subsp. *bernensis* is genetically closer to *A. multicaulis* than to *A. ciliata* s.str. Significantly, despite its very small distribution area, the taxon displays high genetic diversity, and this could also be consistent with the refugial survival of the *A. ciliata* subsp. *bernensis* during the Pleistocene glaciations (nunatak survival) [19].

The possibility of a polyploid speciation origin is supported by the fact that the *A. ciliata* subsp. *bernensis* presents a very high ploidy level (2n = 200), particularly in comparison with *A. multicaulis* (2n = 40), but also with *A. ciliata* s.str. (2n = 40–160) and *A. gothica* (2n = 100) [16,23,28,29]. Its putative hybrid origin and complex polyploidization history is probably one of the reasons researchers have hesitated in attributing it a species status [30], as the traditional species concept is difficult to apply to hybrids and polyploids [31].

The main aim of the present study was to deliver the first evaluation of the genome sizes in the *Arenaria ciliata* species complex using flow cytometry, with special focus on the narrow endemic *A. ciliata* subsp. *bernensis*. The following specific questions have been addressed: (1) What are the differences in genome size among the four taxa belonging to the *A. ciliata* group, occurring in the Alps and neighboring Jura Mountains (*A. ciliata* s.str., *A. ciliata* subsp. *bernensis*, *A. gothica* and *A. multicaulis*)? (2) What is the geographic pattern and stability of the genome size across the whole distribution area of the narrow endemic *A. ciliata* subsp. *bernensis*? (3) Do the obtained results corroborate with the ploidy levels of the four studied taxa? Based on the results for these investigations, we also set out to evaluate the implications of our work for the taxonomy and conservation of *A. ciliata* subsp. *bernensis*.

Figure 1. Habitats and morphology of *Arenaria ciliata* subsp. *bernensis*. (**A**)—Summit area of Gantrisch (2175 m a.s.l., Canton of Bern), view from Leiterenpass, the *locus classicus* of the taxon. (**B**)—Typical habitat on shady and cold slopes with northern exposition at the elevations higher than 2000 m a.s.l. (ca. 2100 m a.s.l., Schafarnisch, Canton of Fribourg). (**C**)—Only few small populations or isolated individuals grow at lower altitudes, at the bases of scree fields with north exposition (Salzmatt next to Kaiseregg, 1650 m a.s.l.). (**D**)—Example of irregular floral morphology of plants with 6 to 9 petals (typical number is 5) (Leiterenpass, 1940 m a.s.l.). Pictures: G. Kozlowski.

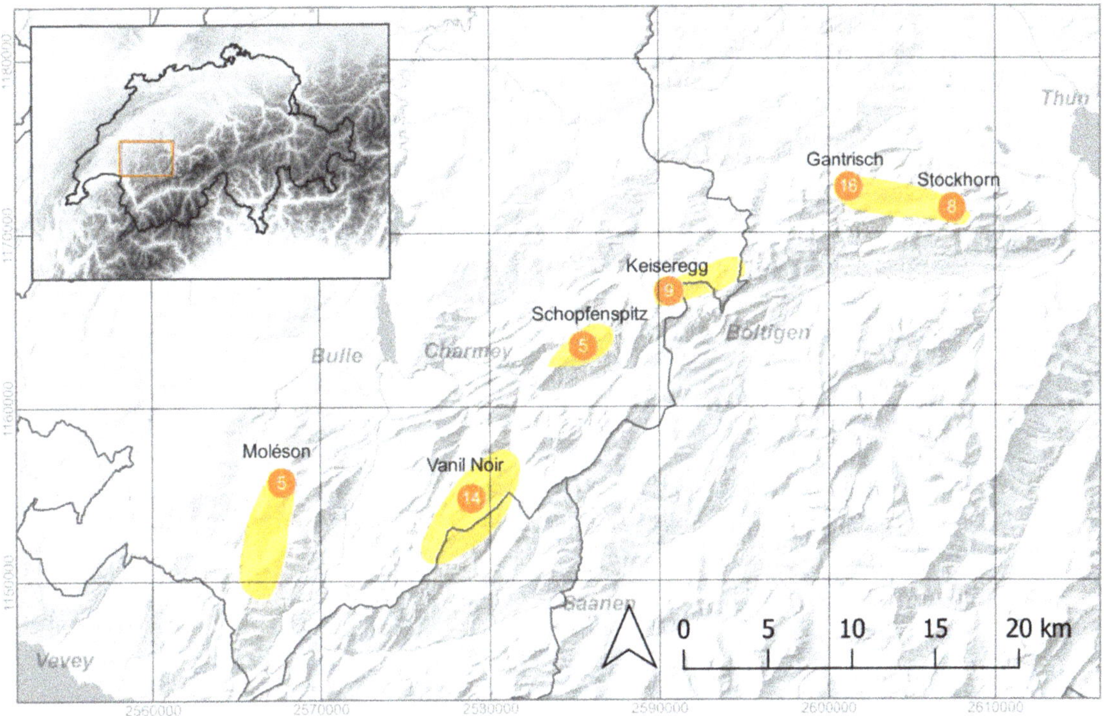

Figure 2. Distribution area of *Arenaria ciliata* subsp. *bernensis* (yellow zones). Orange circles show the position of collection sites with the numbers of individuals sampled in the given summit area.

2. Results

The highest 2c values among all four of the taxa from the *A. ciliata* complex investigated in this study were recorded for *A. ciliata* subsp. *bernensis*, varying between 6.26 pg and 7.75 pg of DNA (Figures 3 and S1, Tables 1 and S1), with a mean 2c value of 6.91 pg. The recorded genome size of *A. gothica* reached approximately half of these 2c values and varied between 3.62 and 3.76 pg of DNA, with a mean 2c value of 3.69 pg. Finally, *A. ciliata* s.str. and *A. multicaulis* both showed similar, but much lower, values, with a mean 2c value of 1.71 for *A. ciliata* s.str. and a mean 2c value of 1.57 for *A. multicaulis*. The higher standard deviation value in the *A. ciliata* subsp. *bernensis* is due to a much larger sample size, of 57 individuals, analyzed for this taxon in comparison to the other three taxa (between five and ten individuals).

Table 1. Genome size (mean ± standard deviation) in *Arenaria ciliata* subsp. *bernensis* in comparison with three other taxa from the *A. ciliata* species complex occurring in Switzerland.

Taxon	Genome Size (pg) Mean (±SD)	Estimated Ploidy Level
A. ciliata subsp. *bernensis* Favarger	6.91 (±0.33)	2n = 20x = 200
A. gothica Fr.	3.69 (±0.07)	2n = 10x = 100
A. ciliata s.str. L.	1.71 (±0.06)	2n = 2x = 40
A. multicaulis L.	1.57 (±0.04)	2n = 2x = 40

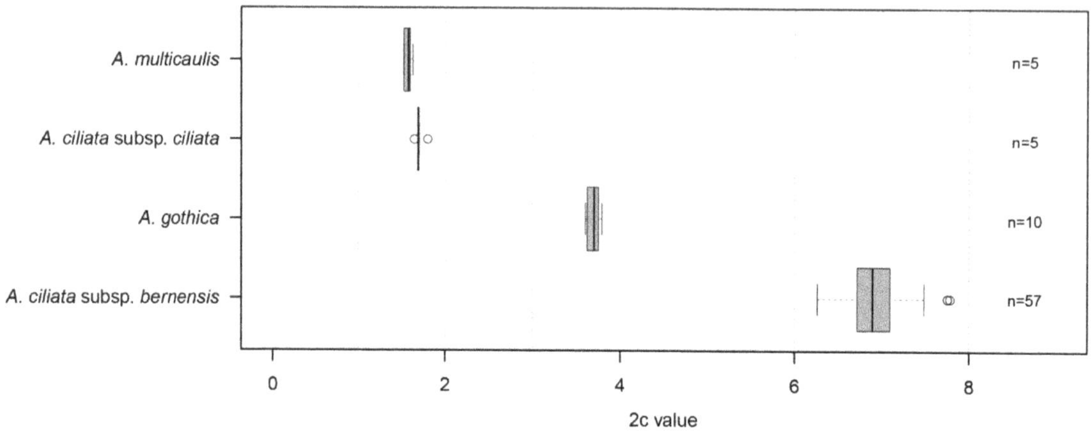

Figure 3. Variation in genome size of four taxa in the *Arenaria ciliata* species complex analyzed in this study, occurring in Swiss Northern Alps and in the Jura Mountains (2c, pg of DNA).

The results show clearly that the genome size of the *A. ciliata* subsp. *bernensis* is very stable across the whole distribution area of the taxon, thus indicating an invariant ploidy level for all of the investigated individuals (Figure 3, Tables 1 and S1). The direct counting of chromosome numbers in the selected samples from Dent de Brenleire (Figure S1) resulted in 2n = 20x = 200, and combined with the stable 2c values across all sites, indicate that the *A. ciliata* subsp. *bernensis* individuals and populations investigated in this study all show 2n = 200.

3. Discussion

Our study delivers the very first genome size (2c values) estimates for the members of the arctic-alpine *Arenaria ciliata* species complex (Table 1, Figure 3). This new data adds to the observed genome sizes and ploidy evaluations in the highly variable genus *Arenaria*, whose base chromosome number ranges between x = 9 (as observed in e.g., *A. balearica*) and x = 15 (*A. saxifraga*) [14]. The most frequent chromosome numbers observed are x = 11 (ca. 40 *Arenaria* spp.) and x = 10 (ca. 25 spp.), and the present data affirm that it is to this latter group that the *A. ciliata* species complex belongs.

3.1. Genome Size Values in the Genus Arenaria

Based on the Kew Plant DNA C-values Database [9] and the published literature, 2c values are available for just six *Arenaria* taxa (Table 2). The diploid *A. leptoclados* has a recorded 2c value of 0.79 pg (2n = 2x = 20) [13]; for *A. gracilis*, 2c = 1.19 pg (2n = 2x = 24) [32]; and the diploid taxa from the *A. grandiflora* complex possess 2c values ranging between 2.11 and 2.70 pg (2n = 2x = 24) [33]. For the tetraploid *A. serpyllifolia*, 2c values between 1.41 and 1.60 pg (2n = 4x = 40) were recorded [12,13,34]. The *Arenaria tetraquetra* subsp. *amabilis* displays 2c = 1.29 pg (2n = 4x = 40) [35], and tetraploid taxa from the *A. grandiflora* complex have 2c values between 4.24 and 5.27 pg (2n = 4x = 44) [12,33]. For *A. deflexa*, with an unrecorded ploidy level, the 2c value = 2.04 pg [36]. To date, no genome size values are available for the *Arenaria* taxa with a ploidy level higher than 4x. From the available published values for *Arenaria*, it is evident that higher ploidy levels are generally (but not always) associated with higher observed 2c values; however, the 2c values appear to be consistently higher for taxa with higher base chromosome numbers (e.g., x = 12), compared to other taxa (e.g., x = 10) within the same ploidy level. In this context, the results presented here are of special relevance in relation to other *Arenaria* taxa with the same basic chromosome numbers as those observed for the *A. ciliata* complex (x = 10), and also those with the same ploidy level (2x diploid).

Table 2. Recorded ploidy and genome size values for *Arenaria* L. in the published literature. The 2c values are indicated in pg estimated mass for each taxon or range of surveyed taxa within a species complex.

Taxon	Basic Chrom. Number	Ploidy Level	2n Chrom. Count	2c (pg)	References
A. leptoclados (Rchb.) Guss	x = 10	2x	20	0.79	[13]
A. gracilis Waldst. and Kit.	x = 12	2x	24	1.19	[32]
A. grandiflora L. complex	x = 12	2x	24	2.11–2.70	[33]
A. serpyllifolia L.	x = 10	2x	40	1.41–1.60	[12,13,34]
A. tetraquetra subsp. *amabilis* (Bory) H.Lindb.	x = 10	2x	40	1.29	[35]
A. grandiflora L. complex	x = 12	2x	44	4.24–5.27	[12,33]
A. deflexa Decne. *	-	-	-	2.04	[36]

* Chromosome counts not recorded.

3.2. *Arenaria ciliata s.str. and A. multicaulis*

Among the four *A. ciliata* taxa investigated in our study (Table 1), the lowest 2c values were obtained for *A. ciliata* s.str. (mean 2c = 1.71 pg) and *A. multicaulis* (mean 2c = 1.57 pg). These values are similar to the two tetraploid taxa *A. serpyllifolia* and *A. tetraquetra* (both with basic chromosome number x = 10) (Table 2). Therefore, our results corroborate with the published chromosome numbers for *A. multicaulis* (2n = 4x = 40, Lauber et al. 2018). According to Favarger [14] and Abukrees et al. [16], the chromosomal variability of *A. ciliata* s.str. is much higher, when analyzing plants from different alpine regions, with 2n-values between 40 and 160. Interestingly, all plants of this taxon analyzed in our study (Swiss Northern Alps) seem to be locally invariant and tetraploids. Since both taxa (*A. multicaulis* and *A. ciliata* s.str.) possess a relatively large distribution area in the Alps and neighboring mountain chains [30], wider investigations covering the whole distribution range are needed to capture the full level of variation in the genome size values.

3.3. *Arenaria gothica*

This taxon is a European boreal-montane plant element, possessing only few and highly disjunct occurrences in Jura Mountains (Lac de Joux, Switzerland) and in Scandinavia (mainly isle of Gotland, Sweden) [28,37]. In Switzerland, the species is extremely difficult to study, as it appears exclusively on the exondated shores of the Lac de Joux, and only during exceptional drought periods [38]. The most recent observed appearance of the population was in 2003, and the plant accessions used in this study are from an ex situ culture of plants collected at this time and maintained at the Botanical Garden of the University of Fribourg (Switzerland). *Arenaria gothica* is a high polyploid taxon with 2n = 10x = 100 [23,29,37]. The genome size estimation in the present study indirectly confirms this recorded chromosome number (Table 1), with the mean 2c value of *A. gothica* at 3.70 pg being approximatively half of the 2c-value of the *A. ciliata* subsp. *bernensis* (mean 2c = 6.91 pg; 2n = 20x = 200) (Figure 3). The relatively small variation in genome size values for *A. gothica* compared to the other sampled taxa may be an artefact of the long-time cultivation in an ex situ collection of closely related individuals.

3.4. *Arenaria ciliata subsp. bernensis*

This narrow endemic taxon, which was the main focus of the present study, shows the highest known 2c values (between 6.26 and 7.75 pg), not only within the *A. ciliata* species complex (Table 1, Figure 3), but also among all other species investigated thus far in the genus *Arenaria* (Table 2). Given the trend evident across the genus (Table 2), this result is consistent with expectations, due to the known high ploidy level for this taxon (2n = 20x = 200), as affirmed in our own observations (Figure S1) and the published literature [14,16,39,40].

Interestingly, the genome size varies only slightly, indicating that ploidy levels appear to be stable across the whole distribution area of the *A. ciliata* subsp. *bernensis* (Figure 2). This situation presents a contrast to the highly variable chromosome numbers in the closely related *A. ciliata* s.str., with 2n ranging between 40 and 160 [16,23]. In his exhaustive studies on the *A. ciliata* species complex [39,40], and also in the publication describing the *A. ciliata* subsp. *bernensis* [14], Favarger was not able to count the exact chromosome number, giving either 2n = 200 or 2n = 240, or writing "env. 240". The latter value was then adopted and repeated in all standard works of the Swiss flora (e.g., [23]). In contrast, our study supports the conclusion of Abukrees et al. [16], who reported 2n = 200 for the *A. ciliata* subsp. *bernensis*.

Berthouzoz et al. [19] highlighted the presence of at least a few irregular flowers with 4–6 styles, 12–16 stamens and 6–9 petals per population of the *A. ciliata* subsp. *bernensis* (Figure 1D). One possibility arising from this observation was that the variation in floral regularity might be associated with polyploid status. To address this question in the present study, we collected several plants with six and nine petals and analyzed their genome size (Table S1). The 2c values of these plants are not significantly different in comparison to other regular individuals, indicating no correlation between such morphological changes and the ploidy level. This finding would appear to have importance beyond the *A. ciliata* subsp. *bernensis*, as irregular flowers are also occasionally observed in many populations of other taxa from the *A. ciliata* species complex across Europe and the arctic (C. Meade, personal observation).

3.5. *Implications for Taxonomy and Conservation of* A. ciliata *subsp.* bernensis

The present study delivers additional and significant evidence regarding the taxonomic distinctness of the high alpine endemic *A. ciliata* subsp. *bernensis*. Our new evidence soundly aligns with other differences in morphology, phylogeny, phenology, ecology and associated plant communities, etc., as described by Favarger [14] and further explored by Berthouzoz et al. [19]. Importantly, in displaying an elevated but stable ploidy level, strongly differentiated compared to adjacent taxa in the *A. ciliata* species complex occurring in the western Alps and in the Jura Mountains, while maintaining a distinct restricted distribution and habitat ecology, the taxon would appear to merit a separate species status.

The type specimens are conserved in the herbarium of the University of Neuchâtel (Switzerland). Its typification based on the original collection and new results and rising it to the species status is thus long overdue. It is important to note that the current subspecies rank slows down, or even completely hinders, the research and development of targeted protective measures for this taxon, as it is not always accepted and included in regional and national floras (e.g., [16,41]).

Growing mainly between 2000 and 2350 m a.s.l., and due to its preferences and ecological adaptation to the high summits of the Northern Alps [19], the taxon belongs in the group of populations whose habitat faces the most acute threats from accelerating climate change [42]. In this context, although human-mediated global warming may be responsible for population declines of many alpine-arctic plants [43], Körner and Hiltbrunner [44] have postulated that high-altitude species are potentially very resistant to the impacts of climate change, particularly in relation to the exploitation of refugial microhabitats. However, the *A. ciliata* subsp. *bernensis* grows exclusively at the very top of the summits and in a geographically very small and isolated area. For this reason, these populations appear to have very limited scope to exploit the mosaic of micro-environmental conditions that may assist other high-altitude plants and climate relicts, to escape into neighboring microhabitats.

4. Materials and Methods

4.1. Taxon Identification in the A. ciliata Species Complex

The identity of *Arenaria ciliata* subsp. *bernensis* individuals were determined in the field, according to the description in Favarger [14,39,40], Lauber et al. [23] and Berthouzoz et al. [19]. The most important characteristics that facilitate the assignment to this taxon, compared to other *Arenaria* taxa in the Swiss Northern Alps, are: (1) large and solitary flowers, ca. 2 cm in diameter; (2) the loose habit of the whole plant with long shoots; and (3) the presence of (at least a few) irregular flowers per population, with higher numbers of petals, stamens and styles (Figure 1). In comparison, the flowers of *A. multicaulis* are only 1 cm in diameter and their inflorescence usually possesses 5–7 flowers. *Arenaria ciliata* s.str. exhibits a very compact habit with short pulvinate shoots. The fourth taxon used in this study, *A. gothica*, is morphologically similar to the subsp. *bernensis* (loose habit) but its petals are much smaller (4–4.5 mm) [23]. In addition, as in Central Europe *A. gothica* occurs exclusively along the well-studied shores of Lac des Joux in Switzerland, where no other taxa from the *A. ciliata* complex are recorded, the discrimination of this taxon is relatively uncomplicated [38].

4.2. Sampling of Plant Material

The plant materials of *A. ciliata* s.str., *A. ciliata* subsp. *bernensis*, and *A. multicaulis* were collected in August 2022. In total, 57 plants of *A. ciliata* subsp. *bernensis* were collected from 6 summit areas, covering the whole known distribution of this taxon from Stockhorn (Canton of Bern) to Moléson (Canton of Fribourg) (Figure 2, Table S1). Plants with irregular flowers, with 6 and 9 petals (Figure 1D), were also collected in order to test the correlation of such morphological anomaly with the genome size and ploidy level (Table S1). Additionally, five individuals of *A. ciliata* s.str. were sampled in the Moléson and Vanil Noir summit areas, and five individuals of *A. multicaulis* in the Gantrisch and Vanil Noir summit areas (Table S1). The plant material of *A. gothica* (10 individuals) was collected in October 2022 from ex situ culture, grown from seeds collected in 2003 from Lac de Joux, the only population of the taxon in Switzerland [38]. The plant material (small portion of flowering stem with flowers) was silica dried and kept for ca. 4 weeks in plastic bags prior to flow cytometry analyses.

4.3. Flow Cytometry Analysis

Approximately 1 cm^2 of silica dried leaves of the *Arenaria* samples were mixed with 1 cm^2 of fresh leaves of the standard plant (*Allium schoenoprasum*, genome size 2c = 15.03 pg). This was chopped with a sharp razor blade to release the nuclei in 100 μL of Cysrain nuclei extraction buffer (Sysmex, Norderstedt, Germany, https://eu.sysmex-flowcytometry.com, accessed on 9 November 2022). The obtained suspension was then sieved through a 40 μm filter, and 1.5 mL of Cystain Pi (propidium iodide) absolute P staining buffer was added. After one hour, the fluorescence of nuclei in the suspension was measured using Sysmex ploidy analyzer (Sysmex, Norderstedt, Germany). The flow cytometry analyses were carried out by Plant Cytometry Services, Didam, The Netherlands, http://www.plantcytometry.nl.

4.4. Confocal Microscopy

Chromosome counting, using confocal microscopy, was performed only for plants of the *A. ciliata* subsp. *bernensis*, the main focus of the present study, grown from seeds collected in Dent de Brenleire (Vanil Noir summit area, Canton of Fribourg). Newly developed shoots were cut from living plants and stored in distilled sterile water for 24 h at 4 °C. For pretreatment, the axillary and apical buds were then cut from the tissue using a dissecting razor and placed in 1.5 mL tubes containing 0.002 mol/L 8-hydroxyquinoline solution (Sigma, Arklow, Ireland), for 4 h at 20 °C. Fixation was carried out in a mixture of 98% 3:1 absolute ethanol: glacial acetic acid (Carnoy's solution), for at least 1 h, at 4 °C. The buds were then washed with distilled water for 5 min. Bud hydrolysis was completed with a solution of 1N HCl (Sigma) at 60 °C for 5–10 min. Following a 2 min rinse in distilled water, the buds were incubated in 50% Schiff's reagent (Feulgen stain) (VWR Chemicals, Leuven, Belgium), for 20 min at room temperature, and then washed with 45% acetic acid, three times, for 5 min each time. The buds were then transferred to a clean slide and covered with 45% acetic acid to prevent drying, and from this stock, one or two buds were placed on a new glass slide and covered with a small drop of acetic acid. Under a dissecting microscope, the epidermis cells were carefully removed by using forceps and a scalpel blade. Using a teasing needle and scalpel, the exposed meristem cells were then separated out as much as possible to form a single layer in order to enable the clear identification of individual cells upon squashing; then, a cover slip was applied. A piece of filter paper was placed over the cover slip and then pressed firmly with the thumb to flatten the cells and to remove excess acetic acid. Using an Olympus FV1000 confocal microscope (Olympus Europa GMBH, Hamburg, Germany) under standard PI (propidium iodide) excitation settings, the Fuelgen-stained chromosomes were then counted by reviewing the layered three-dimensional cell section images, an approach that minimizes halation-related miscounting.

Supplementary Materials: The following supporting information can be downloaded at: https://www.mdpi.com/article/10.3390/plants11243489/s1, Table S1: Characterization of all collected taxa and samples from the *Arenaria ciliata* species complex, with the corresponding genome sizes (2c values in pg of DNA). Irregular floral morphology: * plants with 6 petals, ** plants with 9 petals. Figure S1: A—Confocal micrograph image of Fuelgen-stained late metaphase chromosomes in *Arenaria ciliata* subsp. *bernensis* from Dent de Brenleire (2n = 200, Fribourg, Switzerland). Image scale 1000×. B—Genome size estimation for *A. ciliata* subsp. *bernensis* from Dent de Brenleire using flow cytometry (for more details see Materials and Methods).

Author Contributions: Conceptualization, G.K.; methodology, G.K. and Y.F.; investigation, G.K. and B.C.; writing—original draft preparation, G.K. and C.M.; writing—review and editing, G.K., Y.F. and C.M.; visualization, Y.F.; project administration, G.K. All authors have read and agreed to the published version of the manuscript.

Funding: This research received no external funding.

Data Availability Statement: Not applicable.

Acknowledgments: We would like to thank Peter Wandeler and Emanuel Gerber from the Natural History Museum in Fribourg (NHMF, Switzerland) for their long-standing support in exploration of the flora of the Canton of Fribourg, as well as the Office of the Forest and Nature of the Canton of Fribourg and Pro Natura Fribourg for the collection permits.

Conflicts of Interest: The authors declare no conflict of interest.

References

1. Bennet, M.D.; Leitch, I.J. Plant genome size research: A field in focus. *Ann. Bot.* **2005**, *95*, 1–6. [CrossRef] [PubMed]
2. Pellicer, J.; Hidalgo, O.; Dodsworth, S.; Leitch, I.J. Genome size diversity and its impact on the evolution of land plants. *Genes* **2018**, *9*, 88. [CrossRef] [PubMed]
3. Doležel, J.; Bartoš, J. Plant DNA flow cytometry and estimation of nuclear genome size. *Ann. Bot.* **2005**, *95*, 99–110. [CrossRef] [PubMed]
4. Bourge, M.; Brown, S.C.; Siljak-Yakovlev, S. Flow cytometry as tool in plant sciences, with emphasis on genome size and ploidy level assessment. *Genet. Applic.* **2018**, *2*, 1–12. [CrossRef]
5. García-Fernández, A.; Iriondo, J.M.; Vallès, J.; Orellana, J.; Escidero, A. Ploidy level and genome size of locally adapted populations of *Silene ciliata* across an altitudinal gradient. *Plant Syst. Evol.* **2012**, *298*, 139–146. [CrossRef]
6. Zhang, J.; Wang, M.; Guo, Z.; Guan, Y.; Guo, Y.; Yan, X. Variation in ploidy level and genome size of *Cynodon dactylon* (L.) Pers. along latitudinal gradient. *Folia Geobot.* **2019**, *54*, 267–278. [CrossRef]
7. Greimler, J.; Temsch, E.M.; Xue, Z.; Weiss-Schneeweiss, H.; Volkova, P.; Peintinger, M.; Wasowicz, P.; Shang, H.; Schanzer, I.; Chiapella, J.O. Genome size variation in *Deschampsia cespitosa* sensu lato (Poaceae) in Eurasia. *Plant Syst. Evol.* **2022**, *308*, 9. [CrossRef]
8. Plaschil, S.; Abel, S.; Klocke, E. The variablility of nuclear DNA content of different Pelargonium species estimated by flow cytometry. *PLoS ONE* **2022**, *17*, e0267496. [CrossRef]
9. Kew Plant DNA C-Values Database. Available online: https://cvalues.science.kew.org/ (accessed on 18 November 2022).

10. Christenhusz, M.J.M.; Byng, J.W. The number of known plants species in the world and its annual increase. *Phytotaxa* **2016**, *261*, 201–217. [CrossRef]
11. GeSDaBaF. Genome Size of Balkan Flora Database. Available online: www.gesdabaf.pmf.unsa.ba (accessed on 18 November 2022).
12. Šmarda, P.; Knápek, O.; Březinová, A.; Horová, L.; Grulich, V.; Danihelka, J.; Veselý, P.; Šmarda, J.; Rotreklová, O.; Bureš, P. Genome sizes and genomic guanine+ cytosine (GC) contents of the Czech vascular flora with new estimates for 1700 species. *Preslia* **2019**, *91*, 117–142. [CrossRef]
13. Zonneveld, B.J.M. The DNA weights per nucleus (genome size) of more than 2350 species of the Flora of The Netherlands, of which 1370 are new to science, including the pattern of their DNA peaks. *Forum Geobot.* **2019**, *8*, 24–78.
14. Favarger, C. Nouvelles recherches sur les populations alpines et carpathiques d'*Arenaria ciliata* L. sens. lat. *Bull. Soc. Bot. Suisse* **1963**, *73*, 161–178.
15. Wyse-Jackson, M.B.; Parnell, J.A.N. A biometric study of the *Arenaria ciliata* L. complex (Caryophyllaceae). *Watsonia* **1987**, *16*, 373–382.
16. Abukrees, F.; Kozlowski, G.; Meade, C. Characterization of diverse ploidy in the arctic-alpine *Arenaria ciliata* species complex (Caryophyllaceae) using shoot meristem staining and flow cytometry analysis of archived frozen tissue. *Plant Spec. Biol.* **2018**, *33*, 144–152. [CrossRef]
17. Christenhusz, M.J.M.; Fay, M.F.; Chase, M.W. *Plants of the world: An Illustrated Encyclopedia of Vascular Plants*; University of Chicago Press: Chicago, IL, USA, 2017; ISBN 0-226-53670-X.
18. Dang, X.D.; Kelleher, C.T.; Howard-Williams, E.; Meade, C. Rapid identification of chloroplast haplotypes using High Resolution Melting analysis. *Mol. Ecol. Resour.* **2012**, *12*, 894–908. [CrossRef]
19. Berthouzoz, M.; Maendly, S.; Bétrisey, S.; Mangili, S.; Prunier, S.; Lexer, C.; Kozlowski, G. Some like it cold: Distribution, ecology and phylogeny of *Arenaria bernensis* Favarger (Caryophyllaceae) from western Prealps in Switzerland. *Alpine Bot.* **2013**, *123*, 65–75. [CrossRef]
20. Walker, K.; Howard-Williams, E.; Meade, C. The distribution and ecology of *Arenaria norvegica* Gunn. in Ireland. *Irish Nat. J.* **2013**, *32*, 1–13.
21. Svalbard Flora. Available online: https://www.svalbardflora.no (accessed on 9 November 2022).
22. Panarctic Flora. Available online: http://panarcticflora.org/3701 (accessed on 9 November 2022).
23. Lauber, K.; Wagner, G.; Gygax, A. *Flora Helvetica. 6. Ausfabe*; Haupt: Bern, Switzerland, 2018.
24. Eggenberg, S.; Landolt, E. Für welche Pflanzenarten hat die Schweiz eine internationale Verantwortung? *Bot. Helv.* **2006**, *116*, 119–133. [CrossRef]
25. Gerber, E.; Kozlowski, G.; Mariéthoz, A.-S. *La Flore des Préalpes du lac de Thun au Léman*; Rossolis: Bussigny, Switzerland, 2010.
26. Parisod, C. Plant speciation in the face of recent climate changes in the Alps. *Alpine Bot.* **2022**, *132*, 21–28. [CrossRef]
27. Favarger, C.; Contandriopoulos, J. Essai sur l'endémisme. *Bull. Soc. Bot. Suisse* **1961**, *71*, 384–408.
28. Tralau, H. On the distribution of *Arenaria gothica* Fries and the significance of postglacial plant migration. *Ber. Schweiz. Bot. Ges.* **1959**, *69*, 342–345.
29. Duckert-Henriod, M.-M. Sur l'*Arenaria gothica* Fries de la Vallée de Joux. *Bull. Soc. Neuch. Sc. Nat.* **1962**, *85*, 97–101.
30. Aeschimann, D.; Lauber, K.; Moser, D.; Theurillat, J. *Flora Alpina*; Haupt: Bern, Switzerland, 2004; Volume 1.
31. Hörandl, E. Novel approaches for species concepts and delimitation in polyploids and hybrids. *Plants* **2022**, *11*, 204. [CrossRef] [PubMed]
32. Siljak-Yakovlev, S.; Pustahija, F.; Šolic, E.M.; Bogunic, F.; Muratovic, E.; Bašic, N.; Catrice, O.; Brown, S.C. Towards a genome size and chromosome number database of Balkan flora: C-values in 343 taxa with novel values for 242. *Adv. Sci. Lett.* **2010**, *3*, 190–213. [CrossRef]
33. Daoud, M. Genetic and Phenotypic Patterns of Variablities in *Arenaria grandiflora* L. Species Complex (Caryophyllaceae). New Elements for Taxonomy and Conservation. Ph.D. Thesis, Museum National d'Histoire Naturelle, Paris, France, 2017; pp. 1–180.
34. Bennett, M.D.; Smith, J.B. Nuclear DNA amounts in Angiosperms. *Philos. T. Roy. Soc. B.* **1991**, *334*, 309–345. [CrossRef]
35. Loureiro, J.; Castro, M.; Cerca de Oliveira, J.; Mota, L.; Torices, R. Genome size variation and polyploidy incidence in the alpine flora from Spain. *Ann. Jard. Bot. Madrid* **2013**, *70*, 39–47. [CrossRef]
36. Bou Dagher-Kharrat, M.; Abdel-Samad, N.; Douaihy, B.; Bourge, M.; Fridlender, A.; Siljak-Yakovlev, S.; Brown, S.C. Nuclear DNA C-values for biodiversity screening: Case of the Lebanese flora. *Plant Biosyst.* **2013**, *147*, 1228–1237. [CrossRef]
37. Halliday, G. The identity of *Arenaria gothica* auct. angl. *Watsonia* **1960**, *4*, 207–210.
38. Delarze, R.; Druart, P.; Kozlowski, G.; Moret, J.-L.; Prunier, P.; Gmür, P. *Arenaria gothica* et autres plantes rares des rives lacustres de la Vallée de Joux (Vaud, Suisse). Quelques observations récentes et premières données autoécologiques. *Bull. Soc. Vaude. Sc. Nat.* **2004**, *89*, 1–11.
39. Favarger, C. Recherches cytotaxonomiques sur les populations alpines d'*Arenaria ciliata* L. (sens. lat.). *Bull. Soc. Bot. Suisse* **1960**, *70*, 126–140.
40. Favarger, C. A striking polyploid complex in the alpine flora: *Arenaria ciliata* L. *Bot. Notiser* **1965**, *118*, 273–280.
41. Aeschimann, D.; Heitz, C. *Index Synonymique de la Flore de Suisse et Territoires Limitrophes (ISFS)*; Documenta Floristicae Helvetiae: Geneva, Switzerland, 2005.
42. Bétrisey, S.; Kozlowski, G. *Arenaria bernensis*: Vers la fin d'une longue controverse. *FloraCH* **2013**, *2013*, 29–37.

43. Watts, S.H.; Mardon, D.K.; Mercer, C.; Watson, D.; Cole, H.; Shaw, R.F.; Jump, A.S. Riding the elevator to extinction: Disjunct arctic-alpine plants of open habitats decline as their more competitive neighbours expand. *Biol. Conserv.* **2022**, *272*, 109620. [CrossRef]
44. Körner, C.; Hiltbrunner, E. Why is the Alpine flora comparatively robust against climatic warming? *Diversity* **2021**, *13*, 383. [CrossRef]

plants

MDPI

Article

Genome Size in the *Arenaria ciliata* Species Complex (Caryophyllaceae), with Special Focus on Northern Europe and the Arctic

Gregor Kozlowski [1,2,3,*], Yann Fragnière [1], Benoît Clément [1], Olivier Gilg [4,5], Benoît Sittler [5,6], Johannes Lang [5,7], Pernille Bronken Eidesen [8], Simone I. Lang [9], Pawel Wasowicz [10] and Conor Meade [11]

1 Department of Biology and Botanical Garden, University of Fribourg, Chemin du Musée 10, 1700 Fribourg, Switzerland; yann.fragniere@unifr.ch (Y.F.); benoit.clement@unifr.ch (B.C.)
2 Natural History Museum Fribourg, Chemin du Musée 6, 1700 Fribourg, Switzerland
3 Eastern China Conservation Centre for Wild Endangered Plant Resources, Shanghai Chenshan Botanical Garden, 3888 Chenhua Road, Songjiang, Shanghai 201602, China
4 UMR 6249 Chrono-Environment, CNRS, Université de Bourgogne Franche-Comté, 25000 Besançon, France; olivier.gilg@gmail.com
5 Groupe de Recherche en Ecologie Arctique (GREA), 16 rue de Vernot, 21440 Francheville, France; benoit.sittler@nature.uni-freiburg.de (B.S.); johannes.lang@vetmed.uni-giessen.de (J.L.)
6 Nature Conservation and Landscape Ecology, University of Freiburg, Tannenbacherstrasse 4, 79106 Freiburg im Breisgau, Germany
7 Arbeitsgruppe Wildtierforschung, Justus-Liebig-University Giessen, Frankfurter Strasse 114, 35392 Giessen, Germany
8 Department of Biosciences, University of Oslo, 0316 Oslo, Norway; p.b.eidesen@ibv.uio.no
9 Department of Arctic Biology, The University Centre in Svalbard, P.O. Box 156, 9171 Longyearbyen, Norway; simonel@unis.no
10 Icelandic Institute of Natural History, Borgum við Norðurslóð, 600 Akureyri, Iceland; pawel@ni.is
11 Molecular Ecology & Biogeography Laboratory, Biology Department, Maynooth University, W23 F2H6 Maynooth, Ireland; conor.meade@mu.ie
* Correspondence: gregor.kozlowski@unifr.ch; Tel.: +41-26-300-8842

Abstract: The main aim of the present study has been the completion of genome size data for the diverse arctic-alpine *A. ciliata* species complex, with special focus on the unexplored arctic taxon *A. pseudofrigida*, the north-European *A. norvegica*, and *A. gothica* from Gotland (Sweden). Altogether, 46 individuals of these three Nordic taxa have been sampled from seven different regions and their genome size estimated using flow cytometry. Three other alpine taxa in the *A. ciliata* complex (*A. multicaulis*, *A. ciliata* subsp. *ciliata*, and *A. ciliata* subsp. *bernensis*) were also collected and analyzed for standardization purposes, comprising 20 individuals from six regions. A mean 2c value of 1.65 pg of DNA was recorded for *A. pseudofrigida*, 2.80 pg for *A. norvegica*, and 4.14 pg for *A. gothica*, as against the reconfirmed 2c value of 1.63 pg DNA for the type taxon *A. ciliata* subsp. *ciliata*. Our results presenting the first estimations of genome sizes for the newly sampled taxa, corroborate ploidy levels described in the available literature, with *A. pseudofrigida* being tetraploid (2n = 4x = 40), *A. norvegica* possessing predominantly 2n = 8x = 80, and *A. gothica* with 2n = 10x = 100. The present study also reconfirms genome size and ploidy level estimations published previously for the alpine members of this species complex. Reflecting a likely complex recent biogeographic history, the *A. ciliata* species group comprises a polyploid arctic-alpine species complex characterized by reticulate evolution, polyploidizations and hybridizations, probably associated with rapid latitudinal and altitudinal migrations in the Pleistocene–Holocene period.

Keywords: arctic-alpine plants; *Arenaria norvegica*; *Arenaria gothica*; *Arenaria pseudofrigida*; flow cytometry; ploidy

Citation: Kozlowski, G.; Fragnière, Y.; Clément, B.; Gilg, O.; Sittler, B.; Lang, J.; Eidesen, P.B.; Lang, S.I.; Wasowicz, P.; Meade, C. Genome Size in the *Arenaria ciliata* Species Complex (Caryophyllaceae), with Special Focus on Northern Europe and the Arctic. *Plants* **2024**, *13*, 635. https://doi.org/10.3390/plants13050635

Academic Editor: León A. Bravo

Received: 10 January 2024
Revised: 20 February 2024
Accepted: 23 February 2024
Published: 26 February 2024

1. Introduction

Closely related but disjunct arctic–alpine and boreo-montane taxa offer an excellent model system for the study of the influence of polyploidy on evolution and biogeography of plants, especially in the context of historical climate oscillation events [1,2]. These taxa provide a unique opportunity to explore how polyploidy has shaped their evolutionary trajectories and distribution patterns across changing climatic landscapes over time. Investigating their genetic diversity and adaptation strategies can deepen our understanding of how plants respond to environmental shifts and inform conservation efforts in rapidly changing ecosystems.

A classic example of such a group is the *Arenaria ciliata* L. species complex (Caryophyllaceae) comprising six herbaceous taxa with very similar morphology and ecology but with divergent arctic–alpine distribution ranges and ploidy levels [3–6]. In a recent study of this group, Kozlowski et al. [7] used flow-cytometry and genome size estimations to examine ploidy levels in populations across the Alps and Jura Mountains of central Europe, where four taxa occur in relatively close proximity, namely *A. ciliata* subsp. *ciliata* L., *A. ciliata* subsp. *bernensis* Favarger, *A. gothica* Fr., and *A. multicaulis* L. [4,8], with special focus on *A. ciliata* subsp. *bernensis*, an endemic plant occurring in the Swiss Northern Alps. The focus of the present study is to expand and complete genome size investigation for taxa in Northern Europe and the Arctic, where distribution ranges vary much more widely, and often comprise disjunct populations in reproductively isolated locations.

Genome size is known to vary greatly across organism lineages, including within and between plant groups [9]. This variability arises from a combination of processes leading to, on one hand, duplications and increase in DNA amounts in the genome (particularly polyploidization), and on the other, processes that filter duplications and eliminate DNA (primarily via recombination), in response to both selective pressures and neutral drift events [10]. Correspondingly, nuclear DNA amount and ploidy level are important biodiversity characters and play a significant role in the evolution of land plants, especially with respect to processes of speciation [11–13]. As a result, significant recent research efforts have focused on understanding the factors that shape genome size variation [14–16].

Recognizing the importance of this information in the context of understanding recent diversification in the *A. ciliata* complex across the Euro-Arctic biogeographic region, we focused our present work on genome size estimation in high-latitude taxa, using the same flow cytometry method as in our previous work [7], for the three unstudied target taxa: *A. pseudofrigida* (Ostenf. & O.C.Dahl) Juz. Ex Schischk. & Knorring and *A. norvegica* Gunn., as well as the Scandinavian populations of *A. gothica* Fr. [17–19]. When applied rigorously [20,21], flow-cytometry is established as a benchmark method for estimation of genome sizes and DNA ploidy levels in plants (e.g., [14,22–26]), and provides a replicable protocol for comparison of samples with varying collection, storage, preparation and quantity variables [6]—an important consideration for the assembly of material from remote locations. The present work aims to deliver a targeted synthesis of genomic, biogeographic, and ecological information for three Nordic members of the *A. ciliata* species complex, which have never previously been analyzed collectively.

Arenaria pseudofrigida was described as a distinct taxon (*A. ciliata* subsp. *pseudofrigida*) in 1917 from Finnmark (Persfjorden, Vardø) in the far north of Norway, and was raised to a species status by Russian botanists Boris K. Schischkin and Olga E. Knorring in 1936 [27]. Apart of the neighborhood of the *locus classicus* (where the taxon is still well present), there are several isolated populations in Norway and Finland, numerous scattered occurrences in continental European part of the Russian North (e.g., polar Ural), and on several Russian arctic islands (e.g., Novaya Zemlya, Franz Joseph Land). Additionally, the taxon occurs in Svalbard and in Eastern Greenland, where it is relatively frequent (Figure 1) [5,18,28]. In Svalbard, hundreds of occurrences are known from three islands: Spitsbergen (mainly along Isfjorden, Wijdefjorden, and Kongsfjorden), Edge Island, and Prince Charles Foreland [19]. The most northern populations however, are known from north-eastern Greenland, growing as far as 82°30′ N [29].

Figure 1. Habitats and morphology of the two most Nordic members of the *Arenaria ciliata* species complex. (**A**,**B**): *Arenaria norvegica*. (**A**) Reykholahreppur, Iceland; (**B**) Latrabjarg, Iceland. (**C**,**D**): *Arenaria pseudofrigida*, Traill Island, Karupelv, Greenland. Photos: (**A**,**B**): Gregor Kozlowski; (**C**,**D**): Sven Büchner.

Arenaria pseudofrigida shares a common reproductive strategy typical for the *A. ciliata* species complex. It reproduces sexually via insect pollination (e.g., flies) but is also capable of selfing in the absence of pollinators. Dispersal is nominally quite restricted, as its seeds lack any specialized dispersal mechanisms, relying on wind motion and gravity to dislodge the 0.1–0.5 mm diameter seeds through the neck of the mature fruit capsule, which is indehiscent. Long-distance dispersal (without human influence) is most likely via transport in the digestive tract of birds [17,19].

Arenaria pseudofrigida is a specialist of gravelly ground, either along the coast (on raised beaches as on Figure 1D) or on glacial and alluvial deposits in valleys. In Svalbard, it is confined to circumneutral or alkaline soils. The taxon seems to be very tolerant to drought and wind abrasion [19] and is a typical element of the boreal and arctic base-rich scree and block fields (alliance *Arenarion norvegicae* Nordhagen 1935) [30]. The taxon was described as tetraploid, with $2n = 4x = 40$ [18,19,31].

Arenaria norvegica was described by Johann Ernst Gunnerus at the end of the 18th century, from Laskenstad (isle of Steigen, Northern Norway) in his *Flora Norvegica* [18,32]. The taxon occurs in Iceland (Figure 1) and Norway, where it is quite frequent, as well as possesses scattered and partially extremely small and isolated populations in Sweden, Finland, Shetland

Islands, northern Scotland, western Ireland, and northern England [5,17,18]. The English population is sometimes differentiated as a separated subspecies (*A. norvegica* subsp. *anglica* Halliday), endemic to mid-west Yorkshire [17,33]. *Arenaria norvegica* is found on base-rich sandy substrates and in fine scree and on riverside gravels (Figure 1) [5,17]. Similarly to *A. pseudofrigida*, the species is also very resistant to drought and wind, being able to prosper, for example, in the desert of lava, gravel and blown sand in the central plateau of Iceland [34]. According to the literature, it is an octoploid taxon with 2n = 8x = 80 [18,19,31].

Arenaria gothica, described from the isle of Gotland (Baltic Sea, Sweden) by Wahlenberg [35], was raised to a species rank by Fries [36]. Later, a similar form was discovered in Switzerland, along the shore of the Lac des Joux [37] and was judged by Grenier [38] to be conspecific with *A. gothica* Fr. In the second half of the 19th century, an additional occurrence was found on the Swedish mainland (Kinnekulle, Västergötland) [39,40]. *Arenaria gothica* has therefore an extremely disjunct distribution (distribution maps in [39] and [40]), and its taxonomic position, putative origin and its glacial relict status is a subject of controversy [33,40,41]. The ecology, particularity of the Swiss populations of *A. gothica*, has been explored in detail in our recent work [7]. In Sweden the taxon occurs on chalk and limestone, mainly in open habitat called *alvar* [40]. Alvars are almost level areas that are only sparsely covered by vegetation of the order *Alysso-Sedetalia* (basiphilous dry grasslands of shallow, skeletal soils), and are restricted to the Baltic islands of Sweden and to Estonia [39,42]. The chromosome number from both disjunct regions was counted as 2n = 10x = 100 [7,40].

The aim of the present study was to complement our earlier work exploring alpine and Jura Mountain taxa of the *A. ciliata* complex [7] and deliver the first evaluation of genome size variation for yet unstudied taxa from Northern Europe and the Arctic. In this way, the main focus of the present work is on *A. pseudofrigida* and *A. norvegica*, as well as on the Swedish populations of *A. gothica* from Gotland. The following specific questions have been addressed: (1) What are the differences in genome size among the three north-European and arctic species of the complex, in comparison with closely related taxa occurring in the Alps and neighboring mountain ranges? (2) Do the obtained results corroborate the ploidy levels of the studied taxa known from the literature? Based on our results, the influence of the ploidy level on the evolutionary and biogeography history of the *A. ciliata* species complex will be discussed.

2. Results

The 2c values recorded for *A. pseudofrigida* were varying between 1.54 pg and 1.84 pg of DNA (Figure 2, Table S1), with a mean 2c value of 1.65 pg (Table 1, standard deviation, SD ± 0.11). The recorded genome size of *A. norvegica* was higher, with 2c values varying between 2.68 pg and 2.89 pg of DNA, with a mean 2c value of 2.80 pg (SD ± 0.02). The third Nordic taxon, *A. gothica*, displayed the highest recorded values, varying between 3.90 pg and 4.41 pg of DNA, with mean 2c value of 4.14 (SD ± 0.26). The higher SD-values could be attributed to two different storage techniques of the plant material, as explained in the Materials and Methods. Plant tissue desiccated with silica gel and stored at room temperature showed higher 2c values (4.33 and 4.41) than duplicate samples stored for a longer time at −20 °C (3.90 and 3.94). These differences are most probably due to technical issues, and not natural variation among populations, as explained, for example, by Sliwinska et al. [20].

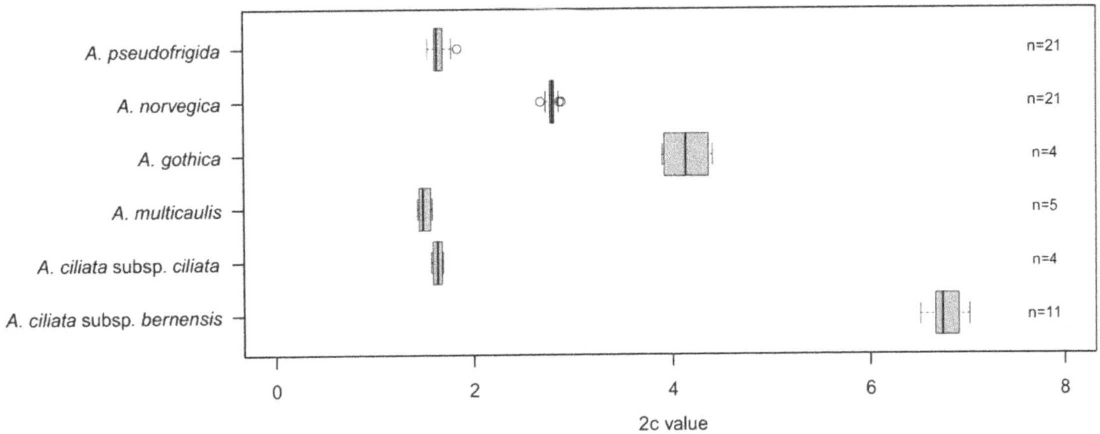

Figure 2. Variation in genome size of four taxa in the *Arenaria ciliata* species complex analyzed in this study (2c values are given in pg of DNA). Box plots showing the quartiles, the 5th and 95th percentiles (whiskers) and the outliers. The number of analyzed individuals is indicated on the right of the plot.

Table 1. Estimated ploidy level and genome size (mean ± standard deviation) in *Arenaria ciliata* species complex in the Arctic, Northern Europe, and the Alps. The genome size (2c values) is given in pg of DNA. The estimated ploidy level is based on comparison against all previously published 2c values and chromosome counts (e.g., [6–8,18,19,33,40,41,43–45]).

Taxon	2c Nuclear DNA Amount (pg DNA), Mean (± SD)	Estimated Ploidy Level
Nordic taxa:		
Arenaria pseudofrigida	1.65 (±0.11)	2n = 4x = 40
Arenaria norvegica	2.80 (±0.02)	2n = 8x = 80
Arenaria gothica (Gotland)	4.14 (±0.26)	2n = 10x = 100
Alpine taxa:		
Arenaria multicaulis	1.50 (±0.06)	2n = 4x = 40
Arenaria ciliata subsp. *ciliata*	1.63 (±0.06)	2n = 4x = 40
Arenaria ciliata subsp. *bernensis*	6.77 (±0.03)	2n = 20x = 200

Arenaria ciliata subsp. *ciliata* and *A. multicaulis*, both showed similar but much lower values, with a mean 2c value of 1.63 (SD ± 0.06) for *A. ciliata* subsp. *ciliata* and a mean 2c value of 1.50 (SD ± 0.06) for *A. multicaulis* (Figure 2, Table 1). The highest 2c values among all the six taxa from the *A. ciliata* complex investigated in this study were recorded for *A. ciliata* subsp. *bernensis*, varying between 6.51 pg and 7.02 pg of DNA, with a mean 2c value of 6.77 pg (SD ± 0.03). The results show that, for all six taxa taken separately, the genome size is very stable, thus indicating an invariant ploidy level of all investigated individuals within a given taxon (Figure 2, Table S1).

3. Discussion

The study identified stable but varying genome sizes across different taxa studied within the *A. ciliata* species complex. It is composed exclusively of polyploid taxa, with no diploids. Polyploid species complexes play an important role in forming local floras [46]. The *A. ciliata* group is one of many polyploid arctic–alpine complexes characterized by reticulate evolution, polyploidizations and hybridizations. Well-studied similar complexes exist elsewhere in the family Caryophyllaceae, for example, in the genera *Cerastium* [47] and *Silene* [48]. Polyploid species complexes are also observed in numerous other well-documented circumpolar, arctic–

alpine, and alpine genera and families, for example, in *Draba* (Brassicaceae, [49]), *Primula* (Primulaceae, [50,51]) and *Calamagrostis* (Poaceae, [18,19,52]).

Our present study, focused on north-European and arctic members of the *A. ciliata* species complex, completes genome size estimations initiated in Kozlowski et al. [7]. Although allowing for the potential limitations of genome size estimations using flow cytometry [20,21], this technique facilitates a synthesis of ploidy level variability within the target arctic–alpine plant group. According to our results, the following three taxa are predominantly tetraploid (2n = 4x = 40): the arctic *A. pseudofrigida* and the two alpine taxa *A. multicaulis* and *A. ciliata* subsp. *ciliata*. Higher ploidy is observed in *A. norvegica* (2n = 8x = 80) and *A. gothica* (2n = 10x = 100), and the highest confirmed ploidy level is for the narrow endemic of Western Alps *A. ciliata* subsp. *bernensis* (2n = 20x = 200). These estimates corroborate chromosome counts available in the published literature [18,19,33,40,41,43–45]. However, for some taxa of the present study, greater variability in ploidy has been documented in previous work, for example, for *A. ciliata* subsp. *ciliata* (2n = 40, 80, 120, 160, 200), but also for *A. norvegica* (2n = 60, 80) [6,8,44]. Abukrees et al. [6] showed that such atypical chromosome counts could be detected in 12.5% of investigated individuals for *A. norvegica* and in as much as 37% of investigated individuals for *A. ciliata* subsp. *ciliata*. In contrast, the available literature and our results demonstrate a very stable ploidy level in the remaining four taxa, namely *A. pseudofrigida* and *A. multicaulis* (both tetraplid), *A. gothica* (decaploid) and *A. ciliata* subsp. *bernensis* (dodecaploid), and was this across the whole investigated distribution area.

Our results thus confirm that there are no extant diploid taxa within the *A. ciliata* species complex. The group comprises an example of a so-called *mature polyploid complex* in the sense of Stebbins [53]. This is not an exception among arctic–alpine disjunct taxa. A similar polyploidy pattern can be observed, for example, in the *Cerastium alpinum/arcticum* species complex [47] and in the *Calamagrostis stricta/neglecta* group [19,52], where the lowest chromosome number known is tetraploid (for *Calamagrostis*) or even octoploid (*Cerastium*). In addition, the *A. ciliata* species complex confirms the conclusion of Brochmann et al. [1], that almost 90% of arctic specialist plant taxa growing in regions that were heavily glaciated during the last ice age, are polyploids.

Polyploidy is thought to infer fitness advantages allowing plants to adapt better to more extreme climatic conditions [54], and in this scenario, it may not be surprising that arctic and northern-latitude habitats include some of the most polyploid-rich floras [1]. In a related inference, it has been postulated that there is a general increasing gradient of polyploidy with latitude (and altitude) [55]. This latter hypothesis is now refuted. Stebbins [56] concluded, for example—by comparing polyploidy along the Pacific Coasts of North America—that the highest frequency of polyploids occurs at between 52° and 54° north, declining to the north and south. Additionally, the majority of typical arctic polyploid complexes (for example, in the genera *Calamagrostis, Campanula, Chamaenerion, Salix*, and *Saxifraga*), reveal the presence of diploids at their northern distributional limits. Similarly, the *A. ciliata* species complex explored in our study does not show a specific latitudinal ploidy gradient, with the lowest chromosome numbers in high arctic *A. pseudofrigida* as well as in the majority of high alpine populations of *A. ciliata* subsp. *ciliata* and *A. multicaulis*. It is the case, however, that the highest ploidy types are only found at high elevation, near the geographic center of diversity for the species complex.

The most plausible hypothesis explaining the high polyploid frequency in arctic–alpine plants, is the so-called *secondary contact hypothesis* [56]. According to this hypothesis, "polyploidy, accompanied by hybridization, is instrumental chiefly for rapid adaptation to new ecological conditions, that become available relatively suddenly" [56]. Migration at the beginning and the end of warm and cold periods is likely to be an important driver of ploidy diversification, particularly for recently evolved polyploids [57], with multiple recurrence of secondary contacts between previously separated closely related taxa [56]. Considering the historical and current geographic ranges of *A. ciliata* complex taxa, including the proliferation of extant populations in recently deglaciated regions of Europe and the Arctic, it is highly likely that secondary contacts played an important role in evolution of high polyploidy in the *A. ciliata* species complex. Our study corroborates conclusions of Abukrees et al. [6], stating that the *A. ciliata* complex arose from a reduced ploidy ancestral stock (2n = 40), probably in the Alps, which after the latitudinal and altitudinal migration gave rise to several polyploidization events. *Arenaria ciliata* subsp. *bernensis* (2n = 200), for example, is proposed to be an allopolyploid neo-endemic taxon resulting from hybridization between different related taxa due to rapid migration events after the last glaciation period (probably, in this case, of *A. multicaulis* and A. *ciliata* subsp. *ciliata*) [4,7,45,58]. The genome size and, thus, also the ploidy level, is stable across the whole distribution area of this taxon. The present study delivers additional support for the taxonomic distinctiveness of the high alpine endemic *A. ciliata* subsp. *bernensis*, which strongly aligns with other differences in morphology, phylogeny, phenology, ecology, and plant communities, described previously. In affirming these differences, further support now exists to re-consider the species status of this taxon. A similar allopolyploid origin after the last glaciation was proposed for *A. gothica* (2n = 100), with *A. norvegica* (2n = 80) and *A. serpyllifolia* subsp. *leptoclados* (2n = 20) as potential parents [33], or between *A. multicaulis* (2n = 40) and *A. serpyllifolia* subsp. *leptoclados* [41]. In this way, ploidy and genome size data from present study confirm that taxonomic and distributional differences among northern and arctic taxa of the *A. ciliata* complex are likely reflecting a discrete genetic origin and migratory history in each case.

4. Materials and Methods

4.1. Sampling of Plant Material

Most of the individuals of the *A. ciliata* species complex were recently collected in the field, explicitly for this study. *Arenaria pseudofrigida* was collected in July 2023 (21 individuals from 2 regions in Northeast Greenland, Figure 3, Table S1). Plant material of *A. norvegica* was sampled in June 2023, mainly in Iceland (21 individuals from 4 populations in western Iceland, Figure 3, Table S1).

Additionally, one individual from Scotland (UK) was included in the analysis from collections of one of the authors (C. Meade, Abukrees et al. [6]). Similarly, four individuals of *A. gothica* collected previously from Gotland in Sweden were also included in the present study. In addition, three alpine taxa of the *A. ciliata* species complex were collected in August and September 2023 in the Swiss Alps: 11 individuals of *A. ciliata* subsp. *bernensis* from 3 summit areas, 4 individuals of *A. ciliata* subsp. *ciliata* from 2 populations, and 5 individuals from one population of *A. multicaulis* (Figure 3, Table S1). The voucher specimens are stored in the herbarium of the Natural History Museum Fribourg (NHMF), Switzerland. None of these alpine populations and individuals were included in our previous study [7]. Plant material (small portion of flowering stem with flowers) was silica dried and kept for ca. 4 weeks in plastic bags prior to flow cytometry analyses. An exception was the sole sample from Scotland (*A. norvegica*) and two samples of *A. gothica* from Sweden which were desiccated and then stored at −20 °C prior to analyses. The whole plant material was sent for analysis to the Plant Cytometry Services (Didam, The Netherlands, www.plantcytometry.nl, accessed on 15 December 2023), a biological research company specialized for ploidy and genome size analysis.

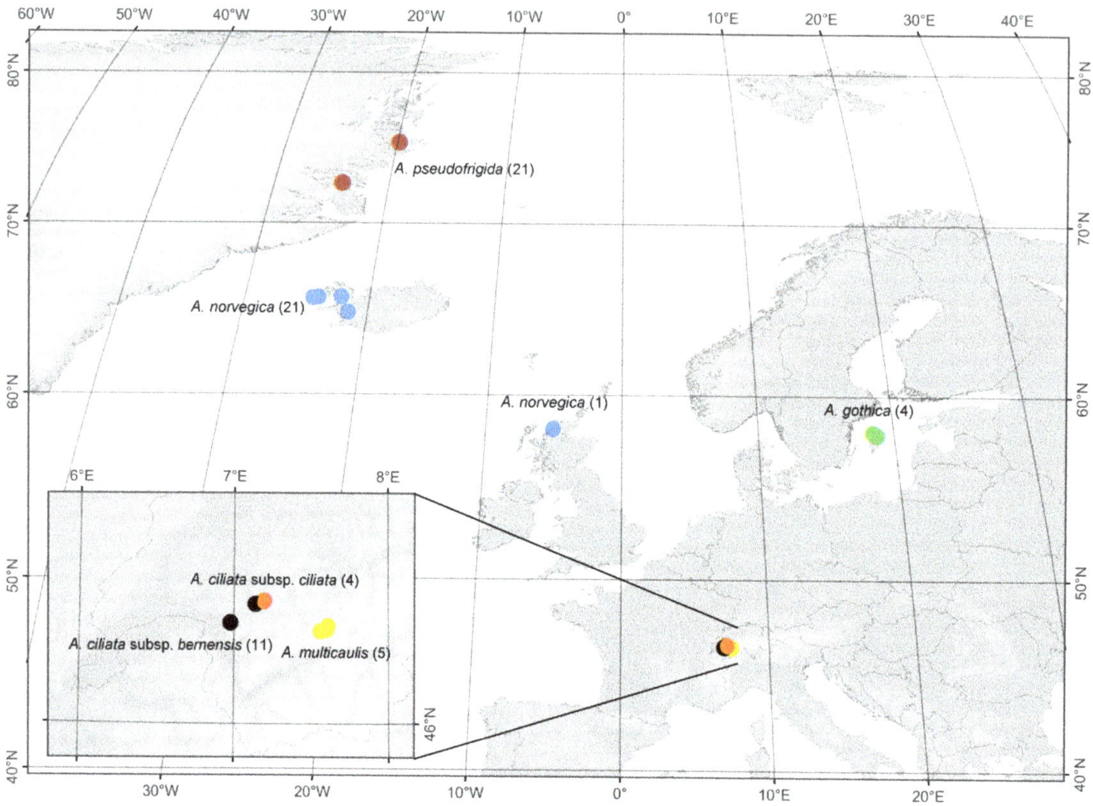

Figure 3. Geographic position of collected sites with *Arenaria ciliata* species complex, with the numbers of individuals sampled in the given area in parentheses. Red: *A. pseudofrigida*; blue: *A. norvegica*; green: *A. gothica*; yellow: *A. multicaulis*; orange: *A. ciliata* subsp. *ciliata*; black: *A. ciliata* subsp. *bernensis*.

4.2. Flow Cytometry Analysis

Approximately 1 cm^2 of leaves of the *Arenaria* samples were mixed with 1 cm^2 of fresh leaves of the standard plants (*Allium schoenoprasum*, genome size 2c = 15.03 pg). This was chopped with a sharp razor blade to release the nuclei in 100 µL of CyStain nuclei extraction buffer (Sysmex, Norderstedt, Germany, https://eu.sysmex-flowcytometry.com, accessed on 25 December 2023). The obtained suspension was then sieved through a 40 µm filter, and 1.5 mL of CyStain PI (propidium iodide) absolute P staining buffer was added. After one hour, the fluorescence of nuclei in the suspension was measured using a Sysmex ploidy analyzer (Sysmex, Norderstedt, Germany). Each individual was analyzed once. The number of nuclei measured for each sample was large enough in order to determine the ploidy and ranged between 100 and 1000 nuclei per sample (Table S1). The use of higher nuclei numbers would influence the genome size estimations by 1–2%. Ploidy level was estimated based on comparison against all previously published 2c values and chromosome counts (e.g., [6–8,18,19,33,40,41,43–45]), as well as on our own chromosome counting in the *A. ciliata* species complex [7].

Supplementary Materials: The following supporting information can be downloaded at: https://www.mdpi.com/article/10.3390/plants13050635/s1, Table S1: Characterization of all collected taxa and samples from the *Arenaria ciliata* species complex, with the corresponding genome sizes (2c values in pg of DNA and the nuclei number per analyzed sample). Figure S1: Examples of histograms of genome size estimations using flow cytometry. A: *Arenaria pseudofrigida* from Traill Island, Greenland. B: *Arenaria norvegica* from Latrabjarg, Iceland (Plant Cytometry Services, Didam, The Netherlands, www.plantcytometry.nl, accessed on 15 December 2023).

Author Contributions: Conceptualization, G.K.; methodology, G.K., Y.F. and C.M.; formal analysis, G.K., Y.F. and C.M.; investigation, G.K., B.C., O.G., B.S., J.L., P.B.E., S.I.L. and P.W.; data curation, G.K., Y.F. and C.M.; writing—original draft preparation, G.K.; writing—review and editing, all authors.; visualization, G.K. and Y.F.; supervision, G.K.; project administration, G.K.; funding acquisition, G.K. and O.G. All authors have read and agreed to the published version of the manuscript.

Funding: This research was funded by the University of Fribourg. APC was funded by the University of Fribourg, Switzerland. Field work at Hochstetter site, NE Greenland (OG), was funded by the French Agence National de la Recherche (Grant: ANR-21-CE02-0024 PACS) and the French Polar Institute-IPEV (Program: Interactions 1036).

Data Availability Statement: Data are contained within the article and supplementary materials.

Acknowledgments: We would like to thank Alain Müller from the Botanical Garden of the University of Fribourg (Switzerland) and Peter Wandeler from the Natural History Museum in Fribourg (NHMF, Switzerland) for their long-standing support, as well as the Office of the Forest and Nature of the Canton of Fribourg and Pro Natura Fribourg for the collection permits. Additionally, we would like to thank Jérémie Guenat from the Cercle vaudois de botanique, Switzerland) for the collection of samples in the Canton of Vaud, and Sven Büchner for the photographs of *A. pseudofrigida* and its habitat. Special thanks go to the Government of Greenland (Ministry of Foreign Affairs) for the issue of permit for utilization of Greenland genetic resources (license no. G23-17). We would also like to thank Bo Göran Johansson and colleagues for collection of material in Gotland (Sweden).

Conflicts of Interest: The authors declare no conflicts of interest.

References

1. Brochmann, C.; Brysting, A.K.; Alsos, I.G.; Borgen, L.; Grundt, H.H.; Scheen, A.-C.; Elven, R. Polyploidy in arctic plants. *Biol. J. Linn. Soc.* **2004**, *82*, 521–536. [CrossRef]
2. Stebbins, G.L. The origin and success of polyploids in the boreal circumpolar flora: A new analysis. *Bot. J. Scot.* **2006**, *58*, 151–164. [CrossRef]
3. Dang, X.D.; Kelleher, C.T.; Howard-Williams, E.; Meade, C. Rapid identification of chloroplast haplotypes using High Resolution Melting analysis. *Mol. Ecol. Resour.* **2012**, *12*, 894–908. [CrossRef] [PubMed]
4. Berthouzoz, M.; Maendly, S.; Bétrisey, S.; Mangili, S.; Prunier, S.; Lexer, C.; Kozlowski, G. Some like it cold: Distribution, ecology and phylogeny of *Arenaria bernensis* Favarger (Caryophyllaceae) from western Prealps in Switzerland. *Alpine Bot.* **2013**, *123*, 65–75. [CrossRef]
5. Walker, K.J.; Howard-Williams, E.; Meade, C. The distribution and ecology of *Arenaria norvegica* Gunn. in Ireland. *Irish Nat. J.* **2013**, *32*, 1–12.
6. Abukrees, F.; Kozlowski, G.; Meade, C. Characterization of diverse ploidy in the arctic-alpine *Arenaria ciliata* species complex (Caryophyllaceae) using shoot meristem staining and flow cytometry analysis of archived frozen tissue. *Plant Spec. Biol.* **2018**, *33*, 144–152. [CrossRef]
7. Kozlowski, G.; Fragnière, Y.; Clément, B.; Meade, C. Genome size in the *Arenaria ciliata* species complex (Caryophyllaceae, with special focus on *A. ciliata* subsp. *bernensis*, a narrow endemic of the Swiss Northern Alps. *Plants* **2022**, *11*, 3489. [CrossRef]
8. Lauber, K.; Wagner, G.; Gygax, A. *Flora Helvetica*, 6th ed.; Haupt: Bern, Switzerland, 2018.
9. Blommaert, J. Genome size evolution: Towards new model systems for old questions. *Proc. R. Soc. B* **2020**, *287*, 20201441. [CrossRef]
10. Bhadra, S.; Leitch, I.J.; Onstein, R.E. From genome size to trait evolution during angiosperm radiation. *Trends Genet.* **2023**, *39*, 728–735. [CrossRef]
11. Bennett, M.D.; Leitch, I.J. Plant genome size research: A field in focus. *Ann. Bot.* **2005**, *95*, 1–6. [CrossRef]
12. Rieseberg, L.; Willis, J.H. Plant Speciation. *Science* **2007**, *317*, 910–914. [CrossRef]
13. Pellicer, J.; Hidalgo, O.; Dodsworth, S.; Leitch, I.J. Genome size diversity and its impact on the evolution of land plants. *Genes* **2018**, *9*, 88. [CrossRef]
14. Loureiro, J.; Castro, M.; Cerca de Oliveira, J.; Mota, L.; Torices, R. Genome size variation and polyploidy incidence in the alpine flora from Spain. *Ann. Jard. Bot. Madrid* **2013**, *70*, 39–47. [CrossRef]

15. Bures, P.; Elliott, T.L.; Vesely, P.; Smarda, P.; Forest, F.; Leitch, I.J.; Lughadha, E.N.; Soto Gomez, S.; Pironon, S.; Brown, M.J.M.; et al. The global distribution of angiosperm genome size is shaped by climate. *bioRxiv* **2023**. [CrossRef]
16. Wang, D.; Zheng, Z.; Li, Y.; Hu, H.; Wang, Z.; Du, X.; Zhang, S.; Zhu, M.; Dong, L.; Ren, G.; et al. Which factors contribute most to genome size variation within angiosperms? *Ecol. Evol.* **2021**, *11*, 2660–2668. [CrossRef]
17. Walker, K.J. The distribution, ecology and conservation of *Arenaria norvegica* subsp. *anglica* Halliday (Caryophyllaecae). *Watsonia* **2000**, *23*, 197–208.
18. Panarctic Flora. Available online: https://www.panarcticflora.org (accessed on 13 December 2023).
19. Svalbard Flora. Available online: https://www.svalbardflora.no (accessed on 13 December 2023).
20. Sliwinska, E.; Loureiro, J.; Leitch, I.J.; Šmarda, P.; Bainard, J.; Bureš, P.; Chumová, Z.; Horová, L.; Koutecký, P.; Lučanová, M.; et al. Application-based guidelines for best practices in plant flow cytometry. *Cytometry Part A* **2022**, *101*, 749–781. [CrossRef] [PubMed]
21. Temsch, E.M.; Koutecký, P.; Urfus, T.; Šmarda, P.; Doležel, J. Reference standards for flow cytometric estimation of absolute nuclear DNA content in plants. *Cytometry Part A* **2021**, *99*, 710–724. [CrossRef] [PubMed]
22. García-Fernández, A.; Iriondo, J.M.; Vallès, J.; Orellana, J.; Escidero, A. Ploidy level and genome size of locally adapted populations of *Silene ciliata* across an altitudinal gradient. *Plant Syst. Evol.* **2012**, *298*, 139–146. [CrossRef]
23. Zhang, J.; Wang, M.; Guo, Z.; Guan, Y.; Guo, Y.; Yan, X. Variation in ploidy level and genome size of *Cynodon dactylon* (L.) Pers. along latitudinal gradient. *Folia Geobot.* **2019**, *54*, 267–278. [CrossRef]
24. Greimler, J.; Temsch, E.M.; Xue, Z.; Weiss-Schneeweiss, H.; Volkova, P.; Peintinger, M.; Wasowicz, P.; Shang, H.; Schanzer, I.; Chiapella, J.O. Genome size variation in *Deschampsia cespitosa* sensu lato (Poaceae) in Eurasia. *Plant Syst. Evol.* **2022**, *308*, 9. [CrossRef]
25. Plaschil, S.; Abel, S.; Klocke, E. The variability of nuclear DNA content of different *Pelargonium* species estimated by flow cytometry. *PLoS ONE* **2022**, *17*, e0267496. [CrossRef] [PubMed]
26. Loureiro, J.; Čertner, M.; Lučanová, M.; Śliwińska, E.; Kolář, F.; Doležel, J.; Garcia, S.; Castro, S.; Galbraith, D.W. The use of flow cytometry for estimating genome sizes and DNA ploidy levels in plants. In *Plant Cytogenetics and Cytogenomics: Methods and Protocols, Methods in Molecular Biology*; Heithkam, T., Garcia, S., Eds.; Springer: New York, NY, USA, 2023; Volume 2672, pp. 25–64.
27. Komarov, V.L. *Flora URSS (Flora Unionis Rerumpublicarum Sovieticum Socialisticarum)*; Academy of Sciences of the USSR: Moscow, Russia, 1936; Volume 6, p. 537.
28. Mossberg, B.; Stenberg, L. *Nordens Flora*; Livonia Print: Riga, Latvia, 2019.
29. Rune, F. *Wild Flowers of Greenland*; Narayana Press: Gylling, Denmark, 2011.
30. Matveyeva, N.V.; Lavrinenko, O.V.; Lavrinenko, I.A. Plant communities with *Arenaria pseudofrigida* (Ostenf. et Dahl) Juz. Ex Schischk. on islands of the south-eastern part of Barents Sea. *Veg. Russ.* **2013**, *22*, 78–85. [CrossRef]
31. Tutin, T.G.; Heywood, V.H.; Burges, N.A.; Moore, D.M.; Valentine, D.H.; Walters, S.M.; Webb, D.A. (Eds.) Flora Europaea. In *Pslotaceae to Plantanaceae*, 2nd ed.; Cambridge University Press: Cambridge, UK, 1993; Volume 1, p. 587.
32. Gunnerus, J.E. *Flora Norvegica—Observationibus Praesertim Oeconomics Panosque Norvegici Locupletata*; Nidrosiae: Trondheim, Norway, 1766.
33. Halliday, G. The identity of *Arenaria gothica* auct. angl. *Watsonia* **1960**, *4*, 207–210.
34. Anderson, F.W.; Falk, P. Observations on ecology of the central desert of Iceland. *J. Ecol.* **1935**, *23*, 406–421. [CrossRef]
35. Wahlenberg, G. Gothlandiae plantae rariores annis 1816, 1817 et 1818 detecta: A Joh. Petr. Rosén, et jam conjunctim cum eo recensitae. *Nova Acta Regiae Soc. Sci. Upsal.* **1821**, *8*, 203–225.
36. Fries, E. *Novitarium Florae Suecicae Mantissa Altera*; Leffler et Sebell: Uppsala, Sweden, 1839.
37. Grenier, C.; Gordon, D.A. *Flore de France*; Jean-Baptiste Baillière: Paris, France, 1848.
38. Grenier, C. Notes critiques sur quelques plantes jurassiques. *Bull. Soc. Bot.* **1869**, *16*, LXI. [CrossRef]
39. Albertson, N. *Österplana hed ett Alvarområde på Kinnekulle*; Almquist & Wiksells Boktryckeri AB: Uppsala, Sweden, 1946.
40. Tralau, H. On the distribution of *Arenaria gothica* Fries and the significance of postglacial plant migration. *Ber. Schweiz. Bot. Ges.* **1959**, *69*, 342–345.
41. Duckert-Henriod, M.-M. Sur l'*Arenaria gothica* Fries de la Vallée de Joux. *Bull. Soc. Neuch. Sc. Nat.* **1962**, *85*, 97–101.
42. Dengler, J. The basiphilous dry grasslands of shallow, skeletal soils (Alysso-Sedetalia) on the island of Öland (Sweden), in the context of North and Central Europe. *Phytocoenologia* **2006**, *36*, 343–391. [CrossRef]
43. Favarger, C. Recherches cytotaxonomiques sur les populations alpines d'*Arenaria ciliata* L. (sens. lat.). *Bull. Soc. Bot. Suisse* **1960**, *70*, 126–140.
44. Favarger, C. Nouvelles recherches sur les populations alpines et carpathiques d'*Arenaria ciliata* L. *sens. lat. Bull. Soc. Bot. Suisse* **1963**, *73*, 161–178.
45. Favarger, C. A striking polyploid complex in the alpine flora: *Arenaria ciliata* L. *Bot. Notiser* **1965**, *118*, 273–280.
46. Stebbins, G.L. Polyploid complexes in relation to ecology and the history of floras. *Am. Nat.* **1942**, *76*, 36–45. [CrossRef]
47. Brysting, A.K.; Hagen, A. Species in Polyploid Complexes? The *Cerastium alpinum*—*C. arcticum* Complex. 2005. Panarctic Flora Project. 2005. Available online: https://www.binran.ru (accessed on 22 December 2023).
48. Popp, M.; Erixon, P.; Eggens, F.; Oxelman, B. Origin and evolution of a circumpolar polyploid species complex in *Silene* (Caryophyllaceae) inferred from low copy nuclear DNA polymerase introns, rDNA, and chloroplast DNA. *Syst. Bot.* **2005**, *30*, 302–313. [CrossRef]

49. Grundt, H.H.; Popp, M.; Brochmann, C.; Oxelman, B. Polyploid origins in a circumpolar complex in *Draba* (Brassicaceae) inferred from cloned nuclear DNA sequences and fingerprints. *Mol. Phylogenet. Evol.* **2004**, *32*, 695–710. [CrossRef] [PubMed]
50. Guggisberg, A.; Mansion, G.; Conti, E. Disentangling reticulate evolution in an arctic-alpine polyploid complex. *Syst. Biol.* **2009**, *58*, 55–73. [CrossRef]
51. Casazza, G.; Granato, L.; Minuto, L.; Conti, E. Polyploid evolution and Pleistocene glacial cycles: A case study from the alpine primrose *Primula marginata* (Primulaceae). *Evol. Biol.* **2012**, *12*, 56. [CrossRef] [PubMed]
52. Eggenberg, S.; Champoud, L.; Leibundgut, M.; Parisod, C.; Wyss, L.; Kozlowski, G. *Calamagrostis lonana* (Poaceae): A new grass species from the Pennine Alps (Switzerland). *Candollea* **2023**, *78*, 1–9. [CrossRef]
53. Stebbins, G.L. *Chromosomal Evolution in Higher Plants*; Edward Arnold Publishers Ltd.: London, UK, 1971.
54. Schinkel, C.C.F.; Kirchheimer, B.; Dellinger, A.S.; Klatt, S.; Winkler, M.; Dullinger, S.; Hörandl, E. Correlations of polyploidy and apomixis with elevation and associated environmental gradients in an alpine plant. *AoB Plants* **2016**, *8*, plw064. [CrossRef] [PubMed]
55. Packer, J.G. Polyploidy in the Canadian Arctic archipelago. *Arct. Antarct. Alp. Res.* **1969**, *1*, 15–28. [CrossRef]
56. Stebbins, G.L. Polyploidy and the distribution of the arctic-alpine flora: New evidence and a review approach. *Bot. Helv.* **1984**, *94*, 1–13.
57. Brochmann, C.; Brysting, A.K. The Arctic—An evolutionary freezer? *Plant Ecol. Divers.* **2008**, *1*, 181–195. [CrossRef]
58. Parisod, C. Plant speciation in the face of recent climate changes in the Alps. *Alpine Bot.* **2022**, *132*, 21–28. [CrossRef]

plants

MDPI

Article

The Perspective of Arctic–Alpine Species in Southernmost Localities: The Example of *Kalmia procumbens* in the Pyrenees and Carpathians

Łukasz Walas [1], Marcin Pietras [1], Małgorzata Mazur [2], Ángel Romo [3], Lydia Tasenkevich [4], Yakiv Didukh [5] and Adam Boratyński [1,*]

[1] Institute of Dendrology Polish Academy of Sciences, 62-035 Kórnik, Poland; lukaswalas@man.poznan.pl (Ł.W.); mpietras@man.poznan.pl (M.P.)
[2] Faculty of Biological Sciences, Kazimierz Wielki University, 85-064 Bydgoszcz, Poland; mazur@ukw.edu.pl
[3] Botanical Institute of Spanish Research Council, 08038 Barcelona, Spain; angel.romo@gmail.com
[4] Department of Botany, Ivan Franko National University of Lviv, 79005 Lviv, Ukraine; tasenkevich@gmail.com
[5] M.G. Kholodny Institute of Botany, NAS of Ukraine, 01601 Kyiv, Ukraine; ya.didukh@gmail.com
* Correspondence: borata@man.poznan.pl

Abstract: High-mountain and arctic plants are considered especially sensitive to climate change because of their close adaptation to the cold environment. *Kalmia procumbens*, a typical arctic–alpine species, reaches southernmost European localities in the Pyrenees and Carpathians. The aim of this study was the assessment and comparison of the current potential niche areas of *K. procumbens* in the Pyrenees and Carpathians and their possible reduction due to climate change, depending on the scenario. The realized niches of *K. procumbens* in the Pyrenees are compact, while those in the Carpathians are dispersed. In both mountain chains, the species occurs in the alpine and subalpine vegetation belts, going down to elevations of about 1500–1600 m, while the most elevated localities in the Pyrenees are at ca. 3000 m, about 500 m higher than those in the Carpathians. The localities of *K. procumbens* in the Carpathians have a more continental climate than those in the Pyrenees, with lower precipitation and temperatures but higher seasonality of temperature and precipitation. The species covered a larger area of geographic range during the Last Glacial Maximum, but its geographic range was reduced during the mid-Holocene. Due to climate warming, a reduction in the potential area of occurrence could be expected in 2100; this reduction is expected to be strong in the Carpathians and moderate in the Pyrenees.

Keywords: biogeography; climate change; ecological niche modeling; geographic range; relict plant

Citation: Walas, Ł.; Pietras, M.; Mazur, M.; Romo, Á.; Tasenkevich, L.; Didukh, Y.; Boratyński, A. The Perspective of Arctic–Alpine Species in Southernmost Localities: The Example of *Kalmia procumbens* in the Pyrenees and Carpathians. *Plants* **2023**, *12*, 3399. https://doi.org/10.3390/plants12193399

Academic Editors: Pablo García Murillo

Received: 9 August 2023
Revised: 20 September 2023
Accepted: 22 September 2023
Published: 26 September 2023

1. Introduction

The origin of cold-adapted plant species and the formation of tundra in Northern Europe, Asia and North America took place in turn of the Pliocene/Pleistocene [1–3] as a reaction of the plant cover to the climate cooling. The plants adapted to the low temperatures and short vegetation periods in the Arctic zone and in the high mountains in southern regions evolved during approximately similar periods [3]. The Arctic plants reached mountains in Central Europe, Asia and North America, escaping to the south ahead of Pleistocene glaciers [1]. Inversely, the alpine plants could migrate to the north in the periglacial zone during deglaciations [4], as in the case of retreating glaciers, observed during the last few centuries in the Alps [5]. During interglacial periods of the Pleistocene (including the Holocene), the plant species connected with the cold climate could survive only in the Arctic and in the high mountains above the timberline, sometimes also on mires at lower altitudes [6–8], these being the glacial relicts [9,10].

High-mountain and arctic plant species are considered to be especially sensitive to climate change because of their specialized adaptation to the cold environment [11]. High temperatures and low humidity are expected to be the most important threats [12–19]. High risk also results from the forest line shift and expansion of trees, high shrubs and herbs, which could colonize or at least shade the sites of light-demanding tundra and alpine plant species [20–24].

Kalmia procumbens (L.) Gift & Kron & P.F.Stevens is an arctic–alpine, circumpolar, amphi-Atlantic plant [25–27]. It is an evergreen, dwarf, prostrate shrub, frequently creeping along the ground, especially at high elevations. In Northern Europe and in the mountains, it prefers the rocky ridges [24,28]. In Europe, it reaches its southernmost localities in the Pyrenees, Alps, northern Dinaric Alps and Carpathians. This species is one of the typical cold-adapted glacial relicts in the Central European mountains [29,30]. Despite this, it can survive in temperatures above 50 °C with a very high diurnal amplitude of temperatures [31].

In the Pyrenees and in the Carpathians, *K. procumbens* localities are confined predominantly to the alpine and sometimes to the subalpine vegetation belts. At lower locations of the subalpine belt, this species is connected mainly with the north-facing slopes [32–34], with relatively shorter vegetation periods, lower temperatures and temperature amplitudes, when compared with the south expositions [35–38]. It is stress-tolerant, adapted to extreme conditions of existence and characterized by slow growth in a relatively long growing season [39]. The formation of its photosynthetic apparatus is aimed at a minimum return of energy, which contributes to the conservation of resources. The population structure is dominated by plants of mature (generative) stages [40]. It prefers rocky ridges with small snow deposition during winter [28,38]. Such characteristics indicate that the species, although resistant to external influences, recovers rather slowly when it loses its position. The sites of *K. procumbens* are susceptible to influence of global climate change [28,38,41]. Despite that, recent observations indicated its more intense growth in the mountains of the Japanese islands [42] and even expansion in the Ukrainian Carpathians resulting from a reduction in snow cover [21]. The species had a broader potential ecological niche area at the global scale during the Last Glacial Maximum (LGM) than at present [41].

This study aimed to assess and compare the current potential niche areas of *K. procumbens* in the Pyrenees and Carpathians and their possible reduction due to climate change depending on the scenario. Additionally, its local, Carpathian and Pyrenean geographic ranges during the LGM were analyzed retrospectively based on the most probable high level of ecological niche conservatism in plants [43–45]. We expected the Carpathian populations would be more prone to the temperature rise and extension of the vegetation period than the Pyrenean ones.

2. Results

2.1. Realized Geographic Niches

The distribution of *K. procumbens* in the Pyrenees and in the Carpathians is connected with the most elevated mountain massifs (Figure 1). The species has a more compact distribution of its localities in the Pyrenees than in the Carpathians. The altitudinal range in the Pyrenees is broader than that in the Carpathians. The altitudinal minima in both mountain systems are similar, but the most elevated localities in the Pyrenees have ascended about 400–500 m higher than those in the Carpathians (Figure 2). The current range of species differs from the potential range in the LGM (Figure S1), when the Carpathian arc was more suitable for the species, whereas conditions in the Pyrenees were less suitable.

The current potential niches of *K. procumbens* in the Pyrenees and Carpathians are determined first of all by elevation, whose influence on the current distribution of the species reaches nearly 80% in the Pyrenees and 60% in the Carpathians (Table 1). From the climatic variables, only precipitation of the driest month (bio 14) has a relatively high degree of influence in both mountain chains (Table 1). Additional important bioclimatic factors, with an influence on the potential niche of *K. procumbens* reaching 1% or more, are the temperature mean diurnal range (bio2), temperature annual range (bio7), and precipitation of the driest quarter (bio17) in the Pyrenees and the annual mean temperature (bio1), temperature seasonality (bio4), precipitation seasonality (bio15), and precipitation of warmest quarter (bio18) in the Carpathians (Table 1).

The average values of every bioclimatic factor differed at a statistically significant level ($p < 0.05$) between localities of *K. procumbens* in the South and East Carpathians (the latter also includes one locality from the West Carpathians) (Table 2). The East Carpathian region of *K. procumbens* occurrence has generally higher precipitation (bio12–bio19) and is characterized by temperature factors (bio1, bio5, bio6, bio8–bio 11) that are slightly but statistically significantly higher than those in the South Carpathian region (Table 2). Populations from the Carpathians occur most often in the northeastern exposition, whereas in the Pyrenees, they occur in the northern exposition (Figure S2).

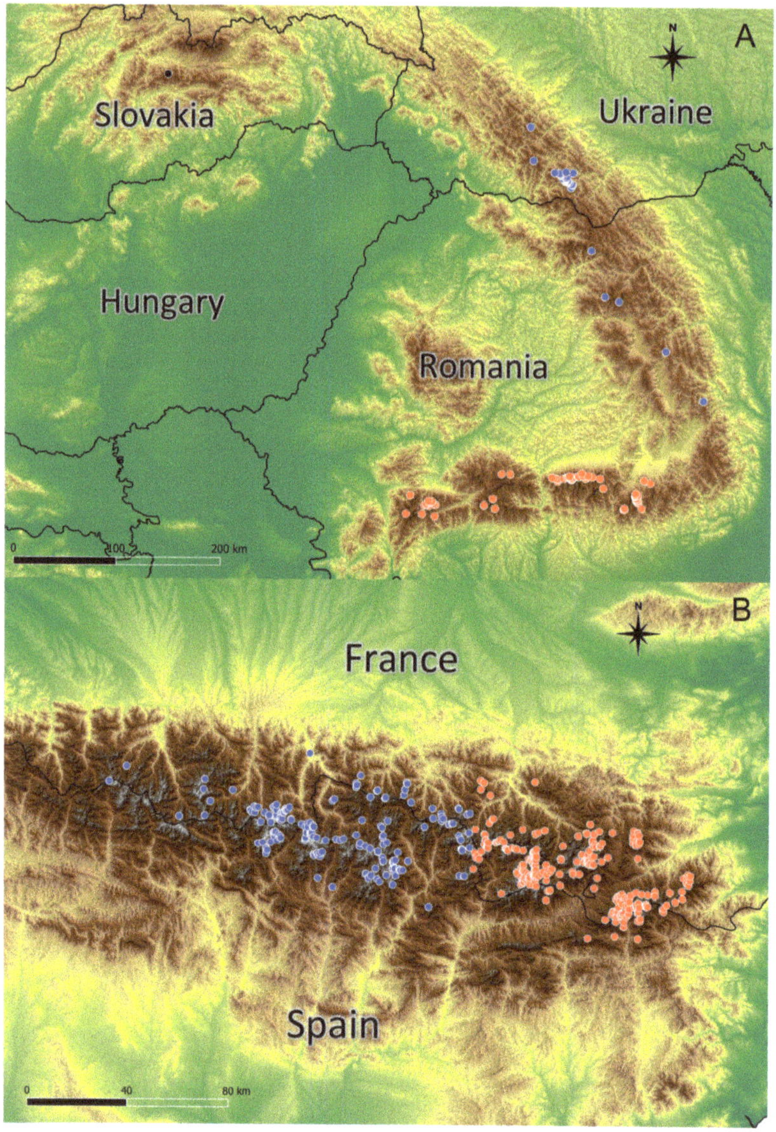

Figure 1. Geographical distribution of *Kalmia procumbens* on the basis of georeferenced data in the Carpathians (**A**) (blue dots—East Carpathians; red dots—South Carpathians) and in the Pyrenees (**B**) (blue dots—West Pyrenees; red dots—East Pyrenees).

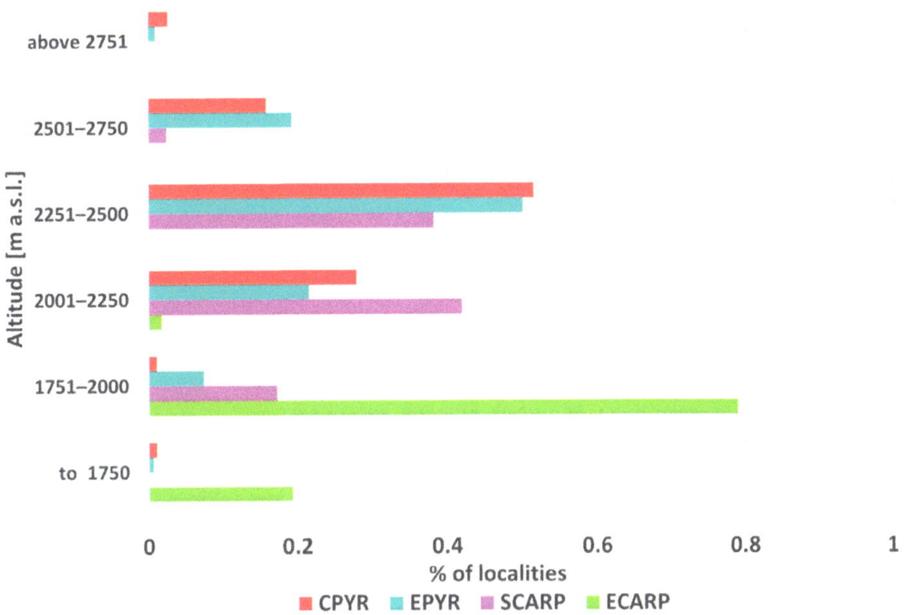

Figure 2. Vertical distribution of *Kalmia procumbens* localities in the Pyrenees (CPYR—Central Pyrenees; EPYR—East Pyrenees) and in the Carpathians (SCARP—South Carpathians; ECARP—East Carpathians).

Table 1. Contribution (%) of bioclimatic variables and altitude to the realized habitats suitable for *Kalmia procumbens* in the Pyrenees (**PYR**), Central Pyrenees (CPYR), East Pyrenees (EPYR), Carpathians (**CARP**), East Carpathians (ECARP) and South Carpathians (SCARP); values of 1.0 and higher **bolded**.

	Bioclimatic Factor	CPYR	EPYR	PYR	ECARP	SCARP	CARP
	AUC	0.994	0.991	0.993	0.999	0.997	0.998
Bio1	Annual Mean Temperature	0.1	0.1	0.1	**2.0**	0.3	**1.3**
Bio2	Mean Diurnal Range	**4.1**	0.9	**2.5**	0.1	0.0	0.1
Bio3	Isothermality	0.6	0.2	0.4	**1.7**	0.0	0.9
Bio4	Temperature Seasonality	0.3	0.2	0.3	**4.7**	0.0	**2.4**
Bio5	Max Temperature of Warmest Month	0.7	0.3	0.5	0.1	0.1	0.1
Bio6	Min Temperature of Coldest Month	0.3	0.0	0.2	0.2	0.2	0.2
Bio7	Temperature Annual Range	**1.0**	**1.0**	**1.0**	0.1	0.0	0.1
Bio8	Mean Temperature of Wettest Quarter	0.4	**2.0**	**1.2**	0.2	0.2	0.2
Bio9	Mean Temperature of Driest Quarter	**3.8**	**6.0**	**4.9**	**5.5**	**8.2**	**6.9**
Bio10	Mean Temperature of Warmest Quarter	0.2	0.1	0.2	0.1	0.1	0.1
Bio11	Mean Temperature of Coldest Quarter	0.1	0.1	0.1	0.2	0.2	0.2
Bio12	Annual Precipitation	0.6	0.2	0.4	0.1	0.0	0.1
Bio13	Precipitation of Wettest Month	0.1	0.3	0.2	0.0	0.0	0.0
Bio14	Precipitation of Driest Month	**1.1**	0.4	0.8	0.0	0.1	0.1
Bio15	Precipitation Seasonality	0.3	0.1	0.2	**3.4**	**10.0**	**6.7**
Bio16	Precipitation of Wettest Quarter	0.3	**1.2**	0.8	0.0	0.0	0.0
Bio17	Precipitation of Driest Quarter	**7.8**	**6.2**	**7.0**	0.1	0.0	0.1
Bio18	Precipitation of Warmest Quarter	0.1	0.1	0.1	**31.1**	**9.9**	**20.5**
Bio19	Precipitation of Coldest Quarter	0.5	0.5	0.5	0.2	0.1	0.2
	Elevation	**77.6**	**80.2**	**78.9**	**50.1**	**70.4**	**60.3**

Table 2. Average values of bioclimatic variables in the studied regions. According to the Mann–Whitney test, all differences between the Pyrenees and the Carpathians (bolded) are significant ($p < 0.05$), as are the differences between the Southern (SCARP) and Western (ECARP) Carpathians. For Eastern (EPYR) and Central (CPYR) Pyrenees, statistically significant differences ($p < 0.05$) were observed for bio2, bio3, bio5, bio7, bio12, bio13, bio14, bio15, bio16, bio17, bio18 and bio19 (shaded).

	Bioclimatic Factor	CPYR	EPYR	Pyrenees	ECARP	SCARP	Carpathians
bio1	Annual Mean Temperature	2.37	2.40	**2.39**	1.07	−0.25	**0.30**
bio2	Mean Diurnal Range	9.27	8.67	**8.90**	6.91	7.16	**7.06**
bio3	Isothermality	33.79	32.62	**33.07**	27.73	29.06	**28.52**
bio4	Temperature Seasonality	649.33	646.56	**647.61**	684.48	657.93	**668.81**
bio5	Max Temperature of Warmest Month	18.69	18.14	**18.35**	13.92	12.52	**13.10**
bio6	Min Temperature of Coldest Month	−8.73	−8.44	**−8.55**	−10.99	−12.08	**−11.62**
bio7	Temperature Annual Range	27.43	26.58	**26.90**	24.91	24.60	**24.73**
bio8	Mean Temperature of Wettest Quarter	−0.62	−0.30	**−0.42**	8.98	6.84	**7.73**
bio9	Mean Temperature of Driest Quarter	10.80	10.78	**10.79**	−6.81	−8.02	**−7.51**
bio10	Mean Temperature of Warmest Quarter	10.93	10.89	**10.90**	9.36	7.83	**8.46**
bio11	Mean Temperature of Coldest Quarter	−4.71	−4.60	**−4.64**	−7.19	−8.14	**−7.74**
bio12	Annual Precipitation	1434.57	1447.54	**1442.62**	1274.90	934.12	**1075.62**
bio13	Precipitation of Wettest Month	154.04	161.53	**158.69**	167.76	132.44	**147.06**
bio14	Precipitation of Driest Month	75.71	68.17	**71.03**	65.56	46.75	**54.58**
bio15	Precipitation Seasonality	20.19	23.53	**22.27**	32.36	39.56	**36.57**
bio16	Precipitation of Wettest Quarter	430.82	457.02	**447.08**	462.96	363.02	**404.52**
bio17	Precipitation of Driest Quarter	263.48	251.75	**256.20**	221.51	149.49	**179.33**
bio18	Precipitation of Warmest Quarter	263.57	251.91	**256.33**	462.67	358.05	**401.49**
bio19	Precipitation of Coldest Quarter	388.06	399.96	**395.45**	228.93	149.75	**182.63**

Principal component analysis (PCA) of the bioclimatic factors for the realized niches of *K. procumbens* indicated a separate grouping of Carpathian and Pyrenean localities. The East Carpathian localities are well separated from the South Carpathian ones, while Central Pyrenean localities are intermixed with East Pyrenean ones to a great extent (Figure 3). The estimation of potential ecological niches on the basis of the species localities in Central and East Pyrenees recognized the realized niches in both Pyrenean regions (Figure 4). Inversely, the estimation of the potential niches of *K. procumbens* in the East Carpathian localities recognized a realized niche in the East Carpathians, but not in the South Carpathians. Additionally, this combination indicated highly suitable conditions in the Tatra Mts. in the West Carpathians, where only one natural locality of the species currently exists. The estimation of potential ecological niches of *K. procumbens* using the South Carpathian localities did not detect suitable environmental conditions for the occurrence of this species either in the East or in the West Carpathians (Figure 4).

2.2. Future Geographic Niches

The potential niches highly suitable for *K. procumbens* in the Carpathians completely disappear in the year 2100, independently of the scenario of climate warming. Only a few South Carpathian populations could persist, mainly in the Fagarash, while the East Carpathian ones would not find suitable conditions (Figure 5). The potential niches in 2100 in the Carpathians would be determined by elevation with a contribution of ca. 50% in the East Carpathians and ca. 70% in the South Carpathians. The next restrictive factor in the East Carpathians would be precipitation of the warmest quarter (bio18), attaining a contribution of 30% or more (Table 3). The other bioclimatic factors influencing the potential niches in 2100 are the same as the current ones (compare Tables 1 and 3), but the values are slightly smaller.

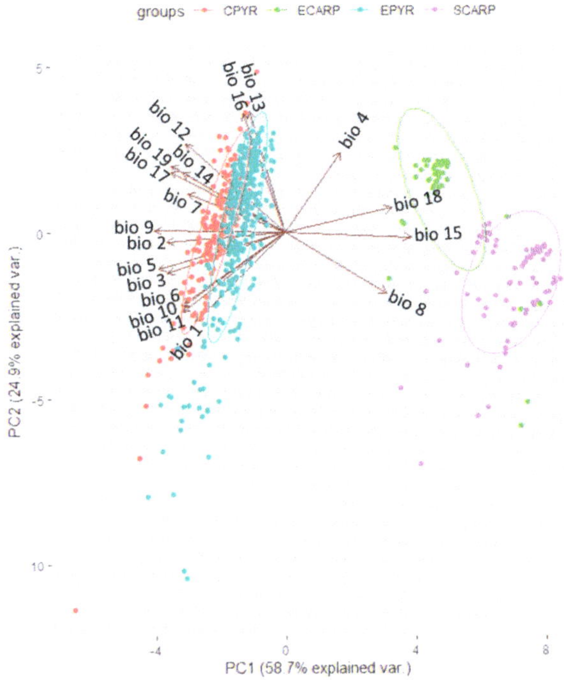

Figure 3. Position of localities of *Kalmia procumbens* from Central Pyrenees (CPYR), East Pyrenees (EPYR), East Carpathians (ECARP) and South Carpathians (SCARP) in PCA on the basis of bioclimatic variables (acronyms as in Table 1); ellipses indicate the 95% confidence intervals.

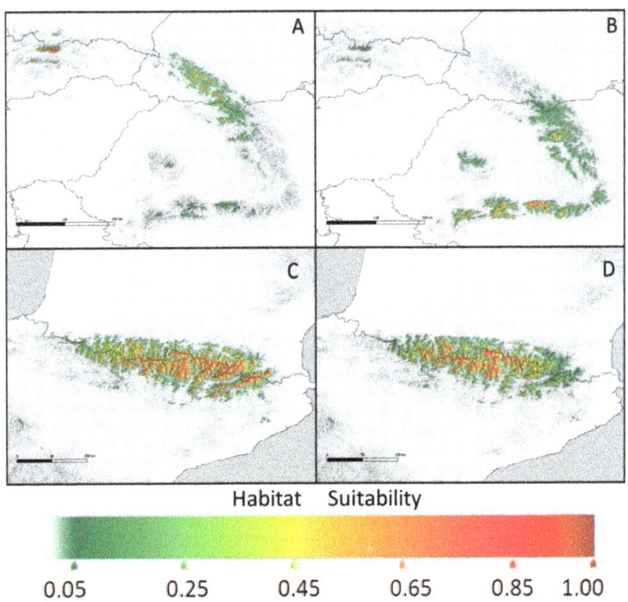

Figure 4. Current potential range of *Kalmia procumbens* in the Carpathians (**A**,**B**), estimated using environmental conditions from (**A**) the East Carpathians and (**B**) the South Carpathians, and in the Pyrenees (c and d), estimated using environmental conditions from (**C**) the East Pyrenees and (**D**) the Central Pyrenees.

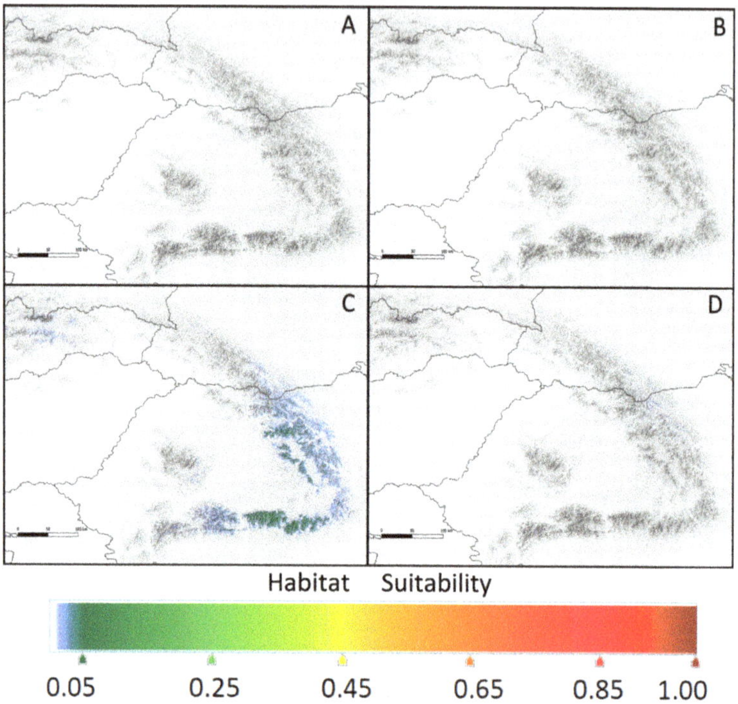

Figure 5. Provided potential range of *Kalmia procumbens* in the Carpathians in 2100 estimated using current environmental conditions from the East Carpathians for (**A**) scenario RCP 2.6 and (**B**) scenario RCP 8.5 and from the South Carpathians (**C**) for scenario RCP 2.6 and (**D**) scenario RCP 8.5.

Table 3. Contribution (%) of bioclimatic variables and altitude to models of the future potential range of *Kalmia procumbens* in the Central Pyrenees (CPYR), East Pyrenees (EPYR), East Carpathians (ECARP) and South Carpathians (SCARP); values of 1.0% and higher **bolded**.

	Bioclimatic Factor	**CPYR**		**EPYR**		**ECARP**		**SCARP**	
		RCP 2.6	**RCP 8.5**	**RCP 2.6**	**RCP 8.5**	**RCP 2.6**	**RCP 8.5**	**RCP 2.6**	**RCP 8.5**
	AUC	0.994	0.994	0.991	0.991	0.999	0.999	0.997	0.997
Bio1	Annual Mean Temperature	0.0	0.1	0.0	0.0	**1.5**	**1.1**	0.6	0.4
Bio2	Mean Diurnal Range	**3.9**	**3.1**	0.9	0.9	0.1	0.1	0.0	0.0
Bio3	Isothermality	0.8	**1.1**	**1.7**	0.1	**2.1**	**2.7**	0.0	0.0
Bio4	Temperature Seasonality	0.3	0.7	0.2	0.2	**4.2**	**4.9**	0.0	0.1
Bio5	Max Temperature of Warmest Month	0.8	0.3	0.2	0.5	0.1	0.1	0.1	0.1
Bio6	Min Temperature of Coldest Month	0.4	0.1	0.1	0.9	0.1	0.1	0.1	0.2
Bio7	Temperature Annual Range	0.9	0.5	0.4	**1.0**	0.1	0.1	0.0	0.0
Bio8	Mean Temperature of Wettest Quarter	**1.2**	0.7	0.4	**4.0**	0.0	0.1	0.1	0.2
Bio9	Mean Temperature of Driest Quarter	**4.3**	**5.5**	**6.4**	**4.0**	**9.5**	**3.8**	**6.6**	**5.9**
Bio10	Mean Temperature of Warmest Quarter	0.2	0.1	0.1	0.2	0.1	0.0	0.1	0.1
Bio11	Mean Temperature of Coldest Quarter	0.1	0.0	0.1	0.1	0.3	0.3	0.3	0.2
Bio12	Annual Precipitation	0.8	0.7	0.4	0.4	0.0	0.1	0.0	0.0
Bio13	Precipitation of Wettest Month	0.3	0.2	0.1	**1.1**	0.0	0.0	0.1	0.0
Bio14	Precipitation of Driest Month	**1.0**	**1.0**	0.4	0.5	0.0	0.1	0.0	0.1
Bio15	Precipitation Seasonality	0.3	0.3	0.0	0.1	**2.0**	**5.4**	**12.1**	**9.8**
Bio16	Precipitation of Wettest Quarter	0.5	0.3	0.2	0.3	0.0	0.0	0.0	0.0
Bio17	Precipitation of Driest Quarter	**7.5**	**8.7**	**7.3**	**5.5**	0.0	0.0	0.0	0.0
Bio18	Precipitation of Warmest Quarter	0.1	0.1	0.2	0.2	**32.0**	**30.3**	**9.6**	**10.2**
Bio19	Precipitation of Coldest Quarter	0.4	0.2	0.3	0.2	0.2	0.2	0.2	0.2
	Elevation	**76.1**	**76.4**	**80.5**	**79.9**	**47.5**	**50.5**	**70.0**	**72.4**

In the Pyrenees, the situation of *K. procumbens* populations in 2100 would not be so drastically worse than that at present (Figure 6). The environmental conditions in the East

Pyrenees allow the species populations to persist and even attain a broader area of potential niche distribution in the more optimistic scenario (Table 4). In the Central Pyrenees, the potential niche area suitable for *K. procumbens* in the high and very high levels would be restricted (Table 4). As in the present, the most influential factor would be elevation, reaching contributions of about 76 and 80% in the Central and East Pyrenees, respectively (Table 3). The remaining bioclimatic factors are the same as those determining the current ecological niches of the species in the Pyrenees (see Tables 1 and 3).

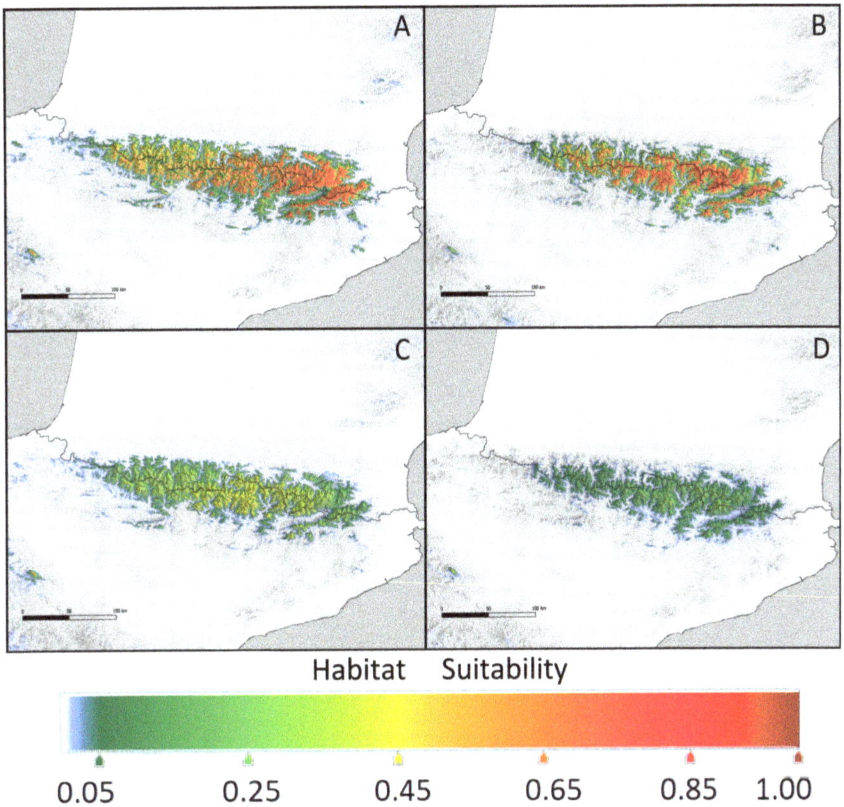

Figure 6. Provided potential range of *Kalmia procumbens* in the Pyrenees in 2100 estimated using current environmental conditions from the East Pyrenees for (**A**) scenario RCP 2.6 and (**B**) scenario RCP 8.5 and from the Central Pyrenees for (**C**) scenario RCP 2.6 and (**D**) scenario RCP 8.5.

The phytoindication method revealed a potential threat to *K. procumbens* occurrence in the East Carpathians within Ukraine. It results mainly from the changes in the hydrological regime of the species sites. The growth in average yearly temperature by 2 °C (pessimistic scenario) put *K. procumbens* at moderate risk of extinction, while that by 3 °C put *K. procumbens* at catastrophic risk of extinction (Figure 7).

Table 4. Area of potential range according to the tested model and four sets of stands: CPYR—Central Pyrenees, EPYR—Eastern Pyrenees, ECARP—Eastern Carpathians, SCARP—Southern Carpathians.

Region	Model	Area in the Probability Levels (Hectares)				Total
		Low (0.1–0.25)	Medium (0.25–0.50)	High (0.5–0.75)	Very High (>0.75)	
CPYR	Current	6099.48	6545.66	4920.93	1288.34	18,854.41
	RCP 2.6	6344.43	10,857.13	420.55	16.35	17,638.46
	RCP 8.5	6698.12	711.52	4.67	0.00	7414.31
EPYR	Current	4825.15	7280.36	8192.59	1338.57	21,636.67
	RCP 2.6	5756.46	5986.58	6773.90	5156.05	23,672.99
	RCP 8.5	4482.04	4812.61	4664.27	4470.36	18,429.28
ECARP	Current	7406.57	2796.66	1059.53	686.48	11,949.24
	RCP 2.6	0.00	0.00	0.00	0.00	0.00
	RCP 8.5	0.00	0.00	0.00	0.00	0.00
SCARP	Current	9808.72	4446.85	2123.17	708.15	17,086.89
	RCP 2.6	771.79	0.00	0.00	0.00	771.79
	RCP 8.5	0.00	0.00	0.00	0.00	0.00

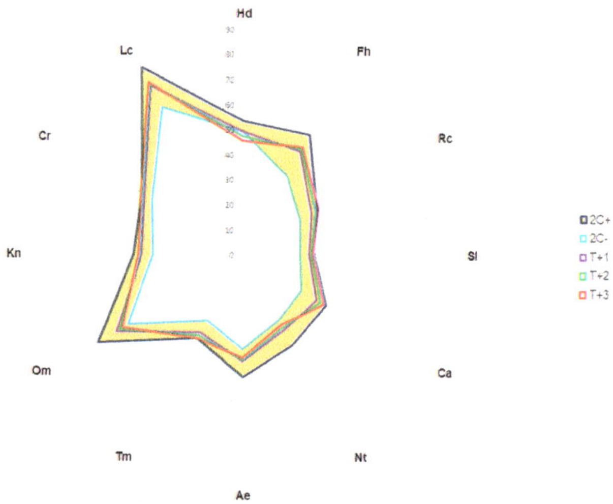

Figure 7. Percentage values of econiche factors of *Kalmia procumbens* characterization in the East Carpathians in Ukraine and potential changes depending on temperature changes: +1 °C (T + 1), +2 °C (T + 2) and +3 °C (T + 3); Hd—soil moisture, Fh—variability in soil moisture, Rc—soil acidity, Sl—soil salinity, Ca—soil carbonate, Nt—content of mineral nitrogen available for assimilation, Ae—soil aeration, Tm—thermal regime, Om—ombro regime, Kn—continental climate, Cr—cryo regime, Lc—illumination; indicator values 2C+ $-$ X + 2σ, 2C− X $-$ 2σ; the range of indicators x ± 2σ marked in orange.

3. Discussion

3.1. Realized Potential Niches in the Carpathians and Pyrenees

During the LGM, the potential range of *K. procumbens* was broader in the Carpathians, and for East Carpathians populations, the model also predicts suitable areas at a lower elevation, closer to the ice sheet. In the Pyrenees during the LGM, the potential range was located at lower elevations, and conditions for Central Pyrenees populations were almost unsuitable. In current conditions, this species attains its southernmost European localities in the Pyrenees and its close to southernmost European localities in the Carpathians [24,27]. *Kalmia procumbens* survived in both mountain chains in the subalpine and alpine vegetation belts due to the high-mountain climate with low temperatures and relatively high precipitation, mostly in the places with restricted snow cover during winters [21,22,28,38].

The low altitudinal borders of *K. procumbens* occurrence in the Carpathians and in the Pyrenees are at similar elevations, as a rule in habitats orographically or edaphically inaccessible for shrubs, tall herbs and grasses. The specific site conditions are mostly on the slopes exposed to the north, in rocky places with very thin layers of soil or on rocks completely without soil, and in places open to the wind. Sometimes, such conditions can be anthropogenic; for example, they could result from over-pasturing. This kind of pressure ceased over the last few decades and could be one of the reasons for the disappearance of the lowest localities of *K. procumbens* in the East Carpathians, as it was reported at 1455 m in the Chornokhora [46] but not found later [22,33]. The reduction in pastoralism in the Pyrenees during the last few decades could also cause the disappearance of the species' lowest localities due to the expansion of the tall herbs and shrubs.

The maximal altitudes of occurrence of *K. procumbens* in the Pyrenees are more elevated than those in the Carpathians. Such a rule was also observed in other subalpine and alpine plants that are common in the Pyrenees and Carpathians, such as *Juniperus communis* L. var. *saxatilis* Pall., *Salix reticulata* L., *Salix herbacea* L., *Salix hastata* L., *Dryas octopetala* L. and *Vaccinium gaultherioides* Bigelow (*V. uliginosum* L.) (Table S1). The differences in the altitudinal maxima of the subalpine and alpine plants between the Carpathians and Pyrenees surely result from the higher elevations of the latter. The Carpathian arc is composed predominantly of medium-sized mountain ridges, with only three or four massifs being sufficiently well revealing, and several others with only a fragmentary developed alpine vegetation belt [47,48]. Inversely, in the Pyrenees, this type of vegetation is more frequent and covers a broader area [28,49–51].

Kalmia procumbens is well adapted to microhabitats with continental climatic conditions, to the extremely high daily amplitude of temperatures during the vegetation season [52,53] and early snowmelt [54], but the species could suffer from frost during the beginning of the vegetation season when development of the generative structures starts [55]. On the other hand, the late frost disturbances and high temperatures in the exposed places of *K. procumbens* occurrence have the effect of reducing most other plant species, promoting the successful regeneration of *Kalmia* [55]. In relation to the surrounding grass plant communities, for which the Index of Continentality is 19.7 or 16.5, after Gorchynsky and Rivas-Martinez, respectively, in *Loiseleurio-Cetrarietum* of the East Carpathians, it attains 21.6 and/or 17.5 [40]. Additionally, the average annual temperatures at the elevation of 1000 m in grass plant communities reach 2.6 °C, but in the rock coenoses dominated by *K. procumbens*, they reach 3.6 °C [19]. Nevertheless, *K. procumbens* does not reach its potential altitudinal maximum in most of the Carpathian ridges, as was determined for some other alpine shrubby plants in the East Carpathians [48].

3.2. Environmental Conditions of the Realized Niches of K. procumbens

Kalmia procumbens is a calcifuge species occurring in mountains composed of metamorphic siliceous rock pH [28,56–58]. The plant communities with dominance of this species are classified as association *Cetrario nivalis-Loiseleurietum procumbentis* Br.-Bl. in Br.-Bl. and Jenny 1926 [59], from alliance *Rhododendro-Vaccinion* Br.-Bl. 1926. In the Pyrenees, the plant community *Cetrario nivalis-Loiseleurietum* is developed on the north-exposed, acid sites where the winds blow out snow, thus reducing snow cover [49,50,57,58,60]. A similar plant community is formed in the South Carpathians [61,62], and a fragmentary community is formed in the East Carpathians [38,56]. In the latter mountains and in the South Carpathians, *K. procumbens* occurs in the grassland communities on siliceous rocks [56,63,64].

The climate in the regions of *K. procumbens* occurrence in the mountains of Central Europe is of an oceanic to sub-continental type, cryo-oro-temperate thermotype and sub-humid to hyper-humid ombrotype [65,66]. Despite this, the average bioclimatic data retrieved from World Clim for *K. procumbens* localities in the Pyrenees revealed slightly milder conditions than those in the Carpathians (Table 2). The climate of the Carpathian localities of. *K. procumbens* appeared more continental with lower factors of temperature and precipitation and higher temperature seasonality. The continental climate of the steppes easterly and southerly from the Carpathians and at a closer distance to the continental climate of the central Euro-Asiatic continent could play some role in lowering positions of *K. procumbens* there, and the alpine and subalpine vegetation belt, in comparison with the Pyrenees.

The average bioclimatic factors of *K. procumbens* localities presented a low level of differences between the Central and Eastern Pyrenees, revealed mostly in the lower values of factors connected with precipitation in the central, more continental parts of this mountain chain (Figure 3). Nevertheless, the more humid climatic conditions in the Eastern Pyrenees can be a reason for the more dispersed and not-so-abundant localities of the species due to prolonged snow cover [28]. Similarly, the more Atlantic climate conditions in the Northern as opposed to the Southern Pyrenees [65.66] can explain a lower abundance and frequency of occurrence of *K. procumbens* on the southern macroslopes [28]. On the other hand, the higher and longer-lasting snow observed in the Eastern Pyrenees could reduce the potential habitats accessible for *K. procumbens*, as the species occurrence is connected mostly with places with thin snow deposits [21,22,28,38,54]. *Kalmia procumbens* occurs mostly on specific microhabitats, mainly the rocky ridges and rocks, where the snow is being blown away and the temperatures reveal high diurnal amplitudes.

The bioclimatic differences between localities of *K. procumbens* in the East versus South Carpathians appeared higher than those between the East and Central Pyrenees (Figure 3). This finding could result from the higher elevations of the South Carpathian mountain massifs as opposed to those in the East Carpathians, and consequently, the greater number of *K. procumbens* populations reported from the higher elevations, which are characterized by lower temperatures and higher mean diurnal amplitude and isothermality.

The snow-free period appeared important for microsites inhabited by small ericaceous shrubs in the high mountains and in the arctic zone [67]. The longer snow-free period positively influenced the wood-ring increment in *Empetrum hermaphroditum* Hagerup [68], a species frequently occurring with *K. procumbens*. At the same time, less snowfall and a shorter duration of snow cover observed during the last decade were a reason for more abundant *K. procumbens* growth [21,69].

The differences in climatic conditions of the current localities of *K. procumbens* in the Pyrenees and Carpathians could result from (1) adaptation to different climates during the Holocene; (2) the origin of the Pyrenean and Carpathian populations from two different regions, namely the Arctic and the Atlantic for the Pyrenees and more continental for the Carpathians; or simultaneous action of both processes.

3.3. Possible Influence of Climate Differences

It is to be expected that the differences detected between the average climatic conditions of *K. procumbens* localities in the Carpathians and Pyrenees influenced the genetic structure of the species. The isolation of *K. procumbens* populations between the Pyrenees, Alps and Carpathians lasting at least during the Holocene [4,70,71] should constitute a reason for genetic and morphological differences. Gene exchange in Ericaceae is limited due to the low rate of seed dispersal [72] and restricted pollen transport, especially between populations from distant mountain chains [73]. These limitations explain the genetic and morphological differences, as described for other subalpine/alpine plant species, such as *Salix herbacea* L. [74], *Ranunculus glacialis* L. [75], *Saxifraga oppositifolia* L. [76], *Soldanella alpina* L. [77] and *Rhododendron ferrugineum* L. [78,79], and even provide a reason for speciation, as in the case of *Rhododendron ferrugineum* and *R. myrtifolim* Schott & Kotschy [80].

The worldwide genetic structure of *K. procumbens* detected using sequences of multiple nuclear loci revealed that southernmost European and East-Asiatic populations are genetically similar but different from the Arctic ones [81,82]. The southern populations of *K. procumbens* in the mountains diverged from the Arctic during the LG [82,83]. This is in contrast with the results of amplified fragment length polymorphism (AFLP) analyses, which indicated the isolation of the Central European mountain clade [84]. The number of verified populations and individuals used in these papers [82,84] was sufficient for the description of the general pattern of geographic differentiation of *K. procumbens* but was rather too small for its genetic differentiation in the Central European mountains. It could be expected that this species reveals differences between Pyrenean, Alpine and Carpathian localities; this hypothesis, however, should be verified in a specific study.

3.4. Future Ecological Niches in the Carpathians and Pyrenees

A climate-change-caused reduction in the potential niches of *K. procumbens* at their southern limit of realized ecological niches in the mountains of Central Europe could be predicted and even expected, as in the case of other subalpine and alpine plants [10,28,29,38,41,48]. In that context, the drastic reduction in potential niches in the Carpathians in general, and the quite complete disappearance of suitable niches in the East Carpathians, is not surprising. The process of reduction in the geographic ranges of cold-adapted, high-mountain plants and their shifts to higher elevations is restricted in the first place by the highest mountain elevations. Extinctions could also result from rather slow uppermost colonization by the plants [85]. The East Carpathians do not possess massifs exceeding an elevation of 2500 m, which are the main centers of occurrence of alpine and subalpine plants [47,86]. It should be stressed that the process of the disappearance of the area of potential niches detected in the case of *K. procumbens* in fact concerns more or less entire alpine and subalpine flora in the East Carpathians [24,48].

In the Pyrenees, the reduction in potential niche area suitable for *K. procumbens* is less drastic than that in the Carpathians. This results from the more "alpine" character of the Pyrenees and the higher elevations of their highest peaks, which form the conditions for the occurrence of alpine flora [28,50]. The more intense reduction in potential niches suitable for *K. procumbens* in the Central as opposed to the East Pyrenees could be explained by the more continental climate of the Central Pyrenees, influenced by the close distance to the very dry and warm central part of the Ebro Basin from the south. This could reduce the influence of the Atlantic climate [58,66]. In fact, the westernmost localities of the species were detected on the northern macroslopes of the Pyrenees, with the more prominent impact of the Atlantic climate (Figure 1). This could also corroborate the amphi-Atlantic biogeographic character of *K. procumbens* proposed by Hultén [25].

The influence of the reduction in the area of potential niches to the currently realized niches of *K. procumbens* could be diminished by the presence of microrefugia suitable for the species within the area of its occurrence [87]. The species in fact settled in such microsites at their lowermost localities. The current localities of *K. procumbens* in the East Carpathians (Figure 1), however, exist in the area of a very low level of suitability of the potential niches, especially at low elevations, which are quite exclusively on the north-facing slopes. It could be expected that *K. procumbens* can exist for a long time outside the optimal environmental conditions on the northern slopes; this makes extinction following climate change more prolonged, especially for plants moderately sensitive to the lack of humidity [88] and heat [31]. Thus, the persistence of *K. procumbens* in environments with a very low level of suitability could result from the specific site conditions of the particular localities eliminating competition of the other plants, but it also could be caused by mycorrhiza. The symbiosis of *K. procumbens* with fungi [89,90] could mitigate the extinction rate, as plants with fungal symbiosis are less vulnerable to harsh environmental conditions [91,92].

The extinction rate of *K. procumbens* due to climate change could also be restricted by the shrub's longevity, found to be more than 100 years [93]. This longevity with tolerance to harsh environments, particularly relatively high temperatures, high diurnal temperature amplitudes, winter frosts [24,52,53,94–96] and episodic summer frosts [97], has the effect of moderating the influence of climate change. *K. procumbens* plants covered with a shallow snow stratum are less vulnerable to spring frosts [98]. On the other hand, the speed of shifting of *K. procumbens* in the mountains after glacier regression could be rather moderate. In the Alps, this species exists together with other ericaceous shrubs, in areas with relatively stabilized plant cover, and colonizes new terrain about a century after glacier regression [5]. Our results suggest that cooler slopes may act as microrefugia, buffering the effects of increases in temperature on plant communities by delaying the extinctions of species with low temperature requirements [88].

The precipitation of the driest month (bio 14) and precipitation of the warmest quarter (bio 18) are the most influential bioclimatic factors for the present realized niche of *K. procumbens* in the East Carpathians in Ukraine. These bioclimatic factors act as limitations due to a lack of water in the vegetation period, moving *K. procumbens* out of the zone of acceptable risk when the average annual temperature rises by 2 °C and deep into the zone of catastrophic risk when the temperature rises by 3 °C. The role of other ecofactors is much lower and concerns only the acidity and salinity of soils, which are stable in the rocks in the localities of *K. procumbens*.

4. Materials and Methods

4.1. Study Areas

The Pyrenees and Carpathians were elevated during alpine orogenesis and currently conserve alpine flora with an abundance of endemic species [4,48,86,99,100]. The Pyrenees include several massifs reaching an elevation of more than 3000 m, with subalpine and alpine vegetation belts harbouring several arctic–alpine plants [28,50,51]. The Pyrenean range is divided into the western portion under Atlantic influence, the central portion with a more continental climate and the eastern portion under the influence of the Mediterranean climate [28,51]. *Kalmia procumbens* occupies elevated parts of the East and Central Pyrenees [28,34].

In comparison to the Pyrenees, the Carpathians cover a broader area and are more fragmented and divided into several mountain chains, but only a few of them are sufficiently high for the development of subalpine and alpine vegetation zones [47,48,86,101,102]. The localities of *K. procumbens* occupy the most elevated sites in the South and East Carpathians [33,103], with only one natural locality in the West Carpathians [104–106].

4.2. Data Sampling and Geographic Analyses

Data on the natural localities of the species were extracted from the Global Biodiversity Information Facility (GBIF) database, the literature, herbaria, and authors' field notes. The geographic coordinates of localities were determined using Google Earth when not reported in the original data. In total, we gathered more than 2000 data, but after verification and exclusion of duplicates and questionable information, we analyzed 641 georeferenced data, 140 for the Carpathians and 501 for the Pyrenees. Maps of the distribution of *K. procumbens* in the Pyrenees and Carpathians were prepared using QGIS 3.16.4 "Hannover" [107]. The altitudinal ranges of the species in both mountain systems are presented in the graphs.

4.3. Environmental Variables

Temperature, precipitation and elevation determine the current (realized) ecological niches of species to the highest degree [108]. In spite of that, we used nineteen bioclimatic variables [109] and altitude (Table 1) to find factors that determine the current potential niche of *K. procumbens*. The usage of all these data could shed new light on the adaptation of the species to specific climatic data, retrieved from WorldClim (WC) database (http://worldclim.org/, accessed on 1 April 2023) [110]. For LGM (21 ka BP), we used data from PaleoClim (PC) (http://www.paleoclim.org/, accessed on 1 April 2023), which is based on the use of the CHELSA algorithm on PMIP3 data [111,112]. The spatial resolution of 30 arc-seconds (~1 km) of climate variables was applied. For the current climate (average for the years 1970–2000), we used WorldClim 2.1 database [110] bioclimatic data. For the future climate, we based our work on the Community Climate System Model (CCSM) [113] and used two representative concentration pathways (RCPs), RCP 2.6 and RCP 8.5 [114]. RCP 2.6

provides an increase in radiation forcing by 2.6 W/m^2 and an increase in temperature by 1 °C before 2070 (average for 2061–2080), and RCP 8.5 provides an increase in radiation forcing by 8.5 W/m^2 and an increase in temperature by 2 °C during the same period. Both are climate projections from GCMs that were downscaled and calibrated using WorldClim 1.4 as the baseline climate.

PC used climate data from the CAPE project [115] and CCSM [113] for the delineation of potential niches during the Eemian Interglacial (LIG, 120–140 ka BP). These climatic data are based on geomorphological and geographical characteristics and do not take into account edaphic features and the structure of the substrate, which somewhat changes the microclimatic conditions.

For retrospective analyses of the climate of the EM (LGM, 21 ka BP), the CHELSA algorithm was used on PMIP3 data. For the Mid-Holocene (MH) climate (ca. 6 ka BP), CCSM4 was used. For the current climate (average for the years 1970–2000), we used WorldClim 2.1 database [110] bioclimatic data. The previsions of future climate changes utilized were scenarios of two representative concentration pathways (RCPs), RCP 2.6 and RCP 8.5 [114].

The average values of bioclimatic variables were compared between the Carpathians and Pyrenees, as well as between regions within these mountain ranges. The Mann–Whitney U test conducted in the R environment was used for this purpose [116]. The influence of particular climatic variables on the current potential niches of *K. procumbens* in the Carpathians and in the Pyrenees was verified by principal component analysis (PCA). In PCA, the data for localities in the North Carpathians and South Carpathians and the data for localities in the Central and East Pyrenees were treated as separate groups.

4.4. Niche Modeling

For the prediction of the potential range of *K. procumbens*, bioclimatic data related to its localities were used. MaxEnt 3.4.1. [117–119] was applied in analyses with maximum entropy modeling for the estimation of a probable distribution of the species outside its realized niche. The model with ENMeval R software [120] for the current climate was evaluated at first. The evaluation procedure followed that described by Salva-Catarineu et al. [121]. For the evaluation of the results of modeling, receiver operating characteristic (ROC) curves were used [122,123], and area under the curve (AUC) values below 0.6 were assessed as nearly random.

QGIS 3.16.4 "Hannover" [107] was applied for mapping the current and predicted potential niches on the climate variables. The potential distribution of *K. procumbens* was calculated for the different classes of suitability [45,121].

The phytoindication method was used for the assessment of econiches, indicators of the leading climatic and edaphic ecofactors, forecasting their changes depending on the climate for the Eastern Carpathians [40,124]. Modeling of econiche change and assessment of habitat loss threats for populations of the Eastern Carpathians were performed on the basis of the phytoindication data. For this purpose, the point values of the amplitude ($x \pm 2\sigma$) were calculated for the leading ecofactors, and their changes were evaluated depending on the increase in average annual temperatures by 1, 2 and 3 °C. The acceptable risk zone is where the average values of the obtained data are inside the confidence intervals of $\pm 2\sigma$, and the catastrophic risk zone is where the amplitudes do not overlap, which means the complete disappearance of the species from this place [40].

5. Conclusions

Kalmia procumbens occurs in the alpine and subalpine vegetation belts, going down to elevations of about 1500–1600 m, while the most elevated localities in the Pyrenees are at ca. 3000 m, about 500 m higher than those in the Carpathians. The localities of *K. procumbens* in the Carpathians have a more continental climate than those in the Pyrenees, with lower precipitation and temperatures but higher seasonality of temperature and precipitation.

Due to climate warming, a strong reduction in the potential area of occurrence by 2100 is expected in the Carpathians, and a moderate reduction is expected in the Pyrenees.

Supplementary Materials: The following supporting information can be downloaded at: https://www.mdpi.com/article/10.3390/plants12193399/s1, Figure S1: Provided potential range of *Kalmia procumbens* during the Last Glacial Maximum in the Carpathians (a and b), estimated using environmental conditions from (a) the East Carpathians and (b) the South Carpathians, and in the Pyrenees (c and d), estimated using conditions from (c) the East Pyrenees and (d) the Central Pyrenees; Figure S2: Occurrence of populations of *Kalmia procumbens* in different expositions in the Carpathians (a and b) (a—East Carpathians; b—South Carpathians) and in the Pyrenees (c and d) (c—East Pyrenees; d—Central Pyrenees); Table S1: Altitudinal maxima of subalpine and alpine species in the Carpathians [48,103,104] and Pyrenees [28]; CEUR–Central European mountain; ARALP—Arctic–Alpine; EUROS—Euro-Siberian mountain.

Author Contributions: Conceptualization, Ł.W. and A.B.; methodology, Ł.W., M.P. and A.B.; validation, M.M., Á.R. and L.T.; formal analysis, A.B; investigation, M.M., Ł.W. and Y.D.; resources, M.M. and A.B.; data curation, M.M. and A.B.; writing—original draft preparation, A.B; writing—review and editing, Ł.W., A.B, L.T. and Y.D.; visualization, Ł.W.; supervision, A.B. All authors have read and agreed to the published version of the manuscript.

Funding: This research was financially supported by the Institute of Dendrology Polish Academy of Sciences under statutory activity.

Data Availability Statement: The data presented in this study are available from the corresponding author upon request.

Acknowledgments: The Institute of Dendrology Polish Academy of Sciences in Kórnik, Kazimierz Wielki University in Bydgoszcz, Botanical Institute of Spanish Research Council in Barcelona, Ivan Franko National University in Lviv and M.G. Kholodny Institute of Botany National Academy of Sciences of Ukraine in Kyiv are acknowledged for the help provided to the authors for this study. Samuel Pyke is greatly acknowledged for linguistic correction.

Conflicts of Interest: The authors declare no conflict of interest.

References

1. Hultén, E. *Outline of the History of Arctic and Boreal Biota during the Quarternary Period*; Bokförlags Aktiebolaget Thule: Stockholm, Sweden, 1937.
2. Billings, W.D. Adaptations and origins of alpine plants. *Arct. Alp. Res.* **1974**, *6*, 129–142. [CrossRef]
3. Hagen, O.; Vaterlaus, L.; Albouy, C.; Brown, A.; Leugger, F.; Onstein, R.E.; Novaes de Santana, C.; Scotese, C.R.; Pellissier, L. Mountain building, climate cooling and the richness of cold-adapted plants in the Northern Hemisphere. *J. Biogeogr.* **2019**, *46*, 1792–1807. [CrossRef]
4. Schmitt, T. Molecular biogeography of the high mountain systems of Europe: An overview. In *High Mountain Conservation in a Changing World*; Catalan, J., Ninot, J.M., Mercè Aniz, M., Eds.; Springer Nature: Berlin/Heidelberg, Germany, 2017; Volume 62, pp. 63–74.
5. Fischer, A.; Fickert, T.; Schwaizer, G.; Patzelt, G.; Groß, G. Vegetation dynamics in Alpine glacier forelands tackled from space. *Sci. Rep.* **2019**, *9*, 13918. [CrossRef]
6. Huntley, B.; Birks, H.J.B. *An Atlas of Past and Present Pollen Maps for Europe: 0–13000 Years Ago*; Cambridge University Press: Cambridge, UK, 1983.
7. Ralska-Jasiewiczowa, M.; Latałowa, M.; Wasylikowa, K.; Tobolski, K.; Madeyska, E.; Wright, H.E.; Turner, C. *Late Glacial and Holocene History of Vegetation in Poland Based on Isopollen Maps*; W. Szafer Institute of Botany Polish Academy of Sciences: Kraków, Poland, 2004.
8. Stivrins, N.; Soininen, J.; Amon, L.; Fontana, S.L.; Gryguc, G.; Heikkilä, M.; Heiri, O.; Kisielienè, D.; Reitalu, T.; Stančikaitè, M.; et al. Biotic turnover rates during the Pleistocene-Holocene transition. *Quat. Sci. Rev.* **2016**, *151*, 100–110. [CrossRef]
9. Kornaś, J.; Medwecka-Kornaś, A. *Geografia Roślin*; Państwowe Wydawnictwo Naukowe: Warszawa, Poland, 2002.
10. Habel, J.C.; Assmann, T. *Relict Species. Phylogeography and Conservation Biology*; Springer: Berlin/Heidelberg, Germany, 2010.
11. Thuiller, W.; Lavorel, S.; Araújo, M.B.; Sykes, M.T.; Prentice, I.C. Climate change threats to plant diversity in Europe. *Proc. Nat. Acad. Sci. USA* **2005**, *102*, 8245–8250. [CrossRef]
12. Lesica, P. Arctic-Alpine plants decline over two decades in Glacier National Park, Montana, USA. *Arctic Antarct. Alp. Res.* **2014**, *46*, 327–332. [CrossRef]

13. Rixen, C.; Wipf, S.; Frei, E.; Stöckli, V. Faster, higher, more? Past, present and future dynamics of alpine and arctic flora under climate change. *Alp. Bot.* **2014**, *124*, 77–79. [CrossRef]
14. Jiménez-Alfaro, B.; García-Calvo, L.; García, P.; Acebes, J.L. Anticipating extinctions of glacial relict populations in mountain refugia. *Biol. Conserv.* **2016**, *201*, 243–251. [CrossRef]
15. Lesica, P.; Crone, E.E. Arctic and boreal plant species decline at their southern range limits in the Rocky Mountains. *Ecol. Let.* **2017**, *20*, 166–174. [CrossRef]
16. Rixen, C.; Wipf, S. Non-equilibrium in alpine plant assemblages: Shifts in Europe's summit floras. In *High Mountain Conservation in a Changing World*; Catalan, J., Ninot, J.M., Mercè Aniz, M., Eds.; Springer Nature: Berlin/Heidelberg, Germany, 2017; Volume 62, pp. 285–303.
17. Pauli, A.; Halloy, S.R.P. High mountain ecosystems under climate change. In *Oxford Research Encyclopedia fo Climate Science*; Oxford University Press: Oxford, UK, 2019. [CrossRef]
18. Fragnière, Y.; Pittet, L.; Clément, B.; Bétrisey, S.; Gerber, E.; Ronikier, M.; Parisod, C.; Kozlowski, G. Climate change and alpine screes: No future for glacial relict *Papaver occidentale* (*Papaveraceae*) in Western Prealps. *Diversity* **2020**, *12*, 346. [CrossRef]
19. Didukh, Y.P.; Chorney, I.I.; Budzhak, V.V.; Tokaryuk, A.I.; Kish, R.Y.; Protopopova, V.V.; Shevera, M.V.; Kozak, O.M.; Kontar, I.S.; Rozenblit, Y.V.; et al. *Climatogenic Changes of Plant Life of the Ukrainian Carpathians*; DrukArt: Chernivtsi, Ukraine, 2016.
20. Trivedi, M.R.; Morecroft, M.D.; Berry, P.M.; Dawson, T.P. Potential effects of climate change on plant communities in three montane nature reserves in Scotland, UK. *Biol. Conserv.* **2008**, *141*, 1665–1675. [CrossRef]
21. Kobiv, Y. Response of rare alpine plant species to climate change in the Ukrainian Carpathians. *Folia Geobot.* **2017**, *52*, 217–226. [CrossRef]
22. Kobiv, Y. Trends in population size of rare plant species in the alpine habitats of the Ukrainian Carpathians under climate change. *Diversity* **2018**, *10*, 62. [CrossRef]
23. Leunda, M.; González-Sampériz, P.; Gil-Romera, G.; Bartolomé, M.; Belmonte-Ribas, Á.; Gómez-García, D.; Kaltenrieder, P.; Rubiales, J.M.; Schwörer, C.; Tinner, W.; et al. Ice cave reveals environmental forcing of long-term Pyrenean tree line dynamics. *J. Ecol.* **2018**, *107*, 814–828. [CrossRef]
24. Löffler, J.; Pape, R. Thermal niche predictors of alpine plant species. *Ecology* **2020**, *101*, e02891. [CrossRef]
25. Hultén, E. *The Amphi-Atlantic Plants and Their Phytogeographic Connections*; Koeltz: Königstein, Germany, 1973.
26. Meusel, H.; Jäger, E.; Rauschert, S.; Weinert, E. *Vergleichende Chorologie der Zentraleuropäischen Flora*; Fischer: Jena, Germany, 1978; Volume 2.
27. Hultén, E.; Fries, M. *Atlas of North European Vascular Plants*; Koelz: Königstein, Germany, 1986; Volume 1–3.
28. Gómez García, D.; Ferrández Palacio, J.V.; Bernal Gálvez, M.; Campo González, A.; López Retamero, J.R.; Ezquerra Rivas, V. *Plantas de las Cumbres del Pirineo*; Instituto Pirenaicode Ecologia: Jaca, Spain; Prames: Zaragoza, Spain, 2020.
29. Abeli, T.; Vamosi, J.C.; Orsenigo, S. The importance of marginal population hotspots of cold-adapted species for research on climate change and conservation. *J. Biogeogr.* **2018**, *45*, 977–985. [CrossRef]
30. Dítě, D.; Hájek, M.; Svitková, I.; Košuthová, A.; Šoltés, R.; Kliment, J. Glacial-relict symptoms in the Western Carpathian flora. *Folia Geobot.* **2018**, *53*, 277–300. [CrossRef]
31. Buchner, O.; Karadar, M.; Bauer, I.; Neuner, G. A novel system for in situ determination of heat tolerance of plants: First results on alpine dwarf shrubs. *Plant Methods* **2013**, *9*, 7. [CrossRef]
32. Chopik, V.I. *Vysokohirna Flora Ukrains'kykh Karpat*; Naukova Dumka: Kyiv, Ukraine, 1976.
33. Boratyński, A.; Didukh, Y.P. *Loiseleuria procumbens* (*Ericaceae*) in the Ukrainian Carpathians. *Dendrobiology* **2002**, *47*, 3–8.
34. Boratyński, A.; Romo, A. *Loiseleuria procumbens* (*Ericaceae*) in the Spanish Pyrenees. *Acta Soc. Bot. Pol.* **2003**, *72*, 125–133. [CrossRef]
35. Graebherr, G.; Gottfried, M.; Pauli, H. Climate effects on mountain plants. *Nature* **1994**, *369*, 448. [CrossRef]
36. Körner, C. *Alpine Plant Life: Functional Plant Ecology of High Mountain Ecosystems*; Springer: Berlin/Heidelberg, Germany, 1999.
37. Körner, C. The green cover of mountains in a changing environment. In *Global Change and Mountain Regions*; Huber, U.M., Bugmann, K.M., Reasoner, M.A., Eds.; Springer: Dordrecht, The Netherlands, 2005; Volume 23, pp. 367–376.
38. Cherepanyn, R.M. Effect of climate changes on the habitat of rare arctic-alpine plant species in the high mountain part of the Ukrainian Carpathians. *Biol. Stud.* **2018**, *12*, 73–86. [CrossRef]
39. Grime, J.P.; Price, S. *The Evolutionary Strategies that Shape Ecosystems*; John Wiley & Sons: West Sussex, UK, 2012.
40. Didukh, Y.P. *World of Plants of Ukraine in Aspect of the Climate Change*; Naukova Dumka: Kyiv, Ukraine, 2023.
41. Pellissier, L.; Eidesen, P.B.; Ehrich, D.; Descombes, P.; Schönswetter, P.; Tribsch, A.; Westergaard, K.B.; Alvarez, N.; Guisan, A.; Zimmermann, N.E.; et al. Past climate-driven range shifts and population genetic diversity in arctic plants. *J. Biogeogr.* **2015**, *43*, 461–470. [CrossRef]
42. Wada, N.; Shimmono, M.; Miyamoto, M.; Kojima, S. Warming effects on shoot developmental growth production in sympatric evergreen alpine dwarf shrubs *Empetrum nigrum* and *Loiseleuria procumbens*. *Ecol. Res.* **2002**, *17*, 125–132. [CrossRef]
43. Rodríguez-Sanchéz, F.; Arroyo, J. Reconstructing the demise of Tethyan plants: Climate-driven range dynamics of *Laurus* since the Pliocene. *Glob. Ecol. Biogeogr.* **2008**, *17*, 685–695. [CrossRef]
44. Svenning, J.-C.; Fløjgaard, C.; Marske, K.A.; Nógues-Bravo, D.; Normand, S. Applications of species distribution modelling to paleobiology. *Quat. Sci. Rev.* **2011**, *30*, 2930–2947. [CrossRef]

45. Walas, Ł.; Sobierajska, K.; Ok, T.; Dönmez, A.A.; Kanoğlu, S.S.; Bou Dagher-Kharrat, M.; Douaihy, B.; Romo, A.; Stephan, J.; Jasińska, A.K.; et al. Past, present and future geographic range of the oro-Mediterranean Tertiary relict: *Juniperus drupacea* case study. *Reg. Environ. Change* **2019**, *19*, 1507–1520. [CrossRef]
46. Zapałowicz, H. *Roślinna Szata Gór Pokucko-Marmaroskich; Sprawozdanie Komisyi Fizyjograficznej 24*; Akademia Umiejętności: Kraków, Poland, 1889.
47. Mráz, P.; Ronikier, M. Biogeography of the Carpathians: Evolutionary and spatial facets of biodiversity. *Biol. J. Linn. Soc.* **2016**, *119*, 528–559. [CrossRef]
48. Tasenkevich, L.; Boratyński, A.; Skrypec, K.; Seniv, M.; Khmil, T.; Walas, Ł. Biodiversity of high-mountain woody plants in the East Carpathians in Ukraine. *Dendrobiology* **2023**, *89*, 1–19. [CrossRef]
49. Braun-Blanquet, L. *La Végétation Alpine des Pyrénées Olrientales, Étude de phytosociologie Compare*; Consejo Superior de Investigaciones Científicas: Barcelona, Spain, 1948.
50. Villar, L.; Sesé, J.A.; Ferrández, J.V. *Atlas de la Flora del Pirineo Aragonés*; Consejo de Protección de la Naturaleza de Aragón e Instituto de Estudios Altoaragoneses: Huesca, Spain, 2001; Volume 2.
51. Gómez García, D.; Ferrández Palacio, J.V.; Tejero, P.; Font, X. Spatial distribution and environmental analysis of the alpine flora in the Pyrenees. *Pirineos* **2017**, *172*, e027.
52. Larcher, W.; Wagner, J. High mountain bioclimate: Temperatures near the ground recorded from the timberline to the nival zone in the Central Alps. *Contrib. Nat. Hist. Mus. Bern* **2009**, *12*, 857–874.
53. Larcher, W.; Wagner, J. Temperatures in the life zones of the Tyrolean Alps. *Sitzungsberichte Abteilung* **2011**, *213*, 31–51. [CrossRef]
54. Wipf, S.; Stoeckli, V.; Bebi, P. Winter climate change in alpine tundra: Plant responses to changes in snow depth and snowmelt timing. *Clim. Change* **2009**, *94*, 105–121. [CrossRef]
55. Ladinig, U.; Hacker, J.; Neuner, G.; Wagner, J. How endangered is sexual reproduction of high-mountain plants by summer frosts? Frost resistance, frequency of frost events and risk assessment. *Oecologia* **2013**, *171*, 743–760. [CrossRef]
56. Tasenkevich, L. *Flora of the Carpathians: Checklist of the Native Vascular Plant Species*; State Museum of Natural History National Academy of Sciences of Ukraine: Lviv, Ukraine, 1998.
57. Ninot, J.M.; Carrillo, E.; Font, X.; Carreras, J.; Ferré, A.; Masalles, R.M.; Soriano, I.; Vigo, J. Altitude zonation in the Pyrenees. A geobotanic interpretation. *Phytocoenologia* **2007**, *37*, 371–398. [CrossRef]
58. Ninot, J.M.; Carrillo, E.; Ferré., A. The Pyrenees. In *The Vegetation of the Iberian Peninsula*; Loidi, J., Ed.; Springer: Cham, Switzerland, 2017; Volume 12, pp. 323–366.
59. HabRef v7. Available online: https://inpn.mnhn.fr/habitat/cd_hab/23233/tab/description (accessed on 17 July 2023).
60. Baudière, A.; Serve, L. Les Landes rases à *Loiseleuria procumbens* en Pyrénées Orientales et leur intérêt phytogéographique. *Coll. Phytosoc.* **1975**, 337–347.
61. Biţă-Nicolae, C.D. The Natural Priority Habitats in the Alpine zone of Bucegi Massif (Romanian Southern Carpathians). *Bot. Serbica* **2011**, *35*, 79–85.
62. Stancu, D.I. The main alpine and subalpine habitats in Râiosu & Buda mountains, Făgăraş massif. *Muzeul Olten. Craiova. Oltenia. Stud. Şi Comunicări. Ştiinţele Nat.* **2013**, *29*, 118–126.
63. Puşcaş, M.; Gafta, D.; Cristea, V. Analyse éco-coenotique des prairies édifiées par *Carex curvala* All. des Carpates Roumaines. *Acta Bot. Gall.* **2005**, *152*, 497–505. [CrossRef]
64. Kricsfalusy, V.V. Mountain grasslands of high conservation value in the Eastern Carpathians: Syntaxonomy, biodiversity, protection and management. *Thaiszia* **2013**, *23*, 67–112.
65. Rivas-Martínez, S.; Rivas-Sáenz, S.; Penas, A. World-wide bioclimatic classification system. *Glob. Geobot.* **2011**, *1*, 1–634 + maps.
66. Rivas-Martínez, S.; Penas, Á.; del Rió, S.; Díaz González, T.E.; Rivas-Sáenz, S. Bioclimatology of the Iberian Peninsula and the Balearic Islands. In *The Vegetation of the Iberian Peninsula*; Loidi, J., Ed.; Springer: Cham, Switzerland, 2017; Volume 12, pp. 29–80.
67. Dobrowski, S.Z. A climatic basis for micro-refugia: The influence of terrain on climate. *Glob. Change Biol.* **2011**, *17*, 1022–1035. [CrossRef]
68. Bär, A.; Pape, R.; Bräuning, A.; Löffler, J. Growth-ring variations of dwarf shrubs reflect regional climate signals in alpine environments rather than topoclimatic differences. *J. Biogeogr.* **2008**, *35*, 625–636. [CrossRef]
69. Steinbauer, M.J.; Grytnes, J.A.; Jurasinski, G.; Kulonen, A.; Lenoir, J.; Pauli, H.; Rixen, C.; Winkler, M.; Bardy-Durchhalter, M.B.; Barni, E.; et al. Accelerated increase in plant species richness on mountain summits is linked to warming. *Nature* **2018**, *556*, 231–234. [CrossRef]
70. Nagy, L.; Grabherr, G.; Körner, C.; Thompson, D.B.A. Alpine biodiversity in space and time. In *Alpine Biodiversity in Europe*; Nagy, L., Grabherr, G., Körner, C., Thompson, D.B.A., Eds.; Springer: Berlin, Germany, 2003; pp. 453–464.
71. Skrede, I.; Eidesen, P.B.; Piñeiro Portela, R.; Brochmann, C. Refugia, differentiation and postglacial migration in arctic-alpine Eurasia, exemplified by the mountain avens (*Dryas octopetala* L.). *Mol. Ecol.* **2006**, *15*, 1827–1840. [CrossRef]
72. Stephenson, C.M.; Kohn, D.D.; Park, K.J.; Atkinson, R.; Edwards, C.; Travis, J.M. Testing mechanistic models of seed dispersal for the invasive *Rhododendron ponticum* (L.). *Persp. Pl. Ecol., Evol. Syst.* **2007**, *9*, 15–28. [CrossRef]
73. Escaravage, N.; Wagner, J. Pollination effectiveness and pollen dispersal in a *Rhododendron ferrugineum* (*Ericaceae*) population. *Pl. Biol.* **2004**, *6*, 606–615. [CrossRef]
74. Alsos, I.G.; Alm, T.; Normand, S.; Brochmann, C. Past and future range shifts and loss of diversity in dwarf willow (*Salix herbacea* L.) inferred from genetics, fossils and modelling. *Glob. Ecol. Biogeogr.* **2009**, *18*, 223–239. [CrossRef]

75. Ronikier, M.; Schneeweiss, G.M.; Schönswetter, P. The extreme disjunction between Beringia and Europe in *Ranunculus glacialis* s. l. (*Ranunculaceae*) does not coincide with the deepest genetic split–a story of the importance of temperate mountain ranges in arctic–alpine phylogeography. *Mol. Ecol.* **2012**, *21*, 5561–5578. [CrossRef]
76. Winkler, M.; Tribsch, A.; Schneeweiss, G.M.; Brodbeck, S.; Gugerli, F.; Holderegger, R.; Abbott, R.J.; Schönswetter, P. Tales of the unexpected: Phylogeography of the arctic-alpine model plant *Saxifraga oppositifolia* (*Saxifragaceae*) revisited. *Mol. Ecol.* **2012**, *21*, 4618–4630. [CrossRef]
77. Kropf, M.; Comes, H.P.; Kadereit, J.W. Past, present and future of mountain species of the French Massif Central–the case of *Soldanella alpina* L. subsp. *alpina* (*Primulaceae*) and a review of other plant and animal studies. *J. Biogeogr.* **2012**, *39*, 799–812. [CrossRef]
78. Suchan, T.; Malicki, M.; Ronikier, M. Relict populations and Central European glacial refugia: The case of *Rhododendron ferrugineum* (*Ericaceae*). *J. Biogeogr.* **2019**, *46*, 392–404. [CrossRef]
79. Lewandowska, A.; Boratyńska, K.; Marcysiak, K.; Gómez, D.; Romo, A.; Malicki, M.; Iszkuło, G.; Boratyński, A. Phenotypic differentiation of *Rhododendron ferrugineum* populations in European mountains. *Dendrobiology* **2022**, *87*, 1–12. [CrossRef]
80. Lewandowska, A.; Marcysiak, K.; Gómez, D.; Jasińska, A.K.; Romo, A.; Didukh, Y.; Sękiewicz, K.; Boratyńska, K.; Boratyński, A. Inference of taxonomic relationships between *Rhododendron ferrugineum* and *R. myrtifolium* (*Ericaceae*) from leaf and fruit morphologies. *Bot. J. Linn. Soc.* **2023**, *201*, 483–497. [CrossRef]
81. Ikeda, H.; Senni, K.; Fujii, N.; Setoguchi, H. High mountains of the Japanese archipelago as refugia for arctic–alpine plants: Phylogeography of *Loiseleuria procumbens* (L.) Desvaux (*Ericaceae*). *Biol. J. Linn. Soc.* **2009**, *97*, 403–412. [CrossRef]
82. Ikeda, H.; Eidesen, P.B.; Yakubov, V.; Barkalov, V.; Brochmann, C.; Setoguchi, H. Late Pleistocene origin of the entire circumarctic range of the arctic-alpine plant *Kalmia procumbens*. *Mol. Ecol.* **2017**, *26*, 5773–5783. [CrossRef]
83. Ikeda, H.; Carlsen, T.; Fujii, N.; Brochmann, C.; Setoguchi, H. Pleistocene climatic oscillations and the speciation history of an alpine endemic and a widespread arctic-alpine plant. *New Phytol.* **2012**, *194*, 583–594. [CrossRef]
84. Eidesen, P.B.; Ehrich, D.; Bakkestuen, V.; Alsos, I.G.; Gilg, O.; Taberlet, P.; Brochmann, C. Genetic roadmap of the Arctic: Plant dispersal highways, traffic barriers and capitals of diversity. *New Phytol.* **2013**, *200*, 898–910. [CrossRef]
85. Pykälä, J. Relation between extinction and assisted colonization of plants in the arctic-alpine and boreal regions. *Conserv. Biol.* **2016**, *31*, 524–530. [CrossRef]
86. Ronikier, M. Biogeography of high-mountain plants in the Carpathians: An emerging phylogeographical perspective. *Taxon* **2011**, *60*, 373–389. [CrossRef]
87. Hülber, K.; Wessely, J.; Gattringer, A.; Moser, D.; Kuttner, M.; Essl, F.; Leitner, M.; Winkler, M.; Ertl, S.; Willner, W.; et al. Uncertainty in predicting range dynamics of endemic alpine plants under climate warming. *Glob. Change Biol.* **2016**, *22*, 2608–2619. [CrossRef]
88. Maclean, I.M.D.; Hopkins, J.J.; Bennie, J.; Lawson, C.R.; Wilson, R.J. Microclimates buffer plant community response to climate change. *Glob. Ecol. Biogeogr.* **2015**, *24*, 1340–1350. [CrossRef]
89. Treu, R.; Laursen, G.A.; Stephenson, S.L.; Landolt, J.C.; Densmore, R. Mycorrhizae from Denali National Park and Preserve, Alaska. *Mycorrhiza* **1996**, *6*, 21–29. [CrossRef]
90. Koizumi, T.; Nara, K. Communities of putative Ericoid Mycorrhizal Fungi isolated from alpine dwarf shrubs in Japan: Effects of host identity and microhabitat. *Microbes Envir.* **2017**, *32*, 147–153. [CrossRef]
91. Haselwandter, K. Mycorrhizal status of ericaceous plants in alpine and subalpine areas. *New Phytol.* **1979**, *83*, 427–431. [CrossRef]
92. Cripps, C.L.; Eddington, L.H. Distribution of mycorrhizal types among alpine vascular plant families on the Beartooth Plateau, Rocky Mountains, U.S.A., in reference to large-scale patterns in Arctic–Alpine habitats. Arctic Antarct. *Alp. Res.* **2005**, *37*, 177–188. [CrossRef]
93. Schweingruber, F.; Poschlod, P. Growth rings in herbs and shrubs: Lifespan, age determination and stem anatomy. *For. Snow Landsc. Res.* **2005**, *79*, 195–415.
94. Scherrer, D.; Körner, C. Topographically controlled thermal-habitat differentiation buffers alpine plant diversity against climate warming. *J. Biogeogr.* **2011**, *38*, 406–416. [CrossRef]
95. Zeidler, M.; Banaš, M. *Vybrané Kapitoly z Ekologie Horských Ekosystémů*; Univerzita Palackého v Olomouci, Přírodovědecká Fakulta: Olomouc, Slovakia, 2013.
96. García, M.B.; Domingo, D.; Pizarro, M.; Font, X.; Gómez-García, D.; Ehrlén, J. Rocky habitats as microclimatic refuges for biodiversity. A close-up thermal approach. *Env. Exp. Bot.* **2020**, *170*, 103886. [CrossRef]
97. Kuprian, E.; Briceño, V.F.; Wagner, J.; Neuner, G. Ice barriers promote supercooling and prevent frost injury in reproductive buds, flowers and fruits of alpine dwarf shrubs throughout the summer. *Environ. Exper. Bot.* **2014**, *106*, 4–12. [CrossRef]
98. Palacio, S.; Lenz, A.; Wipf, S.; Hoch, G.; Rixen, C. Bud freezing resistance in alpine shrubs across snow depth gradients. *Envir. Exper. Bot.* **2015**, *118*, 95–101. [CrossRef]
99. Tasenkevich, L. Endemism in the Carpathian flora–a chorological aspect. In *Geobotanist and Taxonomist. A Volume Dedicated to Professor Adam Zając on the 70th Anniversary of His Birth*; Zemanek, B., Ed.; Istitute of Botany of Jagellonian University: Kraków, Poland, 2011; pp. 157–168.
100. Breman, E.; Hurdu, B.I.; Kliment, J.; Kobiv, Y.; Kučera, J.; Mráz, P.; Pușcaș, M.; Renaud, J.; Ronikier, M.; Šibík, J.; et al. Conserving the endemic flora of the Carpathian Region: An international project to increase and share knowledge of the distribution, evolution and taxonomy of Carpathian endemics and to conserve endangered species. *Pl. Syst. Evol.* **2020**, *306*, 15–59. [CrossRef]

101. Margielewski, W. Geological and geomorphological settings. In *Postglacial History of Vegetation in the Polish Part of the Western Carpathians Based on Isopollen Map*; Obidowicz, A., Madeyska, E., Turner, C., Eds.; W. Szafer Institute of Botany Polish Academy of Sciences: Kraków, Poland, 2013; pp. 11–14.
102. Mirek, Z. Altitudinal vegetation belts of the Western Carpathians. In *Postglacial History of Vegetation in the Polish Part of the Western Carpathians Based on Isopollen Maps*; Obidowicz, A., Madeyska, E., Turner, C., Eds.; W. Szafer Institute of Botany Polish Academy of Sciences: Kraków, Poland, 2013; pp. 15–21.
103. Ţopa, E. Ericaceae. In *Flora Republicii Popularae Romîne*; Săvulescu, T., Ed.; Academia Republicii Populare Romîne: Bucureşti, Romania, 1960; Volume 7, pp. 119–140.
104. Jasičová, M. Ericaceae Juss. In *Flóra Slovenska*; Futák, J., Bertová, L., Eds.; Veda: Bratislava, Slovakia, 1982; Volume 3, pp. 337–359.
105. Turis, P.; Kliment, J.; Feráková, V.; Dítě, D.; Eliáš, P.; Hrivnák, R.; Košťál, J.; Šuvada, R.; Mráz, P.; Bernátová, D. Red List of vascular plants of the Carpathian part of Slovakia. *Thaiszia* **2014**, *24*, 35–87.
106. Mirek, Z. *High Mountain Vascular Plants of the Carpathians. Atlas of Distribution*; W. Szafer Institute of Botany Polish Academy of Sciences: Kraków, Poland, 2020.
107. QGIS.org. QGIS Geographic Information System. QGIS Association. 2022. Available online: http://www.qgis.org (accessed on 1 April 2023).
108. Bradie, L.; Leung, B. A quantitative synthesis of the importance of variables used in Maxent species distribution models. *J. Biogeogr.* **2017**, *44*, 1344–1361. [CrossRef]
109. O'Donnell, M.S.; Ignizio, D.A. Bioclimatic predictors for supporting ecological applications in the conterminous United States. *U.S. Geol. Surv. Data Ser.* **2012**, *691*, 10.
110. Fick, S.E.; Hijmans, R.J. WorldClim 2: New 1 km spatial resolution climate surfaces for global land areas. *Intern. J. Clim.* **2017**, *37*, 4302–4315. [CrossRef]
111. Brown, J.L.; Hill, D.J.; Dolan, A.M.; Carnaval, A.C.; Haywood, A.M. PaleoClim, high spatial resolution paleoclimate surfaces for global land areas. *Nat. Sci. Data* **2018**, *5*, 180–254. [CrossRef]
112. Karger, D.N.; Nobis, M.P.; Normand, S.; Graham, C.H.; Zimmermann, N.E. CHELSA-TraCE21k v1. 0. Downscaled transient temperature and precipitation data since the last glacial maximum. *Clim. Past.* **2021**, *19*, 439–456. [CrossRef]
113. Gent, P.R.; Danabasoglu, G.; Donner, L.J.; Holland, M.M.; Hunke, E.C.; Jayne, S.R.; Lawrence, D.M.; Neale, R.B.; Rasch, P.J.; Vertenstein, M.; et al. The Community Climate System Model Version 4. *J. Clim.* **2011**, *24*, 4973–4991. [CrossRef]
114. Collins, M.; Knutti, R.; Arblaster, J.; Dufresne, J.L.; Fichefet, T.; Friedlingstein, P.; Gao, X.; Gutowski, W.J.; Johns, T.; Krinner, G.; et al. Long-term Climate Change: Projections, Commitments and Irreversibility. In *Climate Change 2013: The Physical Science Basis. Contribution of Working Group I to the Fifth Assessment Report of the Intergovernmental Panel on Climate Change*; Stocker, T.F., Qin, D., Plattner, G.K., Tignor, M., Allen, S.K., Boschung, J., Nauels, A., Xia, Y., Bex, V., Midgley, P.M., Eds.; Cambridge University Press: Cambridge, UK, 2013; pp. 1029–1136.
115. Otto-Bliesner, B.L.; Marshall, S.J.; Overpeck, J.T.; Miller, G.H.; Hu, A. Simulating arctic climate warmth and ice field retreat in the Last Interglaciation. *Science* **2006**, *311*, 1751–1753. [CrossRef]
116. R Core Team. R: A Language and Environment for Statistical Computing. R Foundation for Statistical Computing. Available online: https://www.R-project.org/ (accessed on 20 March 2022).
117. Phillips, S.J. A Brief Tutor. Maxent. 2017. Available online: https://biodiversityinformatics.amnh.org/open_source/maxent/Maxent_tutorial2017.pdf (accessed on 15 March 2020).
118. Phillips, S.J.; Anderson, R.P.; Dudík, M.; Schapire, R.E.; Blair, M.E. Opening the black box: An open-source release of Maxent. *Ecography* **2017**, *40*, 887–893. [CrossRef]
119. Phillips, S.J.; Dudík, M.; Schapire, R.E. Maxent Software for Modeling Species Niches and Distributions (Version 3.4.1). Available online: http://biodiversityinformatics.amnh.org/open_source/maxent/ (accessed on 15 March 2020).
120. Muscarella, R.; Galante, P.J.; Soley-Guardia, M.; Boria, R.A.; Kass, J.M.; Uriarte, M.; Anderson, R.P. ENMeval: An R package for conducting spatially independent evaluations and estimating optimal model complexity for Maxent ecological niche models. *Methods Ecol. Evol.* **2014**, *5*, 1198–1205. [CrossRef]
121. Salvà-Catarineu, M.; Romo, A.; Mazur, M.; Zielińska, M.; Minissale, P.; Dönmez, A.A.; Boratyńska, K.; Boratyński, A. Past, present and future geographic range of the relict Mediterranean and Macaronesian *Juniperus phoenicea* complex. *Ecol. Evol.* **2021**, *11*, 5075–5095. [CrossRef]
122. Wang, Z.; Chang, Y.C.I.; Ying, Z.; Zhu, L.; Yang, Y. A parsimonious threshold-independent protein feature selection method through the area under receiver operating characteristic curve. *Bioinformatics* **2007**, *23*, 2788–2794. [CrossRef]
123. Mas, J.-F.; Soares Filho, B.; Pontius, R.G.; Farfán Gutiérrez, M.; Rodrigues, H. A suite of tools for ROC analysis of spatial models. *ISPRS Intern. J. Geo-Inform.* **2013**, *2*, 869–887. [CrossRef]
124. Didukh, Y. Climate change assessment based on synphytoindication method. In *Handbook of Climate Change Mitigation and Adaptation*, 3rd ed.; Lackner, M., Sajjadi, B., Chen, W.-Y., Eds.; Springer: Heidelberg, Switzerland, 2022; pp. 2759–2814.

plants

Article

Prediction of the Potential Distribution of the Endangered Species *Meconopsis punicea Maxim* under Future Climate Change Based on Four Species Distribution Models

Hao-Tian Zhang and Wen-Ting Wang *

School of Mathematics and Computer Science, Northwest Minzu University, Lanzhou 730030, China
* Correspondence: iamwwt1983@163.com

Abstract: Climate change increases the extinction risk of species, and studying the impact of climate change on endangered species is of great significance to biodiversity conservation. In this study, the endangered plant *Meconopsis punicea Maxim* (*M. punicea*) was selected as the research object. Four species distribution models (SDMs): the generalized linear model, the generalized boosted regression tree model, random forest and flexible discriminant analysis were applied to predict the potential distribution of *M. punicea* under current and future climates scenarios. Among them, two emission scenarios of sharing socio-economic pathways (SSPs; i.e., SSP2-4.5 and SSP5-8.5) and two global circulation models (GCMs) were considered for future climate conditions. Our results showed that temperature seasonality, mean temperature of coldest quarter, precipitation seasonality and precipitation of warmest quarter were the most important factors shaping the potential distribution of *M. punicea*. The prediction of the four SDMs consistently indicated that the current potential distribution area of *M. punicea* is concentrated between 29.02° N–39.06° N and 91.40° E–105.89° E. Under future climate change, the potential distribution of *M. punicea* will expand from the southeast to the northwest, and the expansion area under SSP5-8.5 would be wider than that under SSP2-4.5. In addition, there were significant differences in the potential distribution of *M. punicea* predicted by different SDMs, with slight differences caused by GCMs and emission scenarios. Our study suggests using agreement results from different SDMs as the basis for developing conservation strategies to improve reliability.

Keywords: climate change; endangered plant; potential distribution; species distribution models

Citation: Zhang, H.-T.; Wang, W.-T. Prediction of the Potential Distribution of the Endangered Species *Meconopsis punicea Maxim* under Future Climate Change Based on Four Species Distribution Models. *Plants* **2023**, *12*, 1376. https://doi.org/10.3390/plants12061376

Academic Editor: Gregor Kozlowski

Received: 12 January 2023
Revised: 15 March 2023
Accepted: 15 March 2023
Published: 20 March 2023

1. Introduction

Climate not only plays an important role in the growth of plants, but also determines the distribution of species [1,2]. Future climate change will shift the distribution of species and even cause the extinction of species [3–9]. Some studies have indicated that species will disperse to higher latitudes and/or altitudes under climate change characterized by warming [3,10–12]. At lower altitudes and latitudes, species richness and diversity are therefore predicted to fall, while at higher altitudes and latitudes, endangered species are predicted to face a greater risk of extinction as a result of increased species competition [13]. Thus, alpine ecosystems in response to climate change will be more sensitive and vulnerable [14–16]. Many alpine species, which have ornamental, economic and medicinal values, are at risk from climate change [17–19]. In particular, species in the Qinghai-Tibet Plateau have been adversely affected by climate change [20–22]. Therefore, understanding the impacts of climate change on the potential distribution of species is of great significance for species conservation.

Meconopsis punicea Maxim (*M. punicea*) is a perennial herb belonging to the family *Papaveraceae* and the genus *Meconopsis* [23], mainly distributed in the hillside grassland and alpine shrub at an altitude of 2800–4300 m in northwest Sichuan, northeast Tibet, southeast Qinghai and southwest Gansu, with a flower and fruit period from June to

September [12,23–26]. *M. punicea* is not only a rare resource of Tibetan medicine, but also has a certain ornamental value [23,24]. According to the Red List of Chinese plants database (http://www.chinaplantredlist.org/ (accessed on 12 July 2022)), *M. punicea* is evaluated as a Least Concern (LC) species. Due to global warming, natural ecological degradation and overexploitation of wild resources, the living environment of *M. punicea* has been seriously and repeatedly damaged, and it is facing the danger of exhaustion of natural resources [12,23,27]. Therefore, it is necessary to actively carry out the resource investigation of *M. punicea* and study the response of the potential distribution of *M. punicea* to climate change, which is helpful for the ex situ conservation and the deep resource development of the species, and provides a theoretical basis for the introduction and domestication of the species and resource conservation.

With the development of digital information, species distribution models (SDMs) have been widely used to study the historical geographical distribution of species and their distribution trend under future climate change [28,29]; it provides a reliable theoretical basis for endangered species protection, conservation planning and invasive species control [30–34]. However, there are numerous SDMs algorithms based on statistics and machine learning, and the potential distribution results of species predicted by different algorithms significantly differ [35]. Some studies have shown that being dependent on only one SDM to predict the potential distribution of species would cause the result deviation problem [35,36]. In addition, studies have shown that different global circulation models (GCMs) will bring uncertainty to SDMs prediction [37–39], but the uncertainty caused by emission scenarios is significantly higher than that caused by GCMs [37]. Therefore, the reliability of conservation studies for endangered species could be improved by considering the comprehensive results under the influence of various uncertainties, including SDM algorithms, emission scenarios and GCMs.

In this study, we explored the potential distribution of *M. punicea* under climate change using various SDMs (i.e., generalized linear model, generalized boosted regression tree model, random forest and flexible discriminant analysis). We analyzed the agreement of the prediction results of the four SDMs to avoid the uncertainty brought by the SDM algorithm, and also considered the uncertainty brought by emission scenarios and GCMs. Among them, the emission scenarios are based on the scenarios under the sharing socio-economic pathways (SSPs), which provide more diverse air pollutant emission scenarios, and more scientifically describes future climate change under the mode of socio–economic development [40]. Based on the agreement result of the four SDMs, we aimed to determine the potential distribution of *M. punicea* under current and future climate. In addition, we explored changes in the potential distribution of *M. punicea* by comparing potential current and future distributions, and made conservation recommendations based on the potential distribution changes.

2. Results

2.1. *Evaluation of Model Prediction Accuracy and Significance of Bioclimatic Variables*

According to the evaluation metric AUC, the performance of the four SDMs were "excellent", with the GLM having the highest prediction accuracy, followed by the FDA (Figure 1a). According to the evaluation metric Kappa, the prediction accuracy of FDA was the highest, followed by GLM (Figure 1b). Both evaluation metrics consistently showed higher prediction accuracy of the GLM and FDA than that of the RF and GBM (Figure 1). No matter the current or future climate scenarios, the results consistently showed that the SDM algorithm would generate large uncertainties in predicting the potential distribution of *M. punicea* (Figures 2 and 3). The prediction through GLM model showed a large uncertainty in potential distribution caused by the emission scenarios, while under the other SDMs (i.e., GBM, RF and FDA), the uncertainties caused by the emission scenarios were small (Figure 4). Under different GCMs, the potential distributions predicted by the FDA and GLM models consistently showed large uncertainties, while that predicted by the GBM and RF models were less uncertain (Figure 5). In general, the uncertainty generated by GCMs was smaller than that generated by the SDM algorithms and larger than that generated by the emission scenarios.

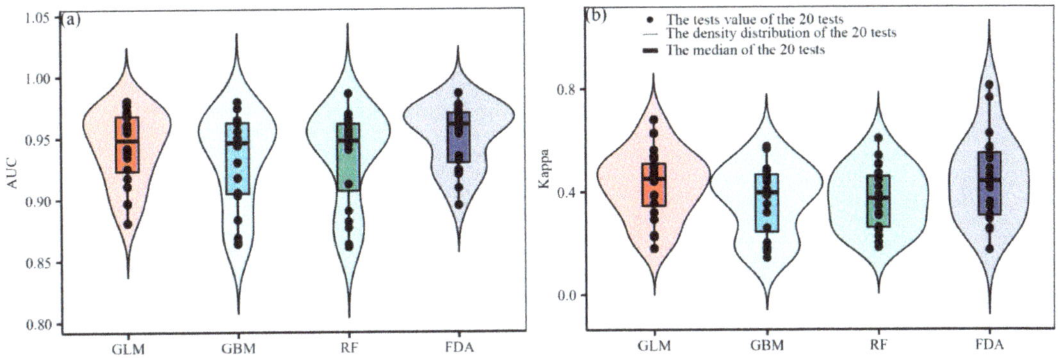

Figure 1. Comparison of prediction accuracy of four species distribution models under (**a**) AUC and (**b**) Kappa. The black dots represent the AUC (Kappa) of each test, the thin outline represents the density distribution of the AUC (Kappa) of 20 random tests, and the thick horizontal line represents the median AUC (Kappa) of 20 random tests.

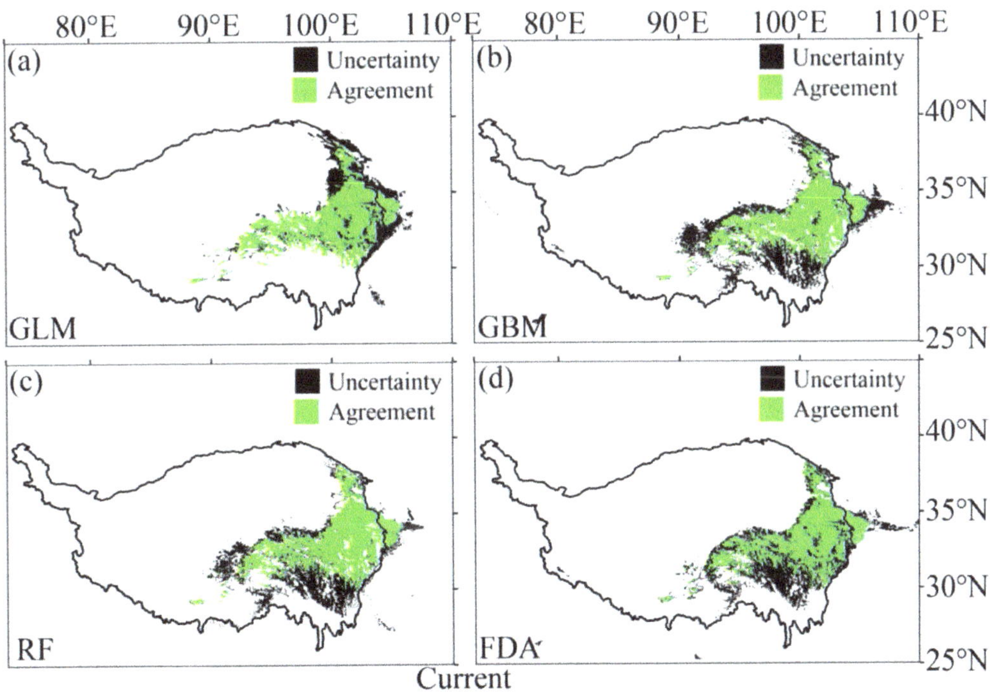

Figure 2. The uncertainty generated by the SDM algorithms in predicting the potential distribution of *M. punicea* under the current climate scenarios. (**a**–**d**) represent the agreement and uncertainty of the potential distribution of *M. punicea* predicted through GLM, GBM, RF and FDA, respectively. The green area indicates the agreement, and the black area indicates the uncertainty. Agreement/Uncertainty means the same/different part of the potential distribution predicted through the four SDMs.

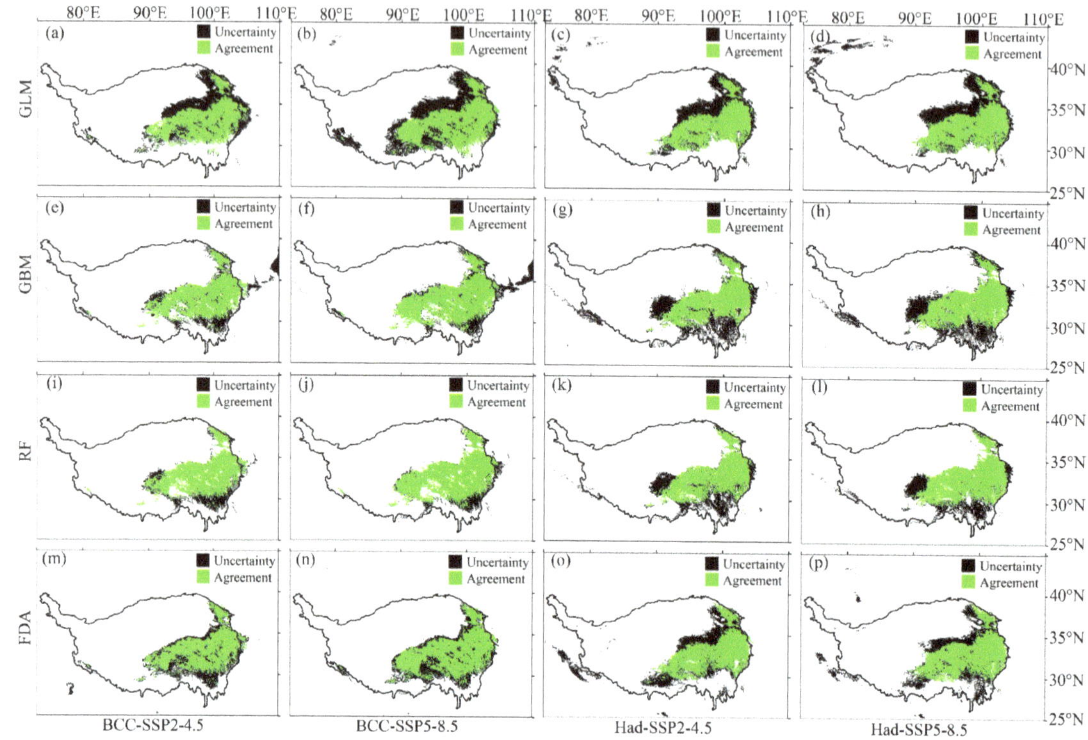

Figure 3. The uncertainty generated by the SDM algorithms in predicting the potential distribution of *M. punicea* under the future climate scenarios. Here, the future global circulation models (GCMs) include BCC-CSM2-MR (BCC) and HadGEM3-GC31-LL (Had), and the emission scenarios include SSP2-4.5 and SSP5-8.5. The first to fourth rows represent the agreement and uncertainty of the predicted potential distribution of *M. punicea* under GLM (**a–d**), GBM (**e–h**), RF (**i–l**) and FDA (**m–p**), respectively. The green area indicates the agreement, and the black area indicates the uncertainty. Agreement/Uncertainty means the same/different part of the potential distribution predicted through the four SDMs under the same climate scenarios.

The importance score of each bioclimatic variable in the four SDMs showed: temperature seasonality (BIO4), mean temperature of coldest quarter (BIO11) and precipitation of warmest quarter (BIO18) have a major contribution to the prediction of the potential distribution of *M. punicea* in the GLM (Table 1). In the GBM and RF, BIO18, precipitation seasonality (BIO15) and BIO11 play a significant role in determining the potential distribution of *M. punicea* (Table 1). In the FDA, the main bioclimatic variables contributing to the potential distribution of *M. punicea* were BIO4, mean diurnal range (BIO2) and isothermality (BIO3) (Table 1). In summary, the dominant bioclimatic variables shaping the potential distribution of *M. punicea* are BIO4, BIO11, BIO15 and BIO18.

Response curves for the dominant bioclimatic variables and potential distribution probabilities fitted through GLM and FDA show similar trends. Among them, the response curves of the distribution probability of *M. punicea* to BIO4, BIO11, BIO15 and BIO18 showed an oscillating trend in a certain range and then stabilized (Figure 6). The response curve trend fitted by GBM and RF were roughly the same, and the response curve of the distribution probability of *M. punicea* to BIO4, BIO11, BIO15 and BIO18 showed a unimodal pattern (Figure 6). The response curves of the four SDMs showed that the temperature seasonality range is 500–750, the mean temperature in the coldest season is below −5 °C, the precipitation seasonality range is above 80 and the precipitation in the warmest season is above 500 mm, which were the most suitable climate environments for the distribution of *M. punicea* (Figure 6). However, the response curve trend of temperature seasonality was different among the four SDMs (Figure 6).

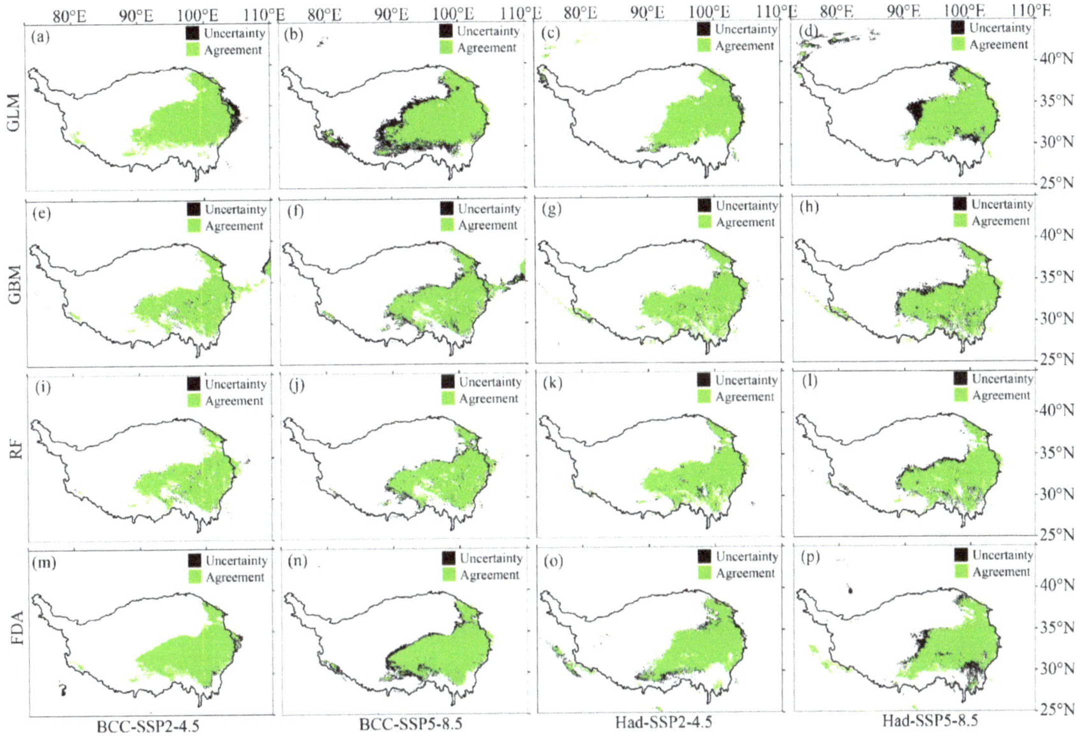

Figure 4. The uncertainty generated by the global circulation models (GCMs) in predicting the future potential distribution of *M. punicea*. Here, the GCMs include BCC-CSM2-MR (BCC) and HadGEM3-GC31-LL (Had), and the emission scenarios include SSP2-4.5 and SSP5-8.5. The first to fourth rows represent the agreement and uncertainty of the predicted potential distribution of *M. punicea* under GLM (**a–d**), GBM (**e–h**), RF (**i–l**) and FDA (**m–p**), respectively. The green area indicates the agreement, and the black area indicates the uncertainty. Agreement/Uncertainty means the same/different part of the potential distribution between BCC and Had under the same SDMs and emission scenarios.

2.2. Potential Distribution of *M. punicea* in Current Climate

The current potential distribution predicted by the four SDMs showed a high consistency, and all the models showed that the potential distribution of *M. punicea* was roughly distributed in Sichuan, Qinghai, Gansu and Qinghai-Tibet (Figure 7). To be specific, the main potential distribution areas were located in Aba Tibetan and Qiang Autonomous Prefecture and Ganzi Tibetan Autonomous Prefecture on the western Sichuan Plateau, Gannan Tibetan Autonomous Prefecture in the south of Gansu Province and Huangnan Tibetan Autonomous Prefecture and Guoluo Tibetan Autonomous Prefecture in the southeast of Qinghai Province, which are concentrated between 29.02° N–39.06° N and 91.40° E–105.89° E (Figure 7). The main potential distribution ranges predicted (29.02° N–39.06° N and 91.40° E–105.89° E) are basically consistent with those of the records (29.53° N–38.32° N and 94.30° E–104.24° E). However, the potential distribution area predicted by each model was slightly different, and the potential distribution area predicted by GLM and FDA was smaller than that predicted by RF and GBM.

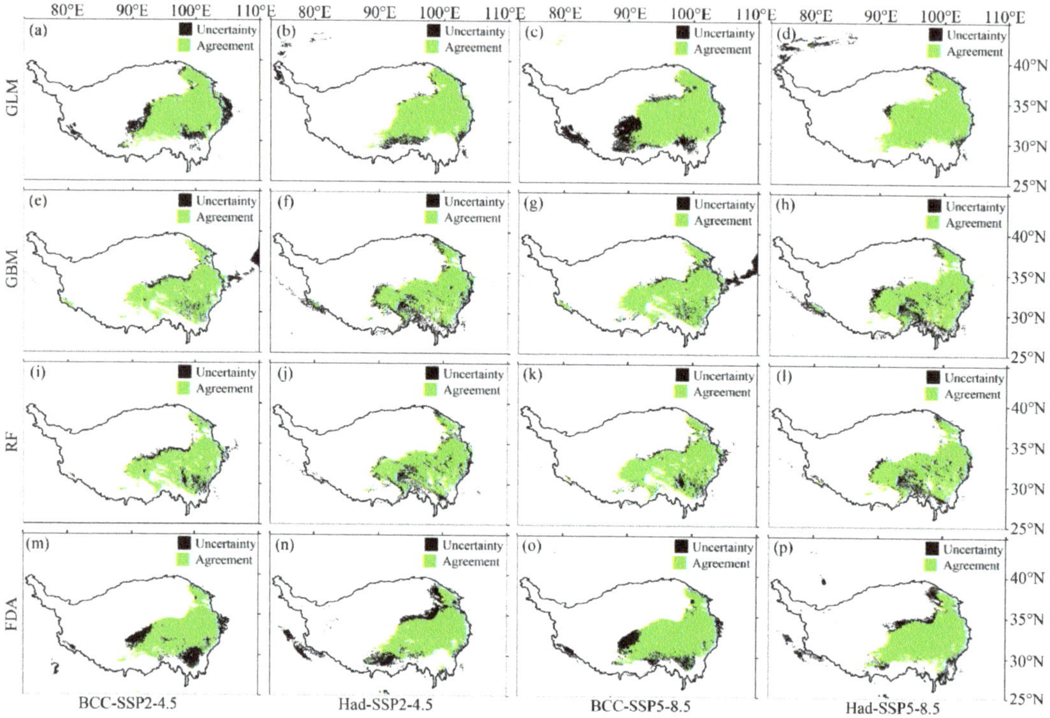

Figure 5. The uncertainty generated by the emission scenarios in predicting the future potential distribution of *M. punicea*. Here, the future global circulation models (GCMs) include BCC-CSM2-MR (BCC) and HadGEM3-GC31-LL (Had), and the emission scenarios include SSP2-4.5 and SSP5-8.5. The first to fourth rows represent the agreement and uncertainty of the predicted potential distribution of *M. punicea* under GLM (**a**–**d**), GBM (**e**–**h**), RF (**i**–**l**) and FDA (**m**–**p**), respectively. The green area indicates the agreement, and the black area indicates the uncertainty. Agreement/Uncertainty means the same part of the potential distribution between SSP2-4.5 and SSP5-8.5 under same SDMs and GCMs.

Table 1. The importance scores of each bioclimatic variable in four species distribution models (SDMs). The top three most important variables in each SDM are bolded.

	GLM	GBM	RF	FDA
BIO2	0.1502	0.0452	0.0198	**0.8284**
BIO3	0.1276	0.0254	0.0194	**0.6464**
BIO4	**0.9842**	0.0092	0.0468	**0.9972**
BIO5	0.5024	0.0282	0.0398	0.0620
BIO11	**0.6760**	**0.1458**	**0.0742**	0.0096
BIO15	0.48186	**0.2244**	**0.1282**	0.1318
BIO17	0.4210	0.0084	0.0244	0.1826
BIO18	**0.7922**	**0.6958**	**0.3162**	0.3958

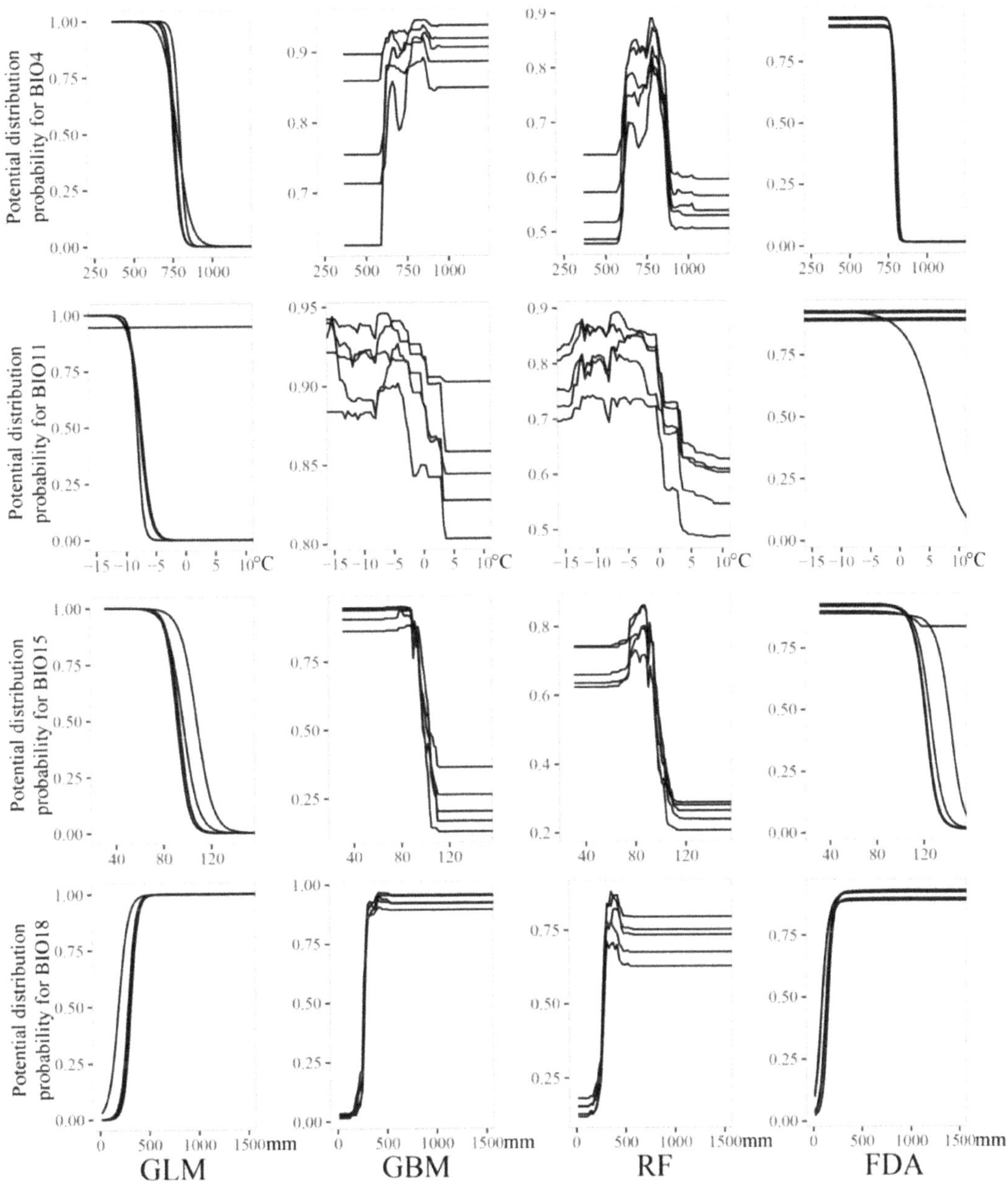

Figure 6. Response curves of potential distribution probability for dominant variables based on four species distribution models for *M. punicea*. The first to the fifth row represent the response curves for BIO4 (temperature seasonality), BIO11 (mean temperature of coldest quarter), BIO15 (precipitation seasonality) and BIO18 (precipitation of warmest quarter), respectively. The first to fourth columns correspond to GLM (generalized linear model), GBM (generalized boosted regression tree model), RF (random forest) and FDA (flexible discriminant analysis), respectively.

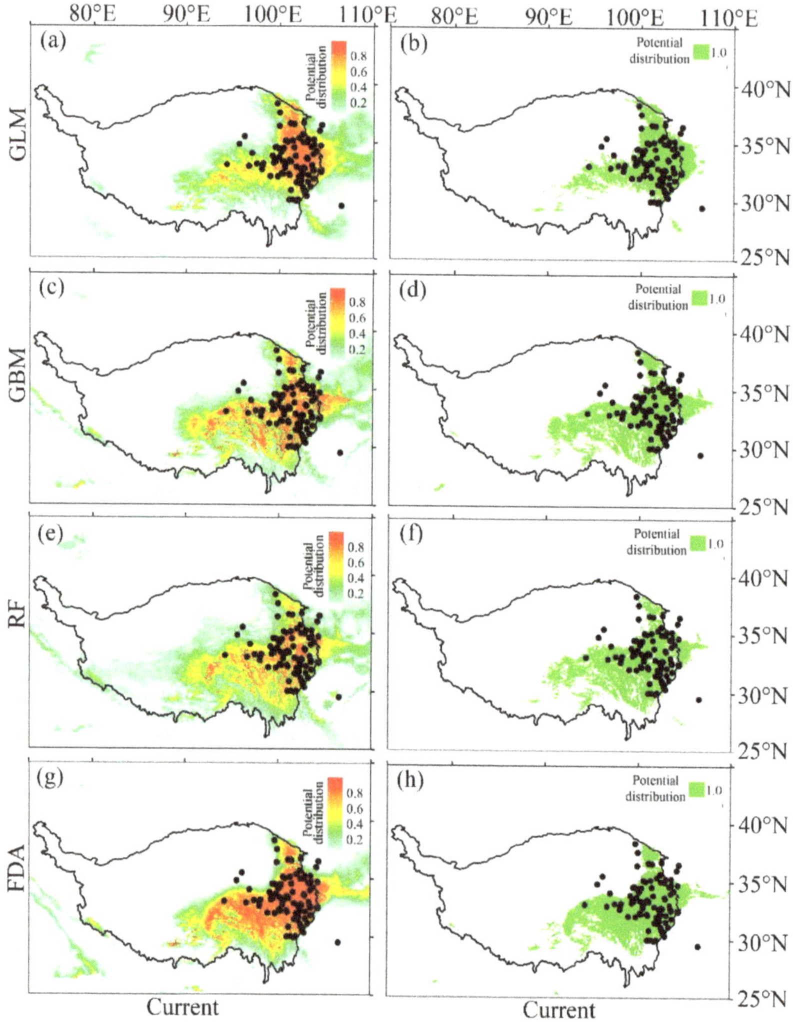

Figure 7. Current potential distribution of *M. punicea* based on four species distribution models. The left column is the prediction of the continuous distribution of *M. punicea* under the current environment. The column on the right shows the 0–1 distribution predictions of *M. punicea* under the current environment. The first to fourth rows represent the predicted potential distribution of *M. punicea* under GLM (**a**,**b**), GBM (**c**,**d**), RF (**e**,**f**) and FDA (**g**,**h**), respectively.

2.3. Changes in Distribution of *M. punicea* under Future Climate Change

Under future climate change, the four SDMs all show that the potential distribution area of *M. punicea* expands to the northwest, while a few areas in the eastern part of the current potential distribution area will not be suitable for *M. punicea* in the future (Figure 8). GLM and FDA predicted that by 2050 (2040–2060 average), the potential distribution area of *M. punicea* in the Qinghai-Tibet Plateau would widely extend to northwest China (Figure 8a–d,m–p). However, GBM and RF predicted that the potential distribution area of *M. punicea* will expand less to the northwest by 2050 (Figure 8e–l). In addition, under the SSP5-8.5 scenario (Figure 8b,d,f,h,j,l,n,p), the potential distribution of *M. punicea* by 2050 expanded more to the northwest than that under the SSP2-4.5 scenario (Figure 8a,c,e,g,i,k,m,o). In addition, no matter what the GCM, the main change trend in the potential distribution of *M. punicea* was basically the same. The minor difference was that the loss of potential distribution under the Had was slightly more than that under the BCC (Figure 8).

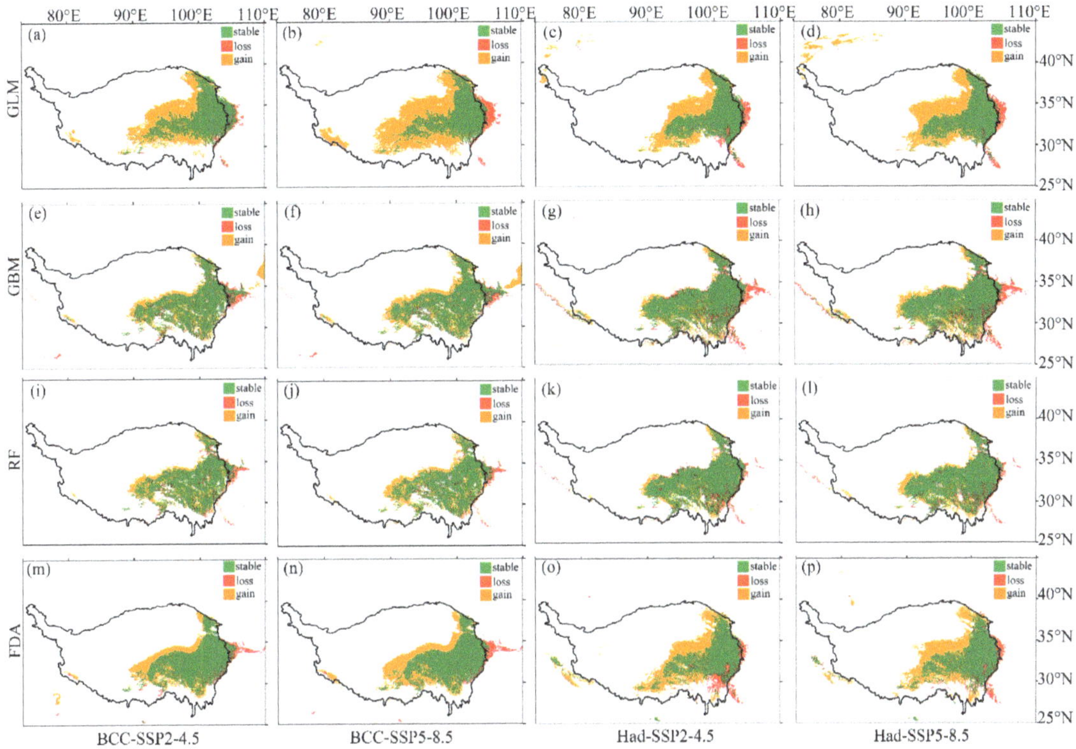

Figure 8. The changes in potential distribution under the future climate scenarios compared to the current climate scenarios based on four species distribution models. Here, the future global circulation models (GCMs) include BCC-CSM2-MR (BCC) and HadGEM3-GC31-LL (Had), and the emission scenarios include SSP2-4.5 and SSP5-8.5. The first to fourth rows show the results of potential distribution changes for GLM (**a**–**d**), GBM (**e**–**h**), RF (**i**–**l**) and FDA (**m**–**p**), respectively. The green area indicates the current and future species presence, denoted as "stable". The red area indicates the current species presence, but with no future presence, denoted as "loss". The orange area indicates no current presence, but with presence in the future, denoted as "gain".

3. Discussion

3.1. Comparison of Prediction Results of Four SDMs

In this study, four SDMs were used to predict the potential distribution of *M. punicea*. We compared the prediction performance of the four SDMs by evaluation metrics and the agreement/uncertainty of the model prediction results. The predictive performance of GLM and FDA was better, as judged by the evaluation metrics (AUC and Kappa). In terms of the agreement/uncertainty of the prediction results, the model with good performance did not always perform well. GLM and FDA had the highest agreement and lowest uncertainty under the current climate, contrary to the performance under the SSP2-4.5 scenario and the SSP5-8.5 (Figures 2 and 3). In addition, not all models with good agreement/uncertainty performance had high evaluation metrics (AUC and Kappa). GBM and RF had the highest agreement and lowest uncertainty under the SSP2-4.5 scenario and the SSP5-8.5, while the AUC (Kappa) was relatively low. This mismatch may be caused by different algorithms based on different SDMs. Because the four SDMs selected in this study were carried out under the same environmental variables, as well as the environmental variables selected and the occurrence data were of the same spatial resolution, the only difference lay in the different algorithms used based on different SDMs. Specifically, GLM and FDA are based on generalized linear methods, and GBM and RF are based on classification methods. In addition, emission scenarios generated less uncertainty in predicting potential distributions than GCMs (Figures 4 and 5). Our findings were inconsistent with previous studies in that the uncertainty caused by emission scenarios was greater than that caused by GCM [37–39]. The reason for the difference may be the different emission scenarios chosen. Therefore, the reliability of the model is still insufficient when only using model evaluation metrics, and the reliability of the model can be evaluated by calculating the agreement/uncertainty of the predicted results of various models under different emission scenarios and GCMs.

3.2. Significant Variables Affecting the Distribution of M. punicea

The potential distribution of plants is impacted by environmental variables, which play a crucial role in plant growth [41], especially precipitation and temperature [42,43]. In this study, the results of the four SDMs consistently showed that the potential distribution of *M. punicea* was significantly impacted by two temperature-related bioclimatic variables (i.e., temperature seasonality (BIO4) and mean temperature of coldest quarter (BIO11)) and two precipitation-related bioclimatic variables (i.e., precipitation seasonality (BIO15) and precipitation of the warmest quarter (BIO18)). *M. punicea* is a type of perennial plant with a seasonal growth and dormancy cycle [23,44,45], and thus strongly correlated with temperature seasonality and precipitation seasonality. The warmest quarter on the Qinghai-Tibet Plateau is usually from July to September, which is also the flowering and fruiting period of *M. punicea* [23,24,26]. Plants in the flowering and fruiting period need proper precipitation; too little precipitation affects the blossom and fruit, and too much water can lead to flower and fruit drops. This is consistent with our results in that the potential distribution probability of *M. punicea* increases with an increase in precipitation of the warmest quarter, and that the optimal water requirement of *M. punicea* in the warmest quarter is when precipitation reaches about 600 mm. In addition, our results show that the increase in the mean temperature of coldest quarter will lead to a decrease in the potential distribution probability of *M. punicea*. This is probably due to the coldest quarter for the dormancy period of *M. punicea*, during which a temperature rise will affect the sprouting and growing during the following year [45,46]. The optimum temperature in the dormancy period of *M. punicea* is below −10 °C, which is consistent with the cold resistance of *M. punicea* [26].

3.3. Impacts of Climate Change on the Potential Distribution of M. Punicea

The prediction results of the four SDMs all showed that the potential distribution of *M. punicea* ranged between 29.02° N–39.06° N and 91.40° E–105.89° E under current climate scenarios. Under future climate change, the potential distribution areas of *M. punicea* will be expanded, showing a trend of extending from southeast to northwest. A few areas in the eastern part of the current potential distribution area will not be suitable for *M. punicea* in the future. Our findings are consistent with Zhao et al. [47] in that climate change will expand the potential distribution of *M. punicea*. However, Shi et al. [19] and He et al. [12] suggested that climate change will reduce the potential distribution of *M. punicea*. The reason for the difference may be the different future global climate model chosen or different environmental variables chosen. Instead of representative concentration pathways (RCPs), we selected a Shared Socio-economic Pathways (SSPs) scenario, which is more suitable for China [48,49]. Furthermore, we only selected climate variables, and did not select soil type variables, which may also have an impact on the distribution of *M. punicea*. Compared with RCPs, the SSPs scenario provides more diverse air pollutant emission scenarios, and more scientifically describes future climate change under the mode of socio–economic development.

Our study supports that plants will migrate to higher latitudes under future climate change [3,7], but the potential distribution area of species will not necessarily shrink under future climate change [7,8]. By comparing different climate scenarios in the future, it is found that the potential distribution area of *M. punicea* will expand to the northwest more widely under the SSP5-8.5 scenario. At the end of the 21st Century, the temperature will increase by 4.7–5.1 °C under the SSP5-8.5 scenario, while the temperature will increase by 3.8–4.2 °C under the SSP2-4.5 scenario [50]. The temperature increases more under the SSP5-8.5 scenario, which forces the migration of the cold resistant plant *M. punicea* to higher latitudes. At the same time, it faces greater competition during migration to higher latitudes.

Although our predictions indicate that the potential distribution of *M. punicea* may expand under climate change, the actual movement of species in a changing climate may be characterized by many challenges, such as competition, predation, physical barriers and lack of dispersal media [51,52]. *M. punicea* are entomophilous plants, which mainly rely on flies to transmit pollen [53]. Therefore, a loss of the dispersal potential of flying insects [54,55] may lead to disruption of the transmission of *M. punicea* during migration to higher elevations areas under climate change. In addition, based on the dispersal distance of plants not exceeding 100 m per year [56–59], we estimate that the species will disperse by 7 km at most in the next 70 years. The potential distribution in the future, expanding by about 120 km (Figure 8), and the dispersal distance of the species is much smaller than the potential distribution in the future. Therefore, most of the potential distribution in the future is unreachable, and only a small part (at most 6%) can be occupied.

3.4. Protection Strategies for M. punicea

M. punicea, as an endangered plant, is not only of medicinal value, but also of ornamental value [23,24,60]. The potential distribution of *M. punicea* provides a prerequisite for the development of conservation strategies. According to the predictions of various SDMs, the potential distribution area of *M. punicea* will expand to the northwest, while a few areas in the eastern part of the current potential distribution area will not be suitable for *M. punicea* in the future. For places where potential distribution is stable in the future, in situ conservation can be adopted, such as the establishment of nature reserves [19]. In addition, field investigation and assessment should be carried out for places where the potential distribution is gained in the future, and ex situ conservation should be considered [61]. In addition, the habitat of *M. punicea* is being destroyed at an accelerated rate due to overexploitation and climate change [62]. Therefore, it is particularly important to reduce human activities that lead to the loss of biodiversity, strengthen the construction of protected areas and take measures to protect and conserve species in natural habitats.

This study only considered the influence of climate conditions on the suitable areas of *M. punicea*, and did not involve the influence of human activities, terrain and soil, etc. Therefore, the future protection status of this plant needs to be studied in many aspects. In addition, only two future climate scenarios were considered in this study, more climate models and emission scenarios were selected for simulation in subsequent studies, and compared with this study, so as to evaluate the response of *M. punicea*, an endangered medicinal plant, to climate change in a more objective and comprehensive way.

4. Materials and Methods

4.1. Overview

This study took *M. punicea* as the research object and three primary steps were conducted to implement this study. Firstly, a pairwise Pearson's correlation test was used to select bioclimatic variables. Secondly, the generalized linear model (GLM), the generalized boosted model (GBM), random forests (RF) and flexible discriminant analysis (FDA) were used to predict the current and future potential distribution of *M. punicea*. At the same time, we considered two global circulation models (GCMs) and two emission scenarios of sharing socio-economic pathways (SSPs) for future climate conditions. Thirdly, the differences in the prediction of four SDMs and the potential distribution of *M. punicea* in response to climate change were analyzed, and different conservation strategies were discussed.

4.2. Species Occurrence Data

A total of 230 records of *M. punicea* (i.e., longitude and latitude of the samples) were obtained from the Chinese Virtual Herbarium (http://www.cvh.ac.cn/ (accessed on 12 July 2022)), the Global Biodiversity Information Facility (http://www.gbif.org/ (accessed on 12 July 2022)) and the related literature [12,25]. In order to match the current climate data range (1970–2000 average), only records after 1970 were screened. In order to reduce spatial autocorrelation, only one record was retained within 5 km; that is, a circle was drawn with the sample point as the center of the circle and the radius of 5 km, and all points except the center of the circle were deleted. Finally, 113 records of *M. punicea* were obtained, which were mainly distributed in the eastern part of the Qinghai-Tibet Plateau, including southwest Gansu, southeast Qinghai, northwest Sichuan and northeast Tibet (Figure 9). According to the distribution range of the records of *M. punicea*, the study area was defined as 73° E–110° E and 25° N–45° N.

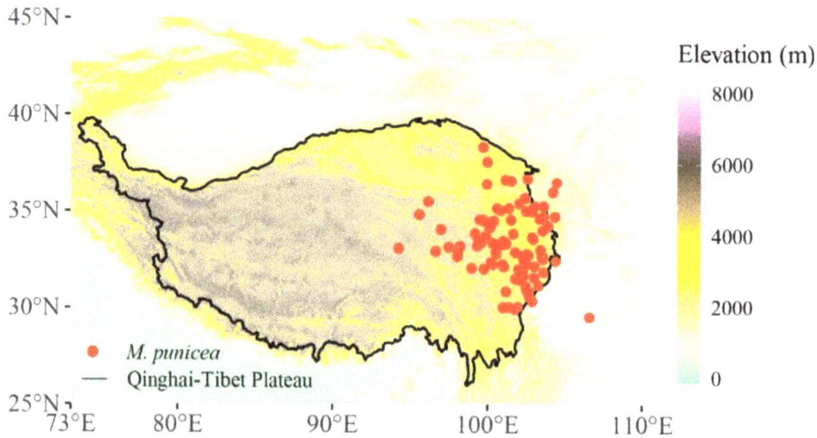

Figure 9. Distribution of occurrence record points of *M. punicea*. The red dots in the figure represent the occurrence record points of *M. punicea*. The polygon data of the Qinghai-Tibet Plateau were obtained through the Global Change Research Data Publishing and Repository (http://www.geodoi.ac.cn/ (accessed on 20 July 2022)).

4.3. Environmental Variables

Bioclimatic variables of current climate (representative of 1970–2000) and future climate (2050: average of 2041–2060) were taken from the WorldClim database version 2.1 (http://www.worldclim.org/ (accessed on 1 June 2022)) with a resolution of 2.5 arc-min [63]. For the GCMs of future climate, we considered BCC-CSM2-MR (BCC) and HadGEM3-GC31-LL (Had), which have been commonly used in studies predicting the potential future distribution of species in the Qinghai-Tibet Plateau [12,19,48,49,64]. We also adopted two emission scenarios, including SSP2-4.5 and SSP5-8.5, which represent diverse air pollutant emission scenarios and more scientific descriptions of future climate change under the mode of socio–economic development [48,65–67]. SSP2-4.5 represents the scenario of moderate social vulnerability and moderate radiation emission (4.5 W/m^2), while SSP5-8.5 simulates the development path of traditional fossil fuels and belongs to the scenario of extremely high radiation emission (8.5 W/m^2) [50]. In order to reduce the correlation between bioclimatic variables; the 113 records of *M. punicea* were first used to extract 19 current climate variables corresponding to geospatial data. Then, we used a pairwise Pearson's correlation test (r) of bioclimatic variables, and eight bioclimatic variables were selected with $|r| < 0.7$ (Table 2) [68].

Table 2. Bioclimatic variables used for species distribution models to predict the potential future distribution of *M. punicea*. These variables were screened by Pearson's correlation test.

Bioclimatic Variables	Meaning of Variables
BIO2	Mean Diurnal Range (mean of monthly (max temp-min temp))/°C
BIO3	Isothermality ((BIO2/BIO7) × 100)
BIO4	Temperature Seasonality (standard deviation ×100)
BIO5	Max Temperature of Warmest Month/°C
BIO11	Mean Temperature of Coldest Quarter/°C
BIO15	Precipitation Seasonality (coefficient of variation)
BIO17	Precipitation of Driest Quarter/mm
BIO18	Precipitation of Warmest Quarter/mm

4.4. Species Distribution Model

4.4.1. Data Preparation

The presence data used to construct SDMs were set as 113 records of *M. punicea*, and the pseudo-absence data were randomly taken from the study area according to three times the amount of presence data (i.e., 113 presence data, 339 pseudo-absence data). Among them, 80% of presence data (pseudo-absence data) were used as the training set, and the remaining 20% were used as the test set. Three climate scenarios (i.e., current, SSP2-4.5 and SSP5-8.5) were cropped according to the scope of the study area, and then stored in stacks, respectively.

4.4.2. Parameter Setting of the Model

The four SDMs were operated in the R environment (version 4.1.1) using the basic packages "mda" (version 0.5–3), "randomForest" (version 4.6-14) and "gbm" (version 2.1.8), as well as the auxiliary package 'MASS' (version 7.3–57), "biomod2" (version 3.5.1). In the GLM model, three forms of variables (i.e., linear term, quadratic term, interaction term) were considered, and the optimal model was determined by stepwise regression according to the AIC value. The stepwise regression was set to bidirectional, while other parameters were kept at default values. In the GBM model, in order to determine the optimal regression tree, it is necessary to consider the number of iterations and learning rate (a smaller rate is better, but the number of iterations should be increased), the complexity of the decision tree (i.e., the depth of the tree), the ratio of resampling and the number of cross validations (used to extract the most appropriate number of regression trees). The number of iterations, the depth of the tree, the learning rate, the resampling ratio and the number of cross validations were respectively set to 10,000, three, 0.01, 0.5 and 10. In the RF model, the number of iterations, which is considered to obtain the optimal model, is set to 1000, and the other parameters were selected as default settings. In the FDA model, the model was adjusted by Multivariate Regression Splines, and the remaining parameters were selected for default settings.

4.5. Data Analysis

4.5.1. Model Evaluation Metrics

In this study, the area under receiver operating characteristic curve (AUC) and Kappa were used to evaluate the prediction accuracy of the four SDMs. AUC is the value of the area under the receiver operating characteristic (ROC) curve, where the ROC curve plots sensitivity (Se) versus 1−specificity (Sp) across all possible thresholds between 0 and 1. Se represents the proportion of presences correctly predicted, and Sp represents the proportion of absences correctly predicted [69,70]. As AUC is not affected by the diagnostic threshold, it is recognized as the best evaluation index at present [70–72]. The evaluation criteria of AUC are: excellent (0.90–1.00); good (0.80–0.90); general (0.70–0.80); worse (0.60–0.70); fail (0.50–0.60) [30,31,70,71]. Kappa is calculated using the following formula: Kappa = $(P_0 - P_e)/(1 - P_e)$, $P_0 = Pr \times Se + (1 - Pr) \times Sp$ and

$P_e = -2 \times (Se + Sp - 1) \times Pr \times (1 - Pr) + P_0$, where Pr is the proportion of presences in the dataset [70]. Kappa is between -1 and 1, usually greater than 0. The larger the value, the higher the accuracy of the model [70,73,74]. Each SDM was randomly simulated 20 times and the AUC (Kappa) was obtained for each simulation. Then, the median of the 20 AUC (Kappa) was used to compare the prediction performance of the model.

In order to further compare the prediction performance of the model, the agreement/uncertainty of the potential distribution predicted by each model was calculated as follows. First, the same part of the potential distribution predicted through each SDM was considered as an agreement. Second, the different part of the potential distribution predicted through each SDM was regarded as an uncertainty.

4.5.2. Comparison of Current and Future Potential Distribution Areas

In species conservation practice, information presented in the form of species presence/absence may be more practical than information presented in terms of probability or suitability. Therefore, a threshold is needed to convert probability or suitability data into presence/absence data [75,76]. Furthermore, the choice of threshold through Kappa maximization is popular in predicting species presence [77,78]. In order to compare the current and future (SSP2-4.5 and SSP5-8.5) changes in potential distribution, the predicted value of the potential distribution was converted as follows. For each model, the operation was repeated 20 times, and the value that made the Kappa value maximum was respectively taken and the average was calculated as the threshold of the model. Then, according to the threshold value, the potential distribution probability of the model was converted into 0–1 data, in which those less than the threshold value were denoted 0 (i.e., absence) and those larger than the threshold value were denoted 1 (i.e., presence). Finally, the calculation was made according to the formula: future potential distribution $\times$ 2 + 1 + current distribution. In the calculation results, the number 4 indicates current and future species presence, denoted as "stable". The number 2 indicates only current species presence, denoted as "loss". The number 3 indicates the species is not currently present, but will be present in the future, denoted as "gain".

Author Contributions: Conceptualization, W.-T.W.; methodology, H.-T.Z.; software, H.-T.Z.; formal analysis, H.-T.Z. and W.-T.W.; data curation, H.-T.Z.; writing—original draft preparation, H.-T.Z.; writing—review and editing, H.-T.Z. and W.-T.W.; funding acquisition, W.-T.W. All authors have read and agreed to the published version of the manuscript.

Funding: This work was supported by the National Natural Science Foundation of China (no. 32260293), the Natural Science Foundation of Gansu Province (no. 21JR11RA023), the Scientific Research Project for Colleges and Universities of Gansu Province (no. 2022QB-017), the Research Fund for Humanities and Social Sciences of the Ministry of Education (no. 20XJAZH006), the Foundation Research Funds for the Central Universities (no. 31920220061; 31920220041).

Data Availability Statement: Not applicable.

Acknowledgments: The authors are grateful to the anonymous reviewers for their constructive suggestions and comments that have helped to improve the quality of this paper, as well as the Innovation Team of Intelligent Computing and Dynamical System Analysis and Application for their support.

Conflicts of Interest: The authors declare no conflict of interest.

References

1. Araújo, M.B.; Rahbek, C. How does climate change affect biodiversity? *Science* **2006**, *313*, 1396–1397. [CrossRef]
2. Beaumont, N.; Austen, M.; Atkins, J.; Burdon, D.; Degraer, S.; Dentinho, T.; Derous, S.; Holm, P.; Horton, T.; Van Ierland, E. Identification, definition and quantification of goods and services provided by marine biodiversity: Implications for the ecosystem approach. *Mar. Pollut. Bull.* **2007**, *54*, 253–265. [CrossRef]
3. Pauli, H.; Gottfried, M.; Dullinger, S.; Abdaladze, O.; Akhalkatsi, M.; Alonso, J.L.B.; Coldea, G.; Dick, J.; Erschbamer, B.; Calzado, R.F. Recent plant diversity changes on Europe's mountain summits. *Science* **2012**, *336*, 353–355. [CrossRef] [PubMed]

4. Flagmeier, M.; Long, D.G.; Genney, D.R.; Hollingsworth, P.M.; Ross, L.C.; Woodin, S.J. Fifty years of vegetation change in oceanic-montane liverwort-rich heath in Scotland. *Plant Ecol. Divers.* **2014**, *7*, 457–470. [CrossRef]
5. Sproull, G.J.; Quigley, M.F.; Sher, A.; González, E. Long-term changes in composition, diversity and distribution patterns in four herbaceous plant communities along an elevational gradient. *J. Veg. Sci.* **2015**, *26*, 552–563. [CrossRef]
6. Christmas, M.J.; Breed, M.F.; Lowe, A.J. Constraints to and conservation implications for climate change adaptation in plants. *Conserv. Genet.* **2016**, *17*, 305–320. [CrossRef]
7. Zhang, K.L.; Yao, L.J.; Meng, J.S.; Tao, J. Maxent modeling for predicting the potential geographical distribution of two peony species under climate change. *Sci. Total Environ.* **2018**, *634*, 1326–1334. [CrossRef]
8. Hu, H.W.; Wei, Y.Q.; Wang, W.Y.; Wang, C.Y. The Influence of climate change on three dominant alpine species under different scenarios on the Qinghai–Tibetan Plateau. *Diversity* **2021**, *13*, 682. [CrossRef]
9. Wei, Y.Q.; Zhang, L.; Wang, J.N.; Wang, W.W.; Niyati, N.; Guo, Y.L.; Wang, X.F. Chinese caterpillar fungus (*Ophiocordyceps sinensis*) in China: Current distribution, trading, and futures under climate change and overexploitation. *Sci. Total Environ.* **2021**, *755*, 142548. [CrossRef]
10. Chen, I.-C.; Hill, J.K.; Ohlemüller, R.; Roy, D.B.; Thomas, C.D. Rapid range shifts of species associated with high levels of climate warming. *Science* **2011**, *333*, 1024–1026. [CrossRef]
11. Xiaodan, W.; Genwei, C.; Xianghao, Z. Assessing potential impacts of climatic change on subalpine forests on the eastern Tibetan Plateau. *Clim. Chang.* **2011**, *108*, 225–241. [CrossRef]
12. He, X.; Burgess, K.S.; Yang, X.F.; Ahrends, A.; Gao, L.M.; Li, D.Z. Upward elevation and northwest range shifts for alpine *Meconopsis* species in the Himalaya–Hengduan Mountains region. *Ecol. Evol.* **2019**, *9*, 4055–4064. [CrossRef] [PubMed]
13. Wang, B.; Chen, T.; Xu, G.; Liu, X.; Wang, W.; Wu, G.; Zhang, Y. Alpine timberline population dynamics under climate change: A comparison between Qilian juniper and Qinghai spruce tree species in the middle Qilian Mountains of northeast Tibetan Plateau. *Boreas* **2016**, *45*, 411–422. [CrossRef]
14. Bugmann, H. A review of forest gap models. *Clim. Chang.* **2001**, *51*, 259–305. [CrossRef]
15. Hörsch, B. Modelling the spatial distribution of montane and subalpine forests in the central Alps using digital elevation models. *Ecol. Model.* **2003**, *168*, 267–282. [CrossRef]
16. Chauchard, S.; Carcaillet, C.; Guibal, F. Patterns of land-use abandonment control tree-recruitment and forest dynamics in Mediterranean mountains. *Ecosystems* **2007**, *10*, 936–948. [CrossRef]
17. Guisan, A.; Theurillat, J.-P. Assessing alpine plant vulnerability to climate change: A modeling perspective. *Integr. Assess.* **2000**, *1*, 307–320. [CrossRef]
18. Halloy, S.R.; Mark, A.F. Climate-change effects on alpine plant biodiversity: A New Zealand perspective on quantifying the threat. *Arct. Antarct. Alp. Res.* **2003**, *35*, 248–254. [CrossRef]
19. Shi, N.; Naudiyal, N.; Wang, J.N.; Gaire, N.P.; Wu, Y.; Wei, Y.Q.; He, J.L.; Wang, C.Y. Assessing the impact of climate change on potential distribution of *Meconopsis punicea* and its influence on ecosystem services supply in the southeastern margin of Qinghai-Tibet Plateau. *Front. Plant Sci.* **2022**, *12*, 3338. [CrossRef]
20. Zhang, Y.; Shao, X.; Wilmking, M. Dynamic relationships between *Picea crassifolia* growth and climate at upper treeline in the Qilian Mts., Northeast Tibetan Plateau, China. *Dendrochronologia* **2011**, *29*, 185–199. [CrossRef]
21. Singh, S.; Sharma, S.; Dhyani, P. Himalayan arc and treeline: Distribution, climate change responses and ecosystem properties. *Biodivers. Conserv.* **2019**, *28*, 1997–2016. [CrossRef]
22. Dorji, T.; Hopping, K.A.; Meng, F.; Wang, S.; Jiang, L.; Klein, J.A. Impacts of climate change on flowering phenology and production in alpine plants: The importance of end of flowering. *Agric. Ecosyst. Environ.* **2020**, *291*, 106795. [CrossRef]
23. Yang, F.S.; Qin, A.L.; Li, Y.F.; Wang, X.Q. Great genetic differentiation among populations of *Meconopsis integrifolia* and its implication for plant speciation in the Qinghai-Tibetan Plateau. *PLoS ONE* **2012**, *7*, e37196. [CrossRef]
24. Wu, H.F.; Wu, Z.J.; Zhu, H.J.; Peng, S.L.; Zhang, X.F. Chemical constituents of *Meconopsis punicea. Nat. Prod. Res. Dev.* **2011**, *23*, 202–207.
25. Wang, W.T.; Guo, W.Y.; Jarvie, S.; Svenning, J.C. The fate of *Meconopsis* species in the Tibeto-Himalayan region under future climate change. *Ecol. Evol.* **2021**, *11*, 887–899. [CrossRef] [PubMed]
26. Shang, X.; Wang, D.; Miao, X.; Wang, Y.; Zhang, J.; Wang, X.; Zhang, Y.; Pan, H. Antinociceptive and anti-tussive activities of the ethanol extract of the flowers of *Meconopsis punicea* Maxim. *BMC Complement. Altern. Med.* **2015**, *15*, 154. [CrossRef]
27. Yasuhara, M.; Hunt, G.; Breitburg, D.; Tsujimoto, A.; Katsuki, K. Human-induced marine ecological degradation: Micropaleontological perspectives. *Ecol. Evol.* **2012**, *2*, 3242–3268. [CrossRef] [PubMed]
28. Kearney, M.R.; Wintle, B.A.; Porter, W.P. Correlative and mechanistic models of species distribution provide congruent forecasts under climate change. *Conserv. Lett.* **2010**, *3*, 203–213. [CrossRef]
29. Baker, D.J.; Hartley, A.J.; Butchart, S.H.; Willis, S.G. Choice of baseline climate data impacts projected species' responses to climate change. *Glob. Chang. Biol.* **2016**, *22*, 2392–2404. [CrossRef]
30. Pouteau, R.; Meyer, J.-Y.; Taputuarai, R.; Stoll, B. Support vector machines to map rare and endangered native plants in Pacific islands forests. *Ecol. Inform.* **2012**, *9*, 37–46. [CrossRef]
31. Ahn, Y.; Lee, D.-K.; Kim, H.G.; Park, C.; Kim, J.; Kim, J.-U. Estimating *Korean Pine* (*Pinus koraiensis*) habitat distribution considering climate change uncertainty-using species distribution models and RCP Scenarios. *J. Korean Soc. Environ. Restor. Technol.* **2015**, *18*, 51–64. [CrossRef]

32. Jeong, S.G.; Lee, D.K.; Ryu, J.E. Riparian connectivity assessment using species distribution model of fish assembly. *J. Korean Soc. Geospat. Inf. Sci.* **2015**, *23*, 17–26.
33. Wellmann, T.; Lausch, A.; Scheuer, S.; Haase, D. Earth observation based indication for avian species distribution models using the spectral trait concept and machine learning in an urban setting. *Ecol. Indic.* **2020**, *111*, 106029. [CrossRef]
34. Yan, H.Y.; Feng, L.; Zhao, Y.F.; Feng, L.; Zhu, C.P.; Qu, Y.F.; Wang, H.Q. Predicting the potential distribution of an invasive species, *Erigeron canadensis L.* in China with a maximum entropy model. *Glob. Ecol. Conserv.* **2020**, *21*, e00822. [CrossRef]
35. Koo, K.A.; Park, S.U.; Kong, W.-S.; Hong, S.; Jang, I.; Seo, C. Potential climate change effects on tree distributions in the Korean Peninsula: Understanding model & climate uncertainties. *Ecol. Model.* **2017**, *353*, 17–27.
36. Li, X.; Wang, Y. Applying various algorithms for species distribution modelling. *Integr. Zool.* **2013**, *8*, 124–135. [CrossRef] [PubMed]
37. Brun, P.; Thuiller, W.; Chauvier, Y.; Pellissier, L.; Wüest, R.O.; Wang, Z.; Zimmermann, N.E. Model complexity affects species distribution projections under climate change. *J. Biogeogr.* **2020**, *47*, 130–142. [CrossRef]
38. Paź-Dyderska, S.; Jagodziński, A.M.; Dyderski, M.K. Possible changes in spatial distribution of walnut (*Juglans regia* L.) in Europe under warming climate. *Reg. Environ. Chang.* **2021**, *21*, 18. [CrossRef]
39. Olszewski, P.; Dyderski, M.K.; Dylewski, Ł.; Bogusch, P.; Schmid-Egger, C.; Ljubomirov, T.; Zimmermann, D.; Le Divelec, R.; Wiśniowski, B.; Twerd, L. European beewolf (*Philanthus triangulum*) will expand its geographic range as a result of climate warming. *Reg. Environ. Chang.* **2022**, *22*, 129. [CrossRef]
40. Riahi, K.; Van Vuuren, D.P.; Kriegler, E.; Edmonds, J.; O'neill, B.C.; Fujimori, S.; Bauer, N.; Calvin, K.; Dellink, R.; Fricko, O. The Shared Socioeconomic Pathways and their energy, land use, and greenhouse gas emissions implications: An overview. *Glob. Environ. Chang.* **2017**, *42*, 153–168. [CrossRef]
41. Beaumont, L.J.; Hughes, L.; Poulsen, M. Predicting species distributions: Use of climatic parameters in BIOCLIM and its impact on predictions of species' current and future distributions. *Ecol. Model.* **2005**, *186*, 251–270. [CrossRef]
42. Molina-Montenegro, M.A.; Atala, C.; Gianoli, E. Phenotypic plasticity and performance of *Taraxacum officinale* (dandelion) in habitats of contrasting environmental heterogeneity. *Biol. Invasions* **2010**, *12*, 2277–2284. [CrossRef]
43. Ganjurjav, H.; Gornish, E.S.; Hu, G.Z.; Schwartz, M.W.; Wan, Y.F.; Li, Y.; Gao, Q.Z. Warming and precipitation addition interact to affect plant spring phenology in alpine meadows on the central Qinghai-Tibetan Plateau. *Agric. For. Meteorol.* **2020**, *287*, 107943. [CrossRef]
44. Iwasa, Y.; Cohen, D. Optimal growth schedule of a perennial plant. *Am. Nat.* **1989**, *133*, 480–505. [CrossRef]
45. Rohde, A.; Bhalerao, R.P. Plant dormancy in the perennial context. *Trends Plant Sci.* **2007**, *12*, 217–223. [CrossRef]
46. Fadón, E.; Rodrigo, J. Unveiling winter dormancy through empirical experiments. *Environ. Exp. Bot.* **2018**, *152*, 28–36. [CrossRef]
47. Zhao, W.; Chen, H.; Liu, L.; Qi, M.; Zhang, J.; Du, T.; Jin, L. The impact of climate change on the distribution pattern of the suitable growing region for endangered Tibetan medicine *Meconopsis punicea*. *Chin. Pharm. J.* **2021**, *56*, 1306–1312.
48. Xin, X.G.; Wu, T.W.; Zhang, J.; Zhang, F.; Li, W.P.; Zhang, Y.W.; Lu, Y.X.; Fang, Y.J.; Jie, W.H.; Zhang, L.; et al. Introduction of BCC models and its participation in CMIP6. *Adv. Clim. Chang. Res.* **2019**, *15*, 533.
49. Gao, T.; Xu, Q.; Liu, Y.; Zhao, J.Q.; Shi, J. Predicting the potential geographic distribution of *Sirex nitobei* in China under climate change using maximum entropy model. *Forests* **2021**, *12*, 151. [CrossRef]
50. Arias, P.; Bellouin, N.; Coppola, E.; Jones, R.; Krinner, G.; Marotzke, J.; Naik, V.; Palmer, M.; Plattner, G.-K.; Rogelj, J. Technical Summary. In *Climate Change 2021: The Physical Science Basis. Contribution of Working Group I to the Sixth Assessment Report of the Intergovernmental Panel on Climate Change*; Cambridge University Press: Cambridge, UK; New York, NY, USA, 2021; pp. 33–144.
51. Wiens, J.A.; Stralberg, D.; Jongsomjit, D.; Howell, C.A.; Snyder, M.A. Niches, models, and climate change: Assessing the assumptions and uncertainties. *Proc. Natl. Acad. Sci. USA* **2009**, *106*, 19729–19736. [CrossRef]
52. Ackerly, D.; Loarie, S.; Cornwell, W.; Weiss, S.; Hamilton, H.; Branciforte, R.; Kraft, N. The geography of climate change: Implications for conservation biogeography. *Divers. Distrib.* **2010**, *16*, 476–487. [CrossRef]
53. Wu, Y.; Liu, Y.; Peng, H.; Yang, Y.; Liu, G.; Cao, G.; Zhang, Q. Pollination ecology of alpine herb *Meconopsis integrifolia* at different altitudes. *Chin. J. Plant Ecol.* **2015**, *39*, 1–13.
54. Roff, D.A. The evolution of flightlessness in insects. *Ecol. Monogr.* **1990**, *60*, 389–421. [CrossRef]
55. McCulloch, G.A.; Foster, B.J.; Dutoit, L.; Ingram, T.; Hay, E.; Veale, A.J.; Dearden, P.K.; Waters, J.M. Ecological gradients drive insect wing loss and speciation: The role of the alpine treeline. *Mol. Ecol.* **2019**, *28*, 3141–3150. [CrossRef]
56. Kinlan, B.P.; Gaines, S.D. Propagule dispersal in marine and terrestrial environments: A community perspective. *Ecology* **2003**, *84*, 2007–2020. [CrossRef]
57. Thomson, F.J.; Moles, A.T.; Auld, T.D.; Kingsford, R.T. Seed dispersal distance is more strongly correlated with plant height than with seed mass. *J. Ecol.* **2011**, *99*, 1299–1307. [CrossRef]
58. Morgan, J.W.; Venn, S. Alpine plant species have limited capacity for long-distance seed dispersal. *Plant Ecol.* **2017**, *218*, 813–819. [CrossRef]
59. Di Musciano, M.; Di Cecco, V.; Bartolucci, F.; Conti, F.; Frattaroli, A.R.; Di Martino, L. Dispersal ability of threatened species affects future distributions. *Plant Ecol.* **2020**, *221*, 265–281. [CrossRef]
60. Qu, Y.; Qu, Z. The research advancement on the genus *Meconpsis*. *North. Hortic.* **2012**, *2*, 191–194.
61. Khoury, C.; Laliberté, B.; Guarino, L. Trends in ex situ conservation of plant genetic resources: A review of global crop and regional conservation strategies. *Genet. Resour. Crop Evol.* **2010**, *57*, 625–639. [CrossRef]

62. Niyati, N.; Wang, J.N.; Wu, N.; Narayan Prasad, G.; Shi, P.L.; Wei, Y.Q.; He, J.L.; Shi, N. Potential distribution of *Abies, Picea,* and *Juniperus* species in the sub-alpine forest of Minjiang headwater region under current and future climate scenarios and its implications on ecosystem services supply. *Ecol. Indic.* **2021**, *121*, 107131.
63. Fick, S.E.; Hijmans, R.J. WorldClim 2: New 1-km spatial resolution climate surfaces for global land areas. *Int. J. Climatol.* **2017**, *37*, 4302–4315. [CrossRef]
64. Kumar, P. Assessment of impact of climate change on Rhododendrons in Sikkim Himalayas using Maxent modelling: Limitations and challenges. *Biodivers. Conserv.* **2012**, *21*, 1251–1266. [CrossRef]
65. Moss, R.H.; Edmonds, J.A.; Hibbard, K.A.; Manning, M.R.; Rose, S.K.; Van Vuuren, D.P.; Carter, T.R.; Emori, S.; Kainuma, M.; Kram, T. The next generation of scenarios for climate change research and assessment. *Nature* **2010**, *463*, 747–756. [CrossRef] [PubMed]
66. Van Vuuren, D.P.; Edmonds, J.; Kainuma, M.; Riahi, K.; Thomson, A.; Hibbard, K.; Hurtt, G.C.; Kram, T.; Krey, V.; Lamarque, J.-F. The representative concentration pathways: An overview. *Clim. Chang.* **2011**, *109*, 5–31. [CrossRef]
67. Fan, X.W.; Miao, C.Y.; Duan, Q.Y.; Shen, C.W.; Wu, Y. The performance of CMIP6 versus CMIP5 in simulating temperature extremes over the global land surface. *J. Geophys. Res. Atmos.* **2020**, *125*, e2020JD033031. [CrossRef]
68. Dormann, C.F.; Elith, J.; Bacher, S.; Buchmann, C.; Carl, G.; Carré, G.; Marquéz, J.R.G.; Gruber, B.; Lafourcade, B.; Leitão, P.J. Collinearity: A review of methods to deal with it and a simulation study evaluating their performance. *Ecography* **2013**, *36*, 27–46. [CrossRef]
69. Hilden, J. The area under the ROC curve and its competitors. *Med. Decis. Mak.* **1991**, *11*, 95–101. [CrossRef] [PubMed]
70. Allouche, O.; Tsoar, A.; Kadmon, R. Assessing the accuracy of species distribution models: Prevalence, kappa and the true skill statistic (TSS). *J. Appl. Ecol.* **2006**, *43*, 1223–1232. [CrossRef]
71. Swets, J.A. Measuring the accuracy of diagnostic systems. *Science* **1988**, *240*, 1285–1293. [CrossRef]
72. Manel, S.; Williams, H.C.; Ormerod, S.J. Evaluating presence-absence models in ecology: The need to account for prevalence. *J. Appl. Ecol.* **2001**, *38*, 921–931. [CrossRef]
73. Cohen, J. A coefficient of agreement for nominal scales. *Educ. Psychol. Meas.* **1960**, *20*, 37–46. [CrossRef]
74. Pearson, R.G.; Dawson, T.P.; Liu, C. Modelling species distributions in Britain: A hierarchical integration of climate and land-cover data. *Ecography* **2004**, *27*, 285–298. [CrossRef]
75. Liu, C.; Berry, P.M.; Dawson, T.P.; Pearson, R.G. Selecting thresholds of occurrence in the prediction of species distributions. *Ecography* **2005**, *28*, 385–393. [CrossRef]
76. Freeman, E.A.; Moisen, G.G. A comparison of the performance of threshold criteria for binary classification in terms of predicted prevalence and kappa. *Ecol. Model.* **2008**, *217*, 48–58. [CrossRef]
77. Guisan, A.; Theurillat, J.P.; Kienast, F. Predicting the potential distribution of plant species in an alpine environment. *J. Veg. Sci.* **1998**, *9*, 65–74. [CrossRef]
78. Pearson, R.G.; Dawson, T.P.; Berry, P.M.; Harrison, P. SPECIES: A spatial evaluation of climate impact on the envelope of species. *Ecol. Model.* **2002**, *154*, 289–300. [CrossRef]

plants

Article

Species-Specific Responses to Human Trampling Indicate Alpine Plant Size Is More Sensitive than Reproduction to Disturbance

Nathalie Isabelle Chardon [1,2,*], Philippa Stone [1,2], Carly Hilbert [3], Teagan Maclachlan [3], Brianna Ragsdale [3], Allen Zhao [3], Katie Goodwin [1,2], Courtney G. Collins [1], Nina Hewitt [4] and Cassandra Elphinstone [1,2]

1 Biodiversity Research Centre, University of British Columbia, Unceded xʷməθkʷəy'əm (Musqueam) Territory, 2212 Main Mall, Vancouver, BC V6T 1Z4, Canada; philippa.stone@botany.ubc.ca (P.S.); katiegoodwin321@gmail.com (K.G.); courtney.collins@botany.ubc.ca (C.G.C.); cassandra.elphinstone@shaw.ca (C.E.)
2 Department of Botany, University of British Columbia, Unceded xʷməθkʷəy'əm (Musqueam) Territory, 3156-6270 University Blvd., Vancouver, BC V6T 1Z4, Canada
3 University of British Columbia, Unceded xʷməθkʷəy'əm (Musqueam) Territory, Vancouver, BC V6T 1Z4, Canada; carlyhilbert02@gmail.com (C.H.); teagmaclachlan@gmail.com (T.M.); bri.ragsdale15@gmail.com (B.R.); allen10to11@gmail.com (A.Z.)
4 Department of Geography, University of British Columbia, Unceded xʷməθkʷəy'əm (Musqueam) Territory, Vancouver, BC V6T 1Z4, Canada; nina.hewitt@ubc.ca
* Correspondence: nathalie.chardon@gmail.com

Abstract: Human disturbance, such as trampling, is an integral component of global change, yet we lack a comprehensive understanding of its effects on alpine ecosystems. Many alpine systems are seeing a rapid increase in recreation and in understudied regions, such as the Coast Mountains of British Columbia, yet disturbance impacts on alpine plants remain unclear. We surveyed disturbed (trail-side) and undisturbed (off-trail) transects along elevational gradients of popular hiking trails in the T'a̲k̲'t'a̲k̲'múy'in tl'a In'inyá̲x̲a7n region (Garibaldi Provincial Park), Canada, focusing on dominant shrubs (*Phyllodoce empetriformis, Cassiope mertensiana, Vaccinium ovalifolium*) and graminoids (*Carex* spp). We used a hierarchical Bayesian framework to test for disturbance by elevation effects on total plant percent cover, maximum plant height and diameter (growth proxies), and buds, flowers, and fruits (reproduction proxies). We found that trampling reduces plant cover and impacts all species, but that effects vary by species and trait, and disturbance effects only vary with elevation for one species' trait. Growth traits are more sensitive to trampling than reproductive traits, which may lead to differential impacts on population persistence and species-level fitness outcomes. Our study highlights that disturbance responses are species-specific, and this knowledge can help land managers minimize disturbance impacts on sensitive vegetation types.

Keywords: alpine plants; human trampling disturbance; global change; recreation; plant traits; trail planning

Citation: Chardon, N.I.; Stone, P.; Hilbert, C.; Maclachlan, T.; Ragsdale, B.; Zhao, A.; Goodwin, K.; Collins, C.G.; Hewitt, N.; Elphinstone, C. Species-Specific Responses to Human Trampling Indicate Alpine Plant Size Is More Sensitive than Reproduction to Disturbance. *Plants* **2023**, *12*, 3040. https://doi.org/10.3390/plants12173040

Academic Editor: Peili Shi

Received: 30 June 2023
Revised: 8 August 2023
Accepted: 16 August 2023
Published: 24 August 2023

1. Introduction

Subalpine and alpine plants are sensitive to disturbances from trampling due to their perennial nature, short growing seasons, and shallow soils [1,2]. Species' responses to trampling vary, with different effects seen in growth and reproduction. These species-specific responses may be related to differences in life history. Grime [3] defined combinations of different life history traits that make certain groups of species more successful along gradients of disturbance, stress, and competition. Ruderals, such as grasses and sedges, are most common in areas with high disturbance, whereas slower-growing, competitive species do not tolerate disturbance well. For example, studies have found that *Phyllodoce* spp. (slow-growing subalpine heathers) are vulnerable to trampling disturbances with reduced heights, whereas *Carex* spp. (sedges) are more resilient [4,5]. The effects of trampling on soils can even have long-term effects on plant community succession, with greater diversity at trampled sites in the Cascade Mountains in Washington [6].

In addition to reducing plant size, trampling can also reduce reproductive output, interrupt succession, and lower species richness [7–15]. Disturbance can also decrease total plant cover, but the severity of effects differs between ecosystems [14,16,17]. Reduced size, reproduction, and survival due to trampling can scale up to negatively affect overall population performance and potentially reduce long-term population viability, especially for hiking trails within small, endangered community types [18]. Population performance may be less sensitive to trampling when this disturbance affects reproduction more than survival [19], and thus the plant tissue that is most affected by trampling will likely influence how trampling impacts overall population performance. However, the extent that trampling might differently affect vegetative growth versus reproduction, and whether this effect is uniform across species, is understudied.

One of the main mechanical effects of trampling on plant habitat is that trampling disturbance induces soil compaction. Increased soil compaction has been found to reduce pore connectivity, which in turn decreases its hydro-conductivity and root permeability [20]. Soil compaction also reduces the relative abundance of soil microbes and mycorrhizal fungi [21]. Such conditions can make trampled habitats inhospitable for many plant species. Elevation may also play an important role in how plants and soils respond to disturbance [22]. Climate, including precipitation and temperature, rates of soil development and hydrology, relative plant canopy cover, and numerous other factors vary consistently across elevation gradients, and thus the response of vegetation to the same intensity of disturbance may vary greatly at different elevation zones [22–24]. Furthermore, species diversity in a plant community can greatly influence sensitivity to human trampling, with mixed communities of subalpine plant species having three times higher survival than the same species in pure stands [25].

The synergistic effects of disturbance and climate further complicate our understanding of trampling impacts on plant communities. Climatic conditions can have varying effects on alpine communities depending on region, and regional climate changes can negatively impact the rate of plant regeneration post-disturbance [2,26]. In other cases, increased precipitation and warmer summers may even increase the richness of alpine plant species in general, although this response is individualistic and regional [27]. These changes could mediate or exacerbate the effects of trampling in unique ways with contrasting responses to trampling across different climatic zones [28,29]. Taken together, the complex interplays between anthropogenic disturbances and climate change emphasize the need to study these effects in alpine regions experiencing rapid changes in climate and human recreation.

One region experiencing rapid changes in climate and human recreation is British Columbia's (BC) Coast Mountains. These ocean-proximate mountains are unique on a global scale as they are among the most southerly glaciated mountains of the northern hemisphere and receive high levels of precipitation year-round [30]. Glacial landscapes and alpine ecosystems, such as those in this region, experience greater warming than other regions in the globe and have seen increasingly rapid glacial retreat in recent years [31]. They are also near Vancouver, Canada, a major population center of 2.64 million people [32]. This large and growing population has led to a rapidly increasing number of outdoor enthusiasts and new trails are being built into the most accessible mountain terrain. For example, visits to Garibaldi Provincial Park within this region increased by 62% (3-year average) between the periods 2006–2009 and 2015–2018 [33–35]. Understanding how these BC ecosystems will respond to warming and increasing trampling due to recreation will help park management determine the least sensitive areas for infrastructure development. However, our understanding of the effects of human trampling disturbance on alpine plant communities in the Coast Mountains of BC is surprisingly limited.

To our knowledge, only two studies have examined the effects of trampling in BC. In a managed subalpine/montane forest stand, one study found negative effects of trampling by livestock (cattle) on lodgepole pine tree plantations primarily in the first 2–3 years of planting in the Kamloops and Merritt Forest Districts [36]. Another study found that disturbance reduces plant percent cover in montane to alpine systems, but does not affect species richness in Mount Robson Provincial Park [37]. Studies in nearby alpine ecosystems in the Colorado Rockies have documented immediate and long-term impacts of trampling disturbance, including decreasing plant cover, lower reproductive output, soil erosion, and changed plant community structure [2,29,38]. However, we expect these findings to differ from trampling responses in the BC Coast Mountains, where a much deeper winter snowpack may buffer trampling impacts through winter but relatively high precipitation and snowmelt in summer may increase mud, resulting in trail widening by hikers.

Given the lack of information on disturbance impacts in the alpine systems of the BC Coastal Mountains, here we study how human trampling disturbance affects characteristic plant species in this region. Our goal is to increase our understanding of how increased human recreation might affect alpine plant communities found across the globe and help to inform future trail designs to minimize disruption locally. We surveyed dominant and common plant communities on and off popular hiking trails in the T'ak̲'t'ak̲'múy'in tl'a In'inyáx̲a7n ("Landing Place of the Thunderbird") region (Garibaldi Provincial Park, British Columbia, Canada) in the Squamish and Lil'wat territories. We quantified the impacts of trampling at community (plant percent cover) and species (growth and reproductive traits) levels over elevational gradients to ask:

1. How does human trampling disturbance affect alpine plant communities and proxy traits for plant growth and reproduction, and does this effect vary with elevation?
2. Which plant species and functional types (evergreen shrubs, deciduous shrubs, sedges) are most sensitive to human trampling disturbance?

We computed plant percent cover from standardized photographs and measured traits on the dominant species in the heath, heather, blueberry, and sedge meadows found frequently throughout this region, encompassing evergreen and deciduous perennial shrubs and graminoids. Our focal species represent common subalpine and alpine plant communities found across arctic and alpine regions worldwide, making our results relevant to other regions around the world. Because of their frequency and broad distribution in the study area, we selected *Phyllodoce empetriformis* (Sm.) D. Don (Ericaceae) and *Cassiope mertensiana* (Bong.) G. Don (Ericaceae), *Vaccinium ovalifolium* Sm. (Ericaceae), and *Carex* spp. (Cyperaceae) (Figure 1). We included a berry species due to their importance to the local Sk̲wx̲x̲wú7mesh (Squamish) and Lil'wat peoples [30,39]. These species provide other important ecosystem services including forage and habitat for wildlife [40], and soil stabilization and high soil carbon storage through fungal mycelium [41]. In addition, the local park authority, BC Parks, aims to better understand how recreation affects common plant communities to help plan for needed infrastructure in areas experiencing an immense increase in recreational traffic, making our study of particular interest from a management perspective.

At a community level, we predict that disturbance will reduce plant percent cover. At a species level, we predict that graminoids (with multiple ramets) will be the least sensitive to disturbance based on their quick recolonization and high survival in disturbed areas [42]. We further predict that the deciduous (*V. ovalifolium*) or evergreen (*P. empetriformis, C. mertensiana*) brittle shrubs will be more sensitive to trampling, with the slow-growing evergreen shrubs most affected by multi-year repeated disturbances [42]. Regarding the effects of trampling on plant size, we hypothesized that trampling would break branches and thus decrease plant size. We expected slow-growing species (*P. empetriformis, C. mertensiana,* and *V. ovalifolium*) to be most sensitive to this. With regard to reproduction, we expected that trampling will either (i) decrease fruit numbers due to smaller plant biomass, but not change the relative number of reproductive structures per biomass of the plant [43], or (ii) increase the density of reproductive structures on trampled plants in response to higher stress [44]. Our results suggest that species have different sensitivities to human trampling, with *V. ovalifolium* being the most sensitive in the T'ak̲'t'ak̲'múy'in tl'a In'inyáx̲a7n region, and elevation might not be as important as widely understood. Future studies will be needed to determine if the responses to trampling that we observed are species-specific in this region or can be generalized for broader groups (e.g., other graminoids, evergreen shrubs, or deciduous shrubs [42]).

(a) *Phyllodoce empetriformis* (b) *Cassiope mertensiana* (c) *Vaccinium ovalifolium* (d) *Carex* spp.

Figure 1. Study species. The four focal species in our studies are common and dominant species of the heath (**a**), heather (**b**), blueberry (**c**), and sedge (**d**) communities in the T'ak̲'t'ak̲'múy'in tl'a In'inyáx̲a7n region (Garibaldi Provincial Park, British Columbia, Canada). They are of cultural importance to the Squamish and Lil'wat Nations and of management concern to the park authority, BC Parks.

2. Results

At the community level, we found that disturbance reduced plant percent cover in quadrats located on trails compared to those off trails (Table 1, Figure 2). At the species level, we found that trampling disturbance reduced size in two out of four species and reproduction in one species, although the intensity of this effect varied by species and some species' traits were not affected by disturbance. Disturbance increased species size in only one case. In particular, disturbance reduced both the maximum height and maximum diameter in the deciduous shrub (*V. ovalifolium*), making this the species with the highest sensitivity to disturbance. However, disturbance only reduced the maximum diameter in one evergreen shrub (*P. empetriformis*) but had no effect on the growth proxies of the other evergreen shrub (*C. mertensiana*). For the sedges (*Carex* spp.), disturbance increased maximum height and had no effect on diameter (Table 1, Figure 2).

Of the three shrub species for which we quantified relative reproductive output (defined as summed buds, flowers, and fruit per plant area, relative to maximum reproductive output for that species), disturbance only affected the reproductive output of the evergreen shrub *C. mertensiana* (Table 1, Figure 2). *Cassiope mertensiana* individuals growing on the trail had a significantly lower reproductive output (per plant area) than those growing off the trail. Individuals of *P. empetriformis* and *V. ovalifolium* showed no difference in reproductive output whether growing on or off the trail. However, total mean and median reproduction were lower on the trails, and since larger plants produce more reproductive structures (Generalized Linear Mixed Model with random effects of species and transect pair: estimate = 0.019, standard error = 0.0009, p-value < 0.001), and this is likely due to smaller plant sizes on trails. The effects of disturbance and elevation only interacted in one case, such that the positive effects of trampling increased with elevation on only *Carex* spp. height (Table 1). However, the effects of trampling did not change with elevation for plant percent cover or other species traits. Elevation alone, within the narrow range we studied, did not affect plant percent cover nor any of the species' traits that we measured (Table 1, Figure S1).

Table 1. Estimated parameters of hierarchical models. Sample size (N), intercept, untransformed parameter estimates, and credible intervals (shown in parentheses) from Bayesian generalized non-linear mixed models testing the effects of disturbance, elevation, and their interaction (Dist. * Elev.) on total plant percent cover, and maximum height, maximum diameter, and relative reproductive output of *Phyllodoce empetriformis, Cassiope mertensiana, Vaccinium ovalifolium*, and *Carex* spp. Estimated parameters whose 95% credible intervals do not overlap zero are in bold, and can be interpreted as having an effect. Note that we report untransformed estimates from negative binomial and beta distribution models, and in the case of height for *Carex* spp., this means that disturbance increased height (see Figure 2d). We fit models with a random intercept of transect pair and fit separate models for plant percent cover and each species by trait combination to account for different responses. We did not measure reproductive output for *Carex* spp. Additional model fitting information is given in Table S1.

Species	Trait	N	Intercept	$ß_{Disturbance}$	$ß_{Elevation}$	$ß_{Dist.*Elev.}$
[All Plants]	Percent Cover	267	6.58 (0.34, 12.61)	**−8.38 (−12.56, −4.4)**	0 (−0.01, 0)	0 (0, 0.01)
P. empetriformis	Height	500	7.04 (2.13, 12.15)	−0.48 (−3.11, 2.12)	0 (0, 0)	0 (0, 0)
	Diameter	500	10.48 (6.66, 14.16)	**−7.15 (−10.53, −3.85)**	0 (0, 0)	0 (0, 0.01)
	Reproduction	500	−4.77 (−7.78, −1.76)	0.39 (−3.47, 4.22)	0 (0, 0)	0 (0, 0)
C. mertensiana	Height	429	7.9 (3.74, 12.03)	0.66 (−1.26, 2.55)	0 (0, 0)	0 (0, 0)
	Diameter	429	9.44 (5.9, 12.98)	−2.51 (−5.13, 0.17)	0 (0, 0)	0 (0, 0)
	Reproduction	429	0.66 (−4.18, 5.38)	**−4.28 (−8.13, −0.52)**	0 (0, 0)	0 (0, 0)
V. ovalifolium	Height	634	2.13 (−4.7, 8.59)	**−4.58 (−7.28, −1.8)**	0 (0, 0.01)	0 (0, 0)
	Diameter	634	6.39 (−0.05, 12.96)	**−9.18 (−12.34, −5.99)**	0 (0, 0)	0.01 (0, 0.01)
	Reproduction	634	−4.5 (−7.93, −0.73)	−3.76 (−7.99, 0.33)	0 (0, 0)	0 (0, 0)
Carex spp.	Height	209	22.62 (5.36, 39.1)	**−19.89 (−30.45, −9.68)**	−0.01 (−0.02, 0)	**0.01 (0.01, 0.02)**
	Diameter	209	−6.29 (−29.7, 16.07)	11.98 (−0.5, 23.86)	0.01 (−0.01, 0.02)	−0.01 (−0.01, 0)

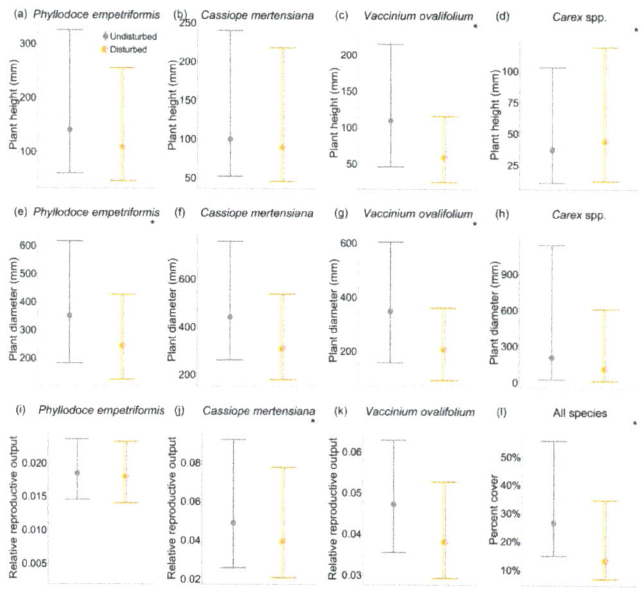

Figure 2. Effects of disturbance on plant traits and plant percent cover. Responses to human trampling disturbance are species-specific, with variable effects on maximum plant height, maximum plant diameter, and relative reproductive output (summed buds, flowers, fruits by total plant area, relative to the maximum per species). Shown are back-transformed parameter estimates (circles) of disturbance with their credible intervals (bars) from Bayesian hierarchical non-linear mixed models. Parameter estimates of disturbance and elevation are shown in Figure S1. Asterisks (*) indicate parameter estimates whose credible intervals do not encompass 0 and can be interpreted as having an effect. Legend for all plots is as in (**a**). Note that y-axes are on different scales.

3. Discussion

This study is one of only a few to document human trampling disturbance in BC's Coastal Mountains. We focused on the dominant plant species heather (*C. mertensiana*), heath (*P. empetriformis*), blueberry (*V. ovalifolium*), and sedge (*Carex* spp.) on- and off-trails in a popular subalpine and alpine recreation region. We quantified community plant percent cover and measured proxies of plant growth and reproduction. We provide detailed measurements for nearly 1800 individuals of four focal plant species that are of particular regional importance, both culturally, in terms of traditional uses (berries), and ecologically, in terms of their importance within the flora (heath, heather, and sedge meadows). Disturbance reduced plant percent cover and decreased proxy traits for growth and reproduction in some species, although the intensity of these effects varied considerably by species. Unexpectedly, disturbance effects increased with elevation for only *Carex* spp. height, whereas disturbance effects on plant percent cover and species traits stayed constant with elevation. Surprisingly, elevation alone did not have an effect at the community or species level.

3.1. Species-Specific Responses

Heather (*C. mertensiana*), heath (*P. empetriformis*), blueberry (*V. ovalifolium*), and sedge (*Carex* spp.) were sensitive to trampling. Human trampling disturbance reduced the size in two out of our four study species (*P. empetriformis, V. ovalifolium*) and reduced reproductive output in just one (*C. mertensiana*). Our empirical findings match those for another study in the nearby Washington Cascades [4], which found reduced height among experimentally trampled, multi-species communities dominated by *P. empetriformis*. Further, their study finding that *Carex* spp. were relatively resilient to trampling disturbance as compared to other taxa corresponds to our finding that *Carex* spp. increased in height with disturbance. Another study in this region found that sedge communities are the most resilient to trampling, and heaths the least resilient [40]. Resilience among sedges and other graminoids may be explained by the presence of adaptations, such as multiple ramets and effective root storage, that buffer against tissue damage [42]. We found *Carex* spp. height, but not diameter, to be sensitive to disturbance impacts, highlighting the need to measure multiple species traits for a better understanding of disturbance effects. While disturbance reduced the diameter of one of our heath species (*P. empetriformis*) and reproductive output in our heather species (*C. mertensiana*), blueberry (*V. ovalifolium*) was most sensitive to disturbance with reduction in both growth proxies (height and diameter).

Other studies have also found differences in disturbance impacts between species and traits [42,45]. Variable responses of flower production to trampling across species might be related to both differences in the sensitivity of flowers to direct trampling and plant response to reduced (i.e., trampled) plant size. Reduced growth from trampling might limit the number of resources available for reproduction, an energetically costly process [46]. Meanwhile, species-specific fruit production responses to disturbance could be related to the effect that disturbance might have on pollinators. For example, insects, which might be negatively affected by human trampling [47], could reduce their pollination services for insect-pollinated species near trails. While it is unclear how negative effects on growth may scale up to influence population persistence in perennial species, decreases in reproduction are more closely tied to plant fitness. However, some work has shown that population performance may be less sensitive to trampling affecting reproduction than survival [19].

3.2. No Effect of Elevation on Species Traits

We found no effects of elevation on species traits in our study, and the effects of disturbance only increased with elevation for one species trait. This is surprising, since plant height and reproduction generally decrease with elevation, as higher areas are regionally cooler and less productive [48,49]. The 400 m elevation gradient of our survey may be too small for macroclimatic differences to override microclimatic effects [50]. Moreover, the particular climatic conditions that characterize this maritime-influenced region, particularly the deep, long-lasting snowpack [51], may protect flora against foot traffic in winter. By comparison, in the nearby drier Colorado Rockies there is evidence that vegetation is vulnerable to trampling on windswept and exposed substrates [2]. Snowpack also suppresses tree recruitment, lowering the elevational transition to alpine tundra by several hundred meters in this region relative to less maritime-influenced regions at similar latitudes [51]. As a result, environmental conditions well above the treeline are likely more similar to those at lower elevations in many other alpine zones. In addition, the strong maritime climate moderate conditions here year-round. Moreover, climate change is altering moisture patterns and producing a thinner snowpack here [51,52] and across western North America [53]. Trampling may thus interact with elevation in the future, amplifying the need for further study.

3.3. Future Research

Although we did not explicitly investigate species richness, other studies have found that decreased species richness occurred alongside growth and reproductive effects of trampling (e.g., [28]), including in heavily experimentally trampled *P. empetriformis*-dominated communities (e.g., [4]). Our results that disturbance reduces plant percent cover may be in line with previous work that found that disturbance reduces species richness and diversity in an alpine system [28]. Community level impacts from trampling are thus likely, but further study will be needed to evaluate these in the present system, and we are conducting ongoing surveys at our transects to quantify the effect of disturbance on species richness and diversity. Higher levels of species richness have also been associated with higher resilience of subalpine plant communities to trampling [25], making this an important consideration of land management to promote highly mixed rather than single species-dominated vegetation alongside trails.

Another major impact of human disturbance, including trampling, is promoting the survival and spread of invasive species in plant communities by the unintentional introduction of seeds carried on clothes and shoes, as well as the intentional introduction of non-natives for ornamental or other purposes [54,55]. Exacerbating this issue is that non-native species can outperform native species in trampled habitats [54,56–58]. However, there is limited data to corroborate such findings for alpine ecosystems. The few studies that have examined disturbance and invasibility find cold-climate ecosystems to be more prone to invasion [59]. While we did not investigate this issue, ongoing studies at our permanent transects examine both changes to species diversity and species invasions.

3.4. Implications for Trail Planning

With interest in outdoor recreation in BC rapidly increasing [34,35], there is a need for more trails to spread out disturbance impacts while still providing public access to outdoor spaces. Our findings suggest that some species are more affected by trampling than others and specific management approaches may need to be tailored toward particular species. These varying species-specific impacts should be considered in the planning of future trail and recreation infrastructure (i.e., avoid trails through regions with species that respond strongly to trampling disturbances and/or exist in very localized areas). We suggest that trails through graminoid-, heath- and heather-dominated meadows may be able to recover more quickly from regular disturbance than trails through deciduous shrubs, such as blueberry, based on our findings and those of previous studies [42]. However, it should be considered that sedge meadows often have significantly moister soil and this can result in trail widening (when hikers avoid getting muddy and wet feet) affecting a much larger area and number of individuals (*pers. obs.*). Based on our results that blueberry (*V. ovalifolium*) was most affected by trampling out of the four species we measured, trails should avoid going through these berry ecosystems since it may decrease berry yields for cultural harvesting and for wildlife. We emphasize that our findings only reflect the four species along the elevational gradient that we studied, and future studies are needed on additional species and in other regions to make our findings generalizable.

Although we studied the common species found in well-established, large subalpine meadows, alpine ecosystems are also filled with less common, sometimes endangered, species found in small microhabitats [18]. Based on our findings that individuals on trails were smaller than those off trails, sometimes resulting in less biomass for reproduction, we suggest that trails should avoid going through microhabitats hosting rare or endangered species for whom a decrease in local reproduction could have population-level consequences. Trail shape and infrastructure (e.g., trail markers, boardwalks, etc.) likely also affect community-level sensitivity to trampling. Small and rare communities are likely much more affected by trails that repeatedly go through them, especially if trampling disturbance impacts decrease in a gradient from the trail edge (e.g., switchbacks). Future studies should investigate how far from trail edges plant size and reproduction are affected. If effects are always directly next to trails, then small, rare plant communities should be avoided for any trail development. Finally, our study highlights the need for public education and outreach efforts aimed to minimize trampling damage. There is some ongoing research into how best to achieve this, for example, by providing the public with short courses and learning materials [60], installing educational signs at park trailheads, and erecting deterrent signage and low fencing along trails [61], strategies that have been effective at reducing or spatially confining trampling impacts in US park settings (e.g., Acadia National Park, Maine; Yosemite National Park, California).

3.5. Conclusions

Given that we monitored these plant communities and their individuals for only a single growing season, it is notable that we find such reduced vegetative growth and reproduction across species, even if this effect is variable. This suggests a need for long-term studies that include more taxa and explore invasive species as well as species diversity. Human recreation in alpine regions will likely continue to grow, increasing the risk of plant trampling by hikers. Furthermore, continued climate change may further exacerbate species-specific plant vulnerability to trampling. For example, trampled trailing edge populations might be more prone to displacement via colonization from lower elevation species expanding their ranges [62]. Given the decreases in size, reproductive capacity, and plant cover observed in our study, it will be important to evaluate possible interactions between trampling and climate change in other systems where displacement by colonization of lower-elevation species is expected.

4. Materials and Methods

4.1. Study Site

We conducted this study in the T'a<u>k</u>'t'a<u>k</u>'múy'in tl'a In'inyá<u>x</u>a7n ("Landing Place of the Thunderbird") region (Garibaldi Provincial Park, British Columbia, Canada) in the Squamish and Lil'wat territories (Figure 3a). The area is classified as Coast Mountain Heather alpine zone within the Biogeoclimatic Ecosystem Classification (BEC) system of BC, where a deep long-lasting winter snowpack limits tree recruitment and the transition to alpine tundra begins at lower elevations than for other alpine zones at comparable latitudes towards the interior [51]. Vegetation is dominated by dwarf shrubs, including *Cassiope* spp. and *Phyllodoce* spp. (mountain heathers and heaths), *Luetkea pectinata* (partridgefoot), and herbaceous species of sedge, grass, and forb. There are patches of stunted trees or "krummholz" at the treeline, comprised mainly of *Abies lasiocarpa* (Subalpine fir), *Picea engelmannii* (Engelmann spruce), and *Tsuga mertensiana* (Mountain hemlock). The mean annual temperatures for the BC alpine have historically been between −4 and 0 °C, with a short growing season and mean annual precipitation of 700–3000 mm, much of which falls as snow [63].

Approximately 100 km from the city of Vancouver, the study area is one of the most popular hiking destinations in southwestern British Columbia [64]. Garibaldi Park attendance has increased substantially, especially since the highway linking Vancouver with the park was upgraded for the 2010 Winter Olympics [33–35]. Popular trailheads (Diamond

Head, Rubble Creek, and Wedgemont Lake) are filled with hikers and backpackers in the summer season (*pers. obs.*). In the last few years, new trail developments have been proposed (e.g., the Darling Lake trail and the hut developments around the Spearhead Range).

Figure 3. Methods schema. (**a**) We quantified the effects of disturbance along elevational gradients on four dominant subalpine and alpine plant species along disturbed (orange) and undisturbed (grey) transects in the T'ak̲'t'ak̲'múy'in tl'a In'inyáx̲a7n ("Landing Place of the Thunderbird") region (Garibaldi Provincial Park, British Columbia, Canada) in the Squamish and Lil'wat territories. We established transects (not all shown) along the popular Taylor Meadows, Black Tusk, and Panorama Ridge trails. (**b**) Along each trail (indicated in brown), we established transects (not all shown) at the edge of the trail (orange = disturbed) and 5 m perpendicular to the disturbed transect (grey = undisturbed) below and above the treeline. We recorded plant data in 100 cm by 50 cm quadrats (photo inserts) along each transect. (**c**) At each quadrat, we randomly selected five individuals of each of our focal species to measure maximum plant height and maximum plant diameter (indicated along picture, not to scale), and counted the number of buds, flowers, and fruits (white circles). We did not count reproductive metrics for *Carex* spp. as it was not possible to accurately distinguish between phenophases. (**d**) At each quadrat, we also took a standardized photograph from which to compute plant percent cover.

We focused our work on four focal species (*Phyllodoce empetriformis*, *Cassiope mertensiana*, *Vaccinium ovalifolium*, *Carex* spp.) that are common and dominant in the region. We sampled all *Carex* species, both clump forming and rhizomatous. Based on the herbarium records for the area, *C. aenea*, *C. albonigra*, *C. aquatalis*, *C. aurea*, *C. canescens*, *C. circinata*,

C. engelmanni, *C. hindsii*, *C. illota*, *C. lenticularis*, *C. luzulina*, *C. macrochaeta*, *C. mertensii*, *C. microptera*, *C. nigricans*, *C. pachystachya*, *C. phaeocephala*, *C. preslii*, *C. pyrenaica*, *C. rossii*, and *C. spectabilis* may have been present in the plots.

4.2. Data Collection

To answer how hiker trampling disturbance affects common subalpine and alpine plant communities (blueberry, heath, heather, and sedge meadows) along elevational gradients, we chose sites at multiple elevations (1580–1980 m a.s.l.) along three different trails (Figure 3b). We set up 14 permanent 10 m × 0.5 m paired trail-side (disturbed) and off-trail (undisturbed) transects in August 2022 along the heavily used Taylor Meadows-Helm Creek, Black Tusk, and Panorama Ridge trails. We established transects directly adjacent to the trail's edge and at least a 5 m perpendicular distance away from the trail to compare trampled and untrampled plant communities, respectively (following [28,29]). To mark the trail's edge, we identified sections of the trails disturbed predominantly by human trampling, but not by water run-off or trail cut construction, thus choosing diffused trail edges with evidence of trampling next to the main trail. We marked the start of each transect with a wooden stake and recorded latitude and longitude in the field with the Gaia mobile application v2022 (WGS 84, ESPG: 4326). This also enabled us to obtain elevation per site (see below).

We recorded plant data in ten continuous 1 m × 0.5 m quadrats within each transect. In each quadrat, we identified all individuals per focal species (*Phyllodoce empetriformis, Cassiope mertensiana, Vaccinium ovalifolium, Carex* spp.) with numbered flags and randomly selected five individuals with a random number generator. If there was a need to eliminate any individuals, it was c. 5 or less. As it is difficult to distinguish ramets from genets of our focal species, we focused on 'individual' plants and defined individuals as plants growing out of a central point with all structures growing in the same direction and with a larger spacing to the next plants than within an individual. We recorded the maximum plant height (height perpendicular to the ground without stretching the plant) and the maximum diameter of each selected plant within the quadrat (Figure 3c, Table 2). To keep the measured plant area consistent across all transects, we only measured the structures growing within a given quadrat. To quantify reproductive output of our focal shrub species, we counted the number of buds, flowers, and fruits on each of the five plants randomly selected. We did not count reproductive metrics for *Carex spp.* as it was not possible to accurately distinguish between phenophases. For image analyses of plant percent cover, we took a standardized photograph 1.5 m directly above each quadrat (Figure 3d). We note that while we had mostly *V. ovalifolium* in our transects, at higher elevations there was likely a bit of *V. uliginosum* and *V. deliciosum* and we do not distinguish these in our analyses.

Table 2. Measured parameters. Description of all the parameters used in this study, and the methodology, unit, spatial scale, and precision of measurement. Precision for plant percent cover is calculated from the pixel size used in our plant percent cover calculation algorithm, and since this varies by camera used, we report the mean pixel size for the lowest resolution camera. Precision for elevation is not available for this region.

Parameter	Method	Unit of Measurement	Spatial Scale	Precision
Maximum plant height	Field measurement	Plant	0.001 – 1 m	0.001 m
Maximum plant diameter	Field measurement	Plant	0.002–1.5 m	0.001 m
Reproductive Structures (buds + flowers + fruits)	Field counts	Plant	NA	1 count
Plant Percent Cover	Computed from field photos	Quadrat	1 × 0.5 m	0.6 mm^2 (0.00001%)
Elevation	Computed from field latitude and longitude	Transect	30 m	NA

In total, we sampled 1772 plants from 280 quadrats along 14 paired trail-off trail transects across a 400 m elevational gradient between 1580–1980 m a.s.l. These plants comprised 500 *P. empetriformis* individuals, 429 *C. mertensiana* individuals, 634 *V. ovalifolium* individuals, and 209 *Carex* spp. individuals.

4.3. Data Processing

We conducted all data processing and analyses in R version 4.2.3 [65]. We used field-recorded latitude and longitude data to obtain elevation per site using the USGS NED1 digital elevation model at 30 m resolution [66]. We used the 'tidyverse' [67] and 'dplyr' [68] packages to check for unusual patterns or outliers in our data. For an area-standardized reproductive output per individual plant, we first calculated a proxy for plant area by multiplying the maximum plant height by the maximum plant diameter. We then summed bud, flower, and fruit counts and divided this by plant area for an area-standardized summed reproductive output. Finally, we calculated relative reproduction for each individual by dividing summed reproductive output by that species' maximum reproductive output (i.e., the individual with the highest reproductive output over all transects). While other studies have shown that reproductive output density increases with plant area [69,70], this is not the case in our study in either disturbed or undisturbed transects (tested with a linear mixed model with package 'lmerTest'; [71]; Figure S2).

To standardize our quadrat photos for percent coverage calculation, we manually flattened and cropped all images to just the quadrat border using Google Drive's photo editing software (https://photos.google.com). To account for any shading in the plot, we increased image saturation to the point where changes in lighting negligibly affected color intensity. Logs and rocks found in quadrats (25% of photos) were counted as non-green coverage and moss was counted as green cover, although very little moss (8% of photos) was seen in the photos. To account for any shading in the plot and highlight any green vegetation, we increased each image's saturation by a factor of 30 with a custom algorithm in Python 3.10 [72]. Using the RGB color space, the algorithm then filters the image to only pixels that have a higher green channel digital number (DN) than both red and green DNs, which through testing on a randomly selected subset of images (n = 20) appeared visually to effectively isolate green vegetation (Figure 3d). The fraction of qualifying pixels out of the entire image is then calculated as an estimate of percent plant coverage in the quadrat.

4.4. Data Analyses

To test how trail disturbance, elevation, and their interaction affect plant percent cover and species traits, we fit Bayesian generalized non-linear mixed models. We fit separate models for plant percent cover and each species by trait to account for potential response differences. These models allow for non-linear relationships and account for the non-independence of data within a transect pair (i.e., on- and off-trail transect pair) by using a random effect of transect pairs. Our model equation is response ~ disturbance * elevation + (1 | transect pair), where 'response' is plant percent cover, maximum plant height, or maximum plant diameter. We account for spatial variation among trails using the random effect of transect pair, which accounts for the unique location of the paired transects within a given trail. Quadrats were simply used to facilitate measurements along the entire length of the transect and measurements between quadrats do not vary more than within quadrats, thus we use transect pair as a random effect instead of trail and quadrat.

To account for the left skew in the diameter and height trait data, we fit negative binomial models (link = 'log', link_shape = 'log') for these response variables and rounded measured diameter and height values to the nearest integer value for this distribution family. The Beta model outperformed the Zero-inflated Beta model (tested on *P. empetriformis*) as checked with a leave-one-out cross-validation ('loo()'), so to best model our left-skewed proportional relative reproduction data, we fit Beta models (link = 'log', link_phi = 'log'). We also modeled plant percent cover with the Beta distribution. We adjusted 0, 1 values by 0.0001 to keep our data within the Beta distribution interval [0,1]. All models and family distributions are described in Table S1.

We fit models with the 'brms' [73–75], 'rjags' [76], and 'R2jags' [77] packages. We fit models using 3 chains, 5000 iterations and 1000 warmup iterations on 4 cores on macOS Version 12.6.5 with a 2.2 GHz Quad-Core Intel Core i7 processor. We set 'init = 0' for plant percent cover and reproduction models. We inferred any parameter estimate's credible interval that did not contain zero as evidence for an effect of the parameter. For each model, we assessed the model fit by checking that (i) Rhat > 1.1 and that Estimated Sample Size (Bulk_ESS and Tail_ESS) > 1000 ('summary()'), and (ii) all MCMC chains converged ('plot()'). We assessed the prior distribution of each model by checking that priors do not overwhelm likelihood ('powerscale_sensitivity()'). We assessed the posterior distribution by checking that (i) predicted values are similar to posterior distribution with 1000 posterior draws ('pp_check(ndraws = 1000)'), (ii) scatterplot matrices of posterior parameter distributions are normally distributed ('pairs()'), (iii) skew is properly modeled using the Fisher-Pearson Skew function ('ppc_stat()'), (iv) Pareto k values < 0.5 ('loo()'), and (v) dispersion is properly modeled ('ppc_loo_pit_overlay()'). The models generally fit well with some exceptions. Skew and dispersion were moderately well to poorly modeled, depending on the species and trait modeled (Table S1). Some models also had one observation where Pareto K values > 0.7, which slightly decreases the reliability of the Monte Carlo error estimates from Bayesian models. The indicated functions are from the 'bayesplot' [78,79], 'loo' [80,81], and 'priorsense' [82] packages. See Table S1 for additional information on model fit.

Supplementary Materials: The following supporting information can be downloaded at: https://www.mdpi.com/article/10.3390/plants12173040/s1, Table S1: Model fitting details; Figure S1: Effects of disturbance and elevation; Figure S2: Reproductive structure density.

Author Contributions: Conceptualization, N.I.C., C.E. and C.G.C.; methodology, N.I.C., C.E. and C.G.C.; software, N.I.C., P.S. and A.Z.; formal analysis, N.I.C. and C.G.C.; investigation, N.I.C., C.H., T.M. and B.R.; data curation, C.H., N.I.C. and P.S.; writing—original draft preparation, N.I.C., P.S., C.H., T.M., B.R., A.Z., K.G., C.G.C., N.H. and C.E.; writing—review and editing, N.I.C., P.S., C.H., T.M., B.R., A.Z., K.G., C.G.C., N.H. and C.E.; visualization, P.S., B.R., C.H., N.I.C. and C.G.C.; supervision, N.I.C., C.E., C.G.C. and N.H.; project administration, N.I.C., C.E. and C.G.C.; funding acquisition, C.E., C.G.C., N.I.C. and N.H. All authors have read and agreed to the published version of the manuscript.

Funding: This research was funded by BC Parks Living Lab for Climate Change and Conservation (2022–Agreement #TP23JHQ003), University of British Columbia's Work Learn program (S22-120640) and Biology Undergraduate Diversity in Research program, and Natural Sciences and Engineering Council (NSERC) of Canada Discovery Grant (to G. Henry). N.I.C. was funded by a Swiss National Science Foundation Postdoc Mobility Fellowship (Grant ID: 194331). C.G.C. was funded by a Biodiversity Research Centre Postdoctoral fellowship. C.E. was funded by the NSERC Vanier. The APC was funded by the Swiss National Science Foundation.

Data Availability Statement: The data supporting reported results and R code for analyses can be found on GitHub at https://github.com/ITEX-sites/Garibaldi/tree/main/trampling_analyses (last updated 18 August 2023).

Acknowledgments: We are guests on the traditional, ancestral, and unceded territory of the xʷməθkʷəy̓əm (Musqueam), Sḵwx̱wú7mesh(Squamish), Stó:lō, and Səl'ílwətaʔ/Selilwitulh (Tsleil Waututh) and the Líl'wat Nations. As such we are communicating and learning from these Nations as much as possible during our studies. This research was supported in part through computational resources and services provided by Advanced Research Computing at the University of British Columbia. We thank the International Tundra Experiment and Blackcomb Helicopters in Squamish for logistical support. Fieldwork was completed by Nathalie Isabelle Chardon, Mackenzie Urquhart-Cronish, Christian Lauber, Carolyn Chong, Teagan Maclachlan, Carly Hilbert, Brianna Ragsdale, and Vicky Lee.

Conflicts of Interest: The authors declare no conflict of interest. The funders had no role in the design of the study; in the collection, analyses, or interpretation of data; in the writing of the manuscript; or in the decision to publish the results.

References and Note

1. Billings, W.D. Arctic and Alpine Vegetations: Similarities, Differences, and Susceptibility to Disturbance. *BioScience* **1973**, *23*, 697–704. [CrossRef]
2. Willard, B.E.; Cooper, D.J.; Forbes, B.C. Natural Regeneration of Alpine Tundra Vegetation after Human Trampling: A 42-Year Data Set from Rocky Mountain National Park, Colorado, U.S.A. *Arct. Antarct. Alp. Res.* **2007**, *39*, 177–183. [CrossRef]
3. Grime, J.P. Vegetation Classification by Reference to Strategies. *Nature* **1974**, *250*, 26–31. [CrossRef]
4. Cole, D.N. *Trampling Effects on Mountain Vegetation in Washington, Colorado, New Hampshire, and North Carolina*; U.S. Department of Agriculture, Forest Service, Intermountain Research Station: Ogden, UT, USA, 1993.
5. Crisfield, V.E.; Macdonald, S.E.; Gould, A.J. Effects of Recreational Traffic on Alpine Plant Communities in the Northern Canadian Rockies. *Arct. Antarct. Alp. Res.* **2012**, *44*, 277–287. [CrossRef]
6. de Gouvenain, R.C. Indirect Impacts of Soil Trampling on Tree Growth and Plant Succession in the North Cascade Mountains of Washington. *Biol. Conserv.* **1996**, *75*, 279–287. [CrossRef]
7. Pickering, C.M.; Hill, W.; Newsome, D.; Leung, Y.-F. Comparing Hiking, Mountain Biking and Horse Riding Impacts on Vegetation and Soils in Australia and the United States of America. *J. Environ. Manag.* **2010**, *91*, 551–562. [CrossRef]
8. Li, W.; He, S.; Cheng, X.; Zhang, M. Functional Diversity Outperforms Taxonomic Diversity in Revealing Short-Term Trampling Effects. *Sci. Rep.* **2021**, *11*, 18889. [CrossRef]
9. Barros, A.; Pickering, C.M. Impacts of Experimental Trampling by Hikers and Pack Animals on a High-Altitude Alpine Sedge Meadow in the Andes. *Plant Ecol. Divers.* **2015**, *8*, 265–276. [CrossRef]
10. Bisht, V.K.; Kuniyal, C.P.; Nautiyal, B.P.; Prasad, P. Integrated Analysis of the Trees and Associated Under-Canopy Species in a Subalpine Forest of Western Himalaya, Uttarakhand, India. *J. Mt. Sci.* **2015**, *12*, 154–165. [CrossRef]
11. Pickering, C.M.; Hill, W. Impacts of Recreation and Tourism on Plant Biodiversity and Vegetation in Protected Areas in Australia. *J. Environ. Manag.* **2007**, *85*, 791–800. [CrossRef]
12. Rossi, G.; Parolo, G.; Ulian, T. Human Trampling as a Threat Factor for the Conservation of Peripheral Plant Populations. *Plant Biosyst. Int. J. Deal. Asp. Plant Biol.* **2009**, *143*, 104–113. [CrossRef]
13. Ballantyne, M.; Pickering, C.M.; McDougall, K.L.; Wright, G.T. Sustained Impacts of a Hiking Trail on Changing Windswept Feldmark Vegetation in the Australian Alps. *Aust. J. Bot.* **2014**, *62*, 263. [CrossRef]
14. Barros, A.; Gonnet, J.; Pickering, C. Impacts of Informal Trails on Vegetation and Soils in the Highest Protected Area in the Southern Hemisphere. *J. Environ. Manag.* **2013**, *127*, 50–60. [CrossRef] [PubMed]
15. Jägerbrand, A.K.; Alatalo, J.M. Effects of Human Trampling on Abundance and Diversity of Vascular Plants, Bryophytes and Lichens in Alpine Heath Vegetation, Northern Sweden. *SpringerPlus* **2015**, *4*, 95. [CrossRef] [PubMed]
16. Yang, M.; Hens, L.; Ou, X.; De Wulf, R. Impacts of Recreational Trampling on Sub-Alpine Vegetation and Soils in Northwest Yunnan, China. *Acta Ecol. Sin.* **2009**, *29*, 171–175. [CrossRef]
17. Cole, D.N. Experimental Trampling of Vegetation. I. Relationship Between Trampling Intensity and Vegetation Response. *J. Appl. Ecol.* **1995**, *32*, 203–214. [CrossRef]
18. Maschinski, J.; Frye, R.; Rutman, S. Demography and Population Viability of an Endangered Plant Species before and after Protection from Trampling. Demografia y Viabilidad Poblacional de Un Especie de Planta En Peligro de Extincion Antes y Despues de Protegerla Contra Pisoteo. *Conserv. Biol.* **2003**, *11*, 990–999. [CrossRef]
19. Pfister, C.A. Patterns of Variance in Stage-Structured Populations: Evolutionary Predictions and Ecological Implications. *Proc. Natl. Acad. Sci. USA* **1998**, *95*, 213–218. [CrossRef]
20. Mooney, S.J.; Nipattasuk, W. Quantification of the Effects of Soil Compaction on Water Flow Using Dye Tracers and Image Analysis. *Soil Use Manag.* **2003**, *19*, 356–363. [CrossRef]
21. Liu, Q.; Li, W.; Nie, H.; Sun, X.; Dong, L.; Xiang, L.; Zhang, J.; Liu, X. The Effect of Human Trampling Activity on a Soil Microbial Community at the Urban Forest Park. *Forests* **2023**, *14*, 692. [CrossRef]
22. Kuss, F.R. A Review of Major Factors Influencing Plant Responses to Recreation Impacts. *Environ. Manag.* **1986**, *10*, 637–650. [CrossRef]
23. Pearcy, R.W.; Ward, R.T. Phenology and Growth of Rocky Mountain Populations of Deschampsia Caespitosa at Three Elevations in Colorado. *Ecology* **1972**, *53*, 1171–1178. [CrossRef]
24. Dykema, J.A. Ecological Impact of Camping upon the Southern Sierra Nevada. Doctoral Thesis, University of California, Los Angeles, CA, USA, 1971. Available online: https://www.proquest.com/openview/8bdb0fa22b4747cbafa2b8f68647dfef/1.pdf?pq-origsite=gscholar&cbl=18750&diss=y (accessed on 29 June 2023).
25. Holmes, D.O.; Dobson, H.E.M. *The Effects of Human Trampling and Urine on Subalpine Vegetation, a Survey of Past and Present Backcountry Use, and the Ecological Carrying Capacity of Wilderness*; Final Report, Ecological Carrying Capacity Research Contract CX8000-4-0026; Department of Plant Pathology, University of California: Davis, CA, USA, 1976.
26. Kycko, M.; Zagajewski, B.; Lavender, S.; Romanowska, E.; Zwijacz-Kozica, M. The Impact of Tourist Traffic on the Condition and Cell Structures of Alpine Swards. *Remote Sens.* **2018**, *10*, 220. [CrossRef]
27. Moen, J.; Lagerström, A. High Species Turnover and Decreasing Plant Species Richness on Mountain Summits in Sweden: Reindeer Grazing Overrides Climate Change. *Arct. Antarct. Alp. Res.* **2008**, *40*, 382–395. [CrossRef]
28. Chardon, N.I.; Wipf, S.; Rixen, C.; Beilstein, A.; Doak, D.F. Local Trampling Disturbance Effects on Alpine Plant Populations and Communities: Negative Implications for Climate Change Vulnerability. *Ecol. Evol.* **2018**, *8*, 7921–7935. [CrossRef]

29. Chardon, N.I.; Rixen, C.; Wipf, S.; Doak, D.F. Human Trampling Disturbance Exerts Different Ecological Effects at Contrasting Elevational Range Limits. *J. Appl. Ecol.* **2019**, *56*, 1389–1399. [CrossRef]
30. Evans, I.S. Climatic Effects on Glacier Distribution Across the Southern Coast Mountains. B.C., Canada. *Ann. Glaciol.* **1990**, *14*, 58–64. [CrossRef]
31. Bevington, A.R.; Menounos, B. Accelerated Change in the Glaciated Environments of Western Canada Revealed through Trend Analysis of Optical Satellite Imagery. *Remote Sens. Environ.* **2022**, *270*, 112862. [CrossRef]
32. Statistics Canada Census of Population 2021. Available online: https://www12.statcan.gc.ca/census-recensement/2021/dp-pd/prof/index.cfm?Lang=E (accessed on 12 June 2023).
33. BC Parks. *2008/09 BC Parks Year End Report*; Ministry of Environment: Victoria, BC, Canada, 2009; p. 12.
34. BC Parks. *BC Parks 2015/16 Statistics Report*; Ministry of Environment: Victoria, BC, Canada, 2016; p. 43.
35. BC Parks. *BC Parks 2017/18 Statistics Report*; Ministry of Environment: Victoria, BC, Canada, 2018; p. 44.
36. Newman, R.; Powell, G.; Cameron, K.; Demaere, C. Effects of Cattle Trampling and Browsing on Lodgepole Pine Plantations. In *Managing the Dry Douglas-Fir Forests of the Southern Interior: Workshop Proceedings*; British Columbia Ministry of Forests Research Program: Kamloops, BC, Canada, 1998; Volume 34.
37. Nepal, S.K.; Way, P. Comparison of Vegetation Conditions along Two Backcountry Trails in Mount Robson Provincial Park, British Columbia (Canada). *J. Environ. Manag.* **2007**, *82*, 240–249. [CrossRef]
38. Curtin, C.G. Can Montane Landscapes Recover from Human Disturbance? Long-Term Evidence from Disturbed Subalpine Communities. *Biol. Conserv.* **1995**, *74*, 49–55. [CrossRef]
39. Squamish Nation Culturally Significant Vegetation. Assembled by Elizabeth Ross Rachel Munger (Squamish Nation). Accessed 30 May 2023
40. Martin, K. Wildlife in Alpine and Sub-Alpine Habitats. In *Wildlife-Habitat Relationships in Oregon and Washington*; Johnson, D.H., O'Neil, T.A., Eds.; Oregon State University Press: Corvallis, OR, USA, 2001; pp. 285–310.
41. Vowles, T.; Björk, R.G. Implications of Evergreen Shrub Expansion in the Arctic. *J. Ecol.* **2019**, *107*, 650–655. [CrossRef]
42. Cole, D.N. Experimental Trampling of Vegetation. II. Predictors of Resistance and Resilience. *J. Appl. Ecol.* **1995**, *32*, 215–224. [CrossRef]
43. Fournier, M.-P.; Paré, M.C.; Buttò, V.; Delagrange, S.; Lafond, J.; Deslauriers, A. How Plant Allometry Influences Bud Phenology and Fruit Yield in Two *Vaccinium* Species. *Ann. Bot.* **2020**, *126*, 825–835. [CrossRef] [PubMed]
44. Wada, K.C.; Takeno, K. Stress-Induced Flowering. *Plant Signal. Behav.* **2010**, *5*, 944–947. [CrossRef]
45. Evju, M.; Hagen, D.; Hofgaard, A. Effects of Disturbance on Plant Regrowth along Snow Pack Gradients in Alpine Habitats. *Plant Ecol.* **2012**, *213*, 1345–1355. [CrossRef]
46. Lee, T.D. Patterns of Fruit and Seed Production. In *Plant Reproductive Ecology*; Oxford University Press: New York, NY, USA, 1988; pp. 179–202.
47. Ciach, M.; Maślanka, B.; Krzus, A.; Wojas, T. Watch Your Step: Insect Mortality on Hiking Trails. *Insect Conserv. Divers.* **2017**, *10*, 129–140. [CrossRef]
48. Macek, P.; Lepš, J. Environmental Correlates of Growth Traits of the Stoloniferous Plant *Potentilla palustris*. *Evol. Ecol.* **2008**, *22*, 419–435. [CrossRef]
49. Mao, L.; Chen, S.; Zhang, J.; Zhou, G. Altitudinal Patterns of Maximum Plant Height on the Tibetan Plateau. *J. Plant Ecol.* **2018**, *11*, 85–91. [CrossRef]
50. Ford, K.R.; Ettinger, A.K.; Lundquist, J.D.; Raleigh, M.S.; Hille Ris Lambers, J. Spatial Heterogeneity in Ecologically Important Climate Variables at Coarse and Fine Scales in a High-Snow Mountain Landscape. *PLoS ONE* **2013**, *8*, e65008. [CrossRef]
51. MacKenzie, W. *The Ecology of the Alpine Zones*; BC Ministry of Forests and Range Research: Victoria, BC, Canada, 2006.
52. Mote, P.W.; Hamlet, A.F.; Clark, M.P.; Lettenmaier, D.P. Declining mountain snowpack in western North America. *Bull. Am. Meteorol. Soc.* **2005**, *86*, 39–50. [CrossRef]
53. Lute, A.C.; Abatzoglou, J.T.; Hegewisch, K.C. Projected Changes in Snowfall Extremes and Interannual Variability of Snowfall in the Western United States. *Water Resour. Res.* **2015**, *51*, 960–972. [CrossRef]
54. Barros, A.; Aschero, V.; Mazzolari, A.; Cavieres, L.A.; Pickering, C.M. Going off Trails: How Dispersed Visitor Use Affects Alpine Vegetation. *J. Environ. Manag.* **2020**, *267*, 110546. [CrossRef] [PubMed]
55. Hulme, P.E.; Pyšek, P.; Pergl, J.; Jarošík, V.; Schaffner, U.; Vilà, M. Greater Focus Needed on Alien Plant Impacts in Protected Areas. *Conserv. Lett.* **2014**, *7*, 459–466. [CrossRef]
56. Luong, J.C.; Villanueva, E.M.; Bauman, T.A. Native and Invasive Bunchgrasses Have Different Responses to Trail Disturbance on California Coastal Prairies. *Plant Ecol.* **2023**, *224*, 125–135. [CrossRef]
57. Urbina, J.C.; Benavides, J.C. Simulated Small Scale Disturbances Increase Decomposition Rates and Facilitates Invasive Species Encroachment in a High Elevation Tropical Andean Peatland. *Biotropica* **2015**, *47*, 143–151. [CrossRef]
58. Bella, E.M. Invasion Prediction on Alaska Trails: Distribution, Habitat, and Trail Use. *Invasive Plant Sci. Manag.* **2011**, *4*, 296–305. [CrossRef]
59. Lembrechts, J.J.; Milbau, A.; Nijs, I. Alien Roadside Species More Easily Invade Alpine than Lowland Plant Communities in a Subarctic Mountain Ecosystem. *PLoS ONE* **2014**, *9*, e89664. [CrossRef]
60. Daniels, M.L.; Marion, J.L. Communicating Leave No Trace Ethics and Practices: Efficacy of Two-Day Trainer Courses. *J. Park Recreat. Adm.* **2005**, *23*, 1–19.

61. USGS. Leave No Trace This Summer as You Explore the Outdoors: The USGS Is Studying How to Enjoy the Wilderness without Impacting the Environment. Available online: https://www.usgs.gov/news/featured-story/leave-no-trace-summer-you-explore-outdoors (accessed on 17 June 2023).
62. Parmesan, C. Ecological and Evolutionary Responses to Recent Climate Change. *Annu. Rev. Ecol. Evol. Syst.* **2006**, *37*, 637–669. [CrossRef]
63. Polar, J.; Stewart, A.C. Alpine Tundra Zone. In *Ecosystems of British Columbia*; Meidinger, D.V., Pojar, J., Eds.; Special Report Series; Research Branch, Ministry of Forests: Victoria, BC, Canada, 1991; pp. 264–274. ISBN 978-0-7718-8997-4.
64. Bryceland, J.; Macaree, D.; Macaree, M. *103 Hikes in Southwestern British Columbia*; Greystone Books: Vancouver, BC, Canada, 2008; ISBN 978-1-926685-02-1.
65. R Core Team 2023. R: A language and environment for statistical computing. R Foundation for Statistical Computing: Vienna, Austria. Available online: https://www.R-project.org/.
66. GPS Visualizer. Available online: https://www.gpsvisualizer.com/ (accessed on 11 June 2023).
67. Wickham, H.; Averick, M.; Bryan, J.; Chang, W.; McGowan, L.D.; François, R.; Grolemund, G.; Hayes, A.; Henry, L.; Hester, J.; et al. Welcome to the Tidyverse. *J. Open Source Softw.* **2019**, *4*, 1686. [CrossRef]
68. Wickham, H.; François, R.; Henry, L.; Müller, K.; Vaughan, D. Dplyr: A Grammar of Data Manipulation version 1.1.0. 2023. Available online: https://cran.r-project.org/web/packages/dplyr/index.html.
69. Bonser, S.P. Size and Phenology Control Plant Reproduction and Agricultural Production. A Commentary on: 'How Plant Allometry Influences Bud Phenology and Fruit Yield in Two *Vaccinium* Species'. *Ann. Bot.* **2020**, *126*, 6–8. [CrossRef] [PubMed]
70. Weiner, J.; Rosenmeier, L.; Massoni, E.S.; Vera, J.N.; Plaza, E.H.; Sebastià, M.-T. Is Reproductive Allocation in *Senecio vulgaris* Plastic? *Botany* **2009**, *87*, 475–481. [CrossRef]
71. Kuznetsova, A.; Brockhoff, P.B.; Christensen, R.H.B. LmerTest Package: Tests in Linear Mixed Effects Models. *J. Stat. Softw.* **2017**, *82*, 1–26. [CrossRef]
72. Python Software Foundation 2021. Python Language Reference, version 3.10. Available online: http://www.python.org.
73. Bürkner, P.-C. Brms: An R Package for Bayesian Multilevel Models Using Stan. *J. Stat. Softw.* **2017**, *80*, 1–28. [CrossRef]
74. Bürkner, P.-C. Advanced Bayesian Multilevel Modeling with the R Package Brms. *R J.* **2018**, *10*, 395–411. [CrossRef]
75. Bürkner, P.-C. Bayesian Item Response Modeling in R with Brms and Stan. *J. Stat. Softw.* **2021**, *100*, 1–54. [CrossRef]
76. Plummer, M.; Stukalov, A.; Denwood, M. Rjags: Bayesian Graphical Models Using MCMC, version 4-13. 2023. Available online: https://cran.r-project.org/web/packages/rjags/index.html.
77. Su, Y.-S.; Yajima, M. R2jags: Using R to Run "JAGS", version 0.7-1. 2021. Available online: https://cran.r-project.org/web/packages/R2jags/index.html.
78. Gabry, J.; Simpson, D.; Vehtari, A.; Betancourt, M.; Gelman, A. Visualization in Bayesian Workflow. *J. R. Stat. Soc. Ser. A Stat. Soc.* **2019**, *182*, 389–402. [CrossRef]
79. Gabry, J.; Mahr, T. Bayesplot: Plotting for Bayesian Models, version 1.10.0. 2022. Available online: http://mc-stan.org/bayesplot/.
80. Vehtari, A.; Gelman, A.; Gabry, J. Practical Bayesian Model Evaluation Using Leave-One-Out Cross-Validation and WAIC. *Stat. Comput.* **2017**, *27*, 1413–1432. [CrossRef]
81. Vehtari, A.; Gabry, J.; Magnusson, M.; Bürkner, P.-C.; Paananen, T.; Gelman, A. Loo: Efficient-Leave-One-Out Cross-Validation and WAIC for Bayesian Models, version 2.5.1. 2022. Available online: https://mc-stan.org/loo/.
82. Kallioinen, N.; Bürkner, P.-C.; Paananen, T.; Vehtari, A. Priorsense: Prior Diagnostics and Sensitivity Analysis, Version 0.0.0.9000. 2023. Available online: https://rdrr.io/github/n-kall/priorsense/.

plants

MDPI

Article

Climate Overrides the Influence of Microsite Conditions on Radial Growth of the Tall Multi-Stemmed Shrub *Alnus alnobetula* at Treeline

Walter Oberhuber *, Anna-Lena Dobler, Tamara Heinzle, Francesca Scandurra, Andreas Gruber and Gerhard Wieser

Department of Botany, University of Innsbruck, A-6020 Innsbruck, Austria; andreas.gruber@uibk.ac.at (A.G.); gerhard.wieser@uibk.ac.at (G.W.)
* Correspondence: walter.oberhuber@uibk.ac.at

Abstract: Green alder (*Alnus alnobetula*), a tall multi-stemmed deciduous shrub, is widespread at high elevations in the Central European Alps. Its growth form frequently leads to asymmetric radial growth and anomalous growth ring patterns, making development of representative ring-width series a challenge. In order to assess the variability among radii of one shoot, among shoots belonging to one stock and among stocks, 60 stem discs were sampled at treeline on Mt. Patscherkofel (Tyrol, Austria). Annual increments were measured along 188 radii and analyzed in terms of their variability by applying dendrochronological techniques. Results revealed a high agreement in ring-width variation among radii of one shoot, among shoots of one stock and largely among stocks from different sites, confirming the pronounced limitation of radial stem growth by climate forcing at the alpine treeline. In contrast to this, a high variability in both absolute growth rates and long-term growth trends was found, which we attribute to different microsite conditions and disturbances. These factors also override climate control of radial growth under growth-limiting environmental conditions. Based on our findings we provide recommendations for the number of samples needed to carry out inter- and intra-annual studies of radial growth in this multi-stemmed clonal shrub.

Keywords: climate forcing; green alder; growth variability; multi-stemmed shrub; radial stem growth; ring width; tree ring

Citation: Oberhuber, W.; Dobler, A.-L.; Heinzle, T.; Scandurra, F.; Gruber, A.; Wieser, G. Climate Overrides the Influence of Microsite Conditions on Radial Growth of the Tall Multi-Stemmed Shrub *Alnus alnobetula* at Treeline. *Plants* **2023**, *12*, 1708. https://doi.org/10.3390/plants12081708

Academic Editor: Gregor Kozlowski

Received: 14 March 2023
Revised: 13 April 2023
Accepted: 17 April 2023
Published: 20 April 2023

1. Introduction

During the last decades, land use change and climate warming favor the spreading of shrubs in arctic and alpine environments (e.g., [1–3]), resulting in widespread impacts on ecosystem services, plant diversity, soil nutrient content and biogeochemical cycles [4–8]. The most expanding shrub species in the European Alps is green alder (*Alnus alnobetula* (Ehrh.) K. Koch former *Alnus viridis* (Chaix) DC; [9–11]), a tall multi-stemmed early successional shrub forming dense 2–4 m tall canopies with shoots showing arched ascending growth [12,13]. The spreading of tall *Alnus* spp. shrubs is also reported for the tundra ecotone [14–16]. Due to clonal growth by layering, stocks are developed from which numerous shoots sprout [17]. Preference for clonal growth over vertical stem growth is part of the successful and rapid expansion strategy within the treeline ecotone triggered primarily by decreasing grazing pressure and land abandonment [18–20].

Due to environmental alterations that occur in the course of shrub expansion, the drivers of shrub growth are being increasingly studied by applying dendrochronological techniques (for a review see Myers-Smith et al. [21]). With regard to tall shrubs, dendroclimatic analyses were applied to quantify the climatic sensitivity of radial growth of, e.g., *Salix lanata*, *Salix glauca*, *Alnus viridis* ssp. *fruticosa* and *Alnus alnobetula* [22–26]. These studies focused on the mean climate response of all individuals (=stocks) at a given site, disregarding variability in radial growth among radii of one shoot, shoots belonging to one stock and stocks. In multi-stemmed *Alnus alnobetula*, ascending growth of stems often leads to a pronounced asymmetric radial growth, forming ellipse-shaped growth rings with eccentric pith in the lower ascending stem region, which is enhanced by heavy snow load occurring in avalanche tracks. Furthermore, above the closed forest, environmental conditions and disturbances (e.g., snow load, wind) frequently change at small spatial scales and temperature gradients, resulting in differences in insolation that can develop within tall multi-stemmed shrubs [27]. For this reason, it can be assumed that radii of one shoot and shoots belonging to one individual may show high variability in radial growth due to varying microclimates and disturbances. Hence, the development of reliable ring-width series necessary to determine, e.g., climate forcing of radial growth or long-term growth trends in response to climate warming, is associated with great difficulties due to frequently poor agreement among ring-width series that are (i) measured along different radii, (ii) developed from different shoots belonging to one individual or (iii) among individuals. In order to cope with these challenges, several authors suggested methodological approaches for dendrochronological studies of especially dwarf shrubs [21,28–30].

The main focus of this study was to evaluate the variability in radial growth among radii of one shoot, among shoots of one individual and among individuals of the tall multi-stemmed shrub *Alnus alnobetula* spreading above 2100 m asl within the alpine treeline ecotone on Mt. Patscherkofel (Tyrol, Austria; [26]). Based on our considerations, we formulated the following three Hypotheses (H):

H1. *Radial growth measured along several radii of one shoot shows high variability in absolute growth and low synchrony in year-to-year ring-width fluctuations.*

H2. *Radial growth compared among radii of different shoots within one individual (stock) shows low variability in absolute growth and high synchrony in year-to-year ring-width fluctuations.*

H3. *Radial growth compared between different individuals exposed to site-specific environmental conditions and disturbances shows high variability in absolute growth and long-term growth trends and low synchrony in year-to-year ring-width fluctuations.*

Recording growth variability among different radii of one shoot, among shoots belonging to one individual and among individuals will allow specifying how many radii and shoots of one multi-stemmed individual and how many individuals must be sampled in order to develop representative ring-width time series of this tall multi-stemmed shrub. An adequate sampling design is a prerequisite to determine climate forcing of radial stem growth and long-term growth trends in this clonal shrub spreading rapidly at the alpine treeline.

2. Results

Collected shoots had a mean stem length of 2.5 m (stocks IM and AG) and 1.6 m (stock WE), and mean stem diameter ranged between 2.7 cm (stock WE) and 4 cm (stock AG; Table 1). Mean cambial age close to the stem base varied between 23 and 29 yrs, and mean ring width was 462 µm (stock WE), 566 µm (stock IM) and 618 µm (stock AG). Recorded growth parameters as well as shoot age of stock WE differed significantly ($p \leq 0.01$) from the other individuals (i.e., stocks IM and AG).

2.1. Variability in Radial Growth among Radii of One Shoot

Ring-width series of all radii (cf. Figure 1) recorded in all shoots taken from one stock are depicted in Figure 2a–c. In all three stocks investigated, radius *a* was significantly wider than radii *b*, *c* and *d* and varied between 508 to 703 µm at the WE and AG sites, respectively (Figure 2d–f). At the AG site, mean ring width of radius *d* was also significantly different ($p \leq 0.01$) from mean ring widths of radii *a* and *c*. Mean ring widths of opposite radii (=*ab* and *cd*) were significantly different ($p \leq 0.05$) from each other at the IM and AG sites, but were not significantly different from mean ring widths of all radii (=*abcd*) in all stocks (Figure 3).

The agreement of inter-annual radial growth among radii of one shoot is depicted in Figure 4a–f. The highly significant agreement ($p \leq 0.001$) in ring width variations of radius *a* and *b* (i.e., *a:b*) and radius *a* and mean of radii *c* and *d* (i.e., *a:cd*), as well as mean of radii *a* and *b* and *c* and *d* (i.e., *ab:cd*), are striking (Figure 4a–c). Mean t_{BP}-scores for all radius comparisons in all individuals were more than twice as high as the value of 3.5, indicating highly secure cross-dating.

2.2. Variability in Radial Growth among Radii of Different Shoots of One Stock

All statistical parameters estimating agreement in year-to-year variation of ring-width series (i.e., EPS, S/N-ratio, EV and Rbar) were high for all single radii, mean values of opposite radii and mean of all radii (Table 2). The highest values for single radii (i.e., radius *a*, *b*, *c* or *d*) were obtained for radius *a* at the WE and AG sites. These statistics increased only slightly when calculated on the basis of the mean value of all radii. MS values were quite similar in all radii (including mean values calculated from different radii) and all stocks, ranging from 36.7% to 48.9%. AC was rather low ($r < 0.26$) in all radii recorded in shoots sampled at the WE and AG sites. AC reached at most a value of $r = 0.430$ at the IM site.

Table 1. Characteristics of selected *Alnus alnobetula* individuals (RW = ring width; SD = standard deviation; ShA = shoot age; ShD = shoot diameter; ShL = shoot length; sites: WE = wind-exposed ridge; IM = intermediate site; AG = avalanche gully). Statistically significant differences of mean values among individuals (=stocks) are indicated by different letters ($p \leq 0.01$; Student's *t*-test).

Site	Shoots [1] (*n*)	ShL (m) Mean ± SD	ShD (cm) Mean ± SD	ShA [2] (yrs) Min/Max/Mean ± SD	RW [3] (μm) Mean ± SD
WE	15	1.6 ± 0.2^{a}	2.7 ± 0.3^{a}	$19/25/23.5 \pm 1.8^{a}$	462 ± 59^{a}
IM	16	2.5 ± 0.1^{b}	3.8 ± 0.6^{b}	$23/32/28.6 \pm 2.6^{b}$	566 ± 107^{b}
AG	16	2.5 ± 0.3^{b}	4.0 ± 0.7^{b}	$19/35/29.2 \pm 5.5^{b}$	618 ± 134^{b}

[1] Number of shoots taken from one individual (total number of shoots varied between 30–40 for each stock). [2] Mean cambial age of shoots (stem discs were taken close to the stem base). [3] Mean of all four radii measured in all shoots belonging to one stock.

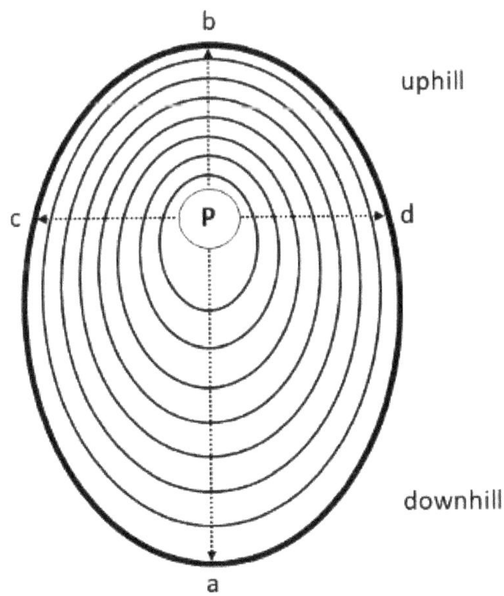

Figure 1. Schematic representation of measured radii. Radii *a* and *b* are oriented on the lower and upper side of the stem, respectively. Radii *c* and *d* are aligned perpendicular to these radii (*p* = pith).

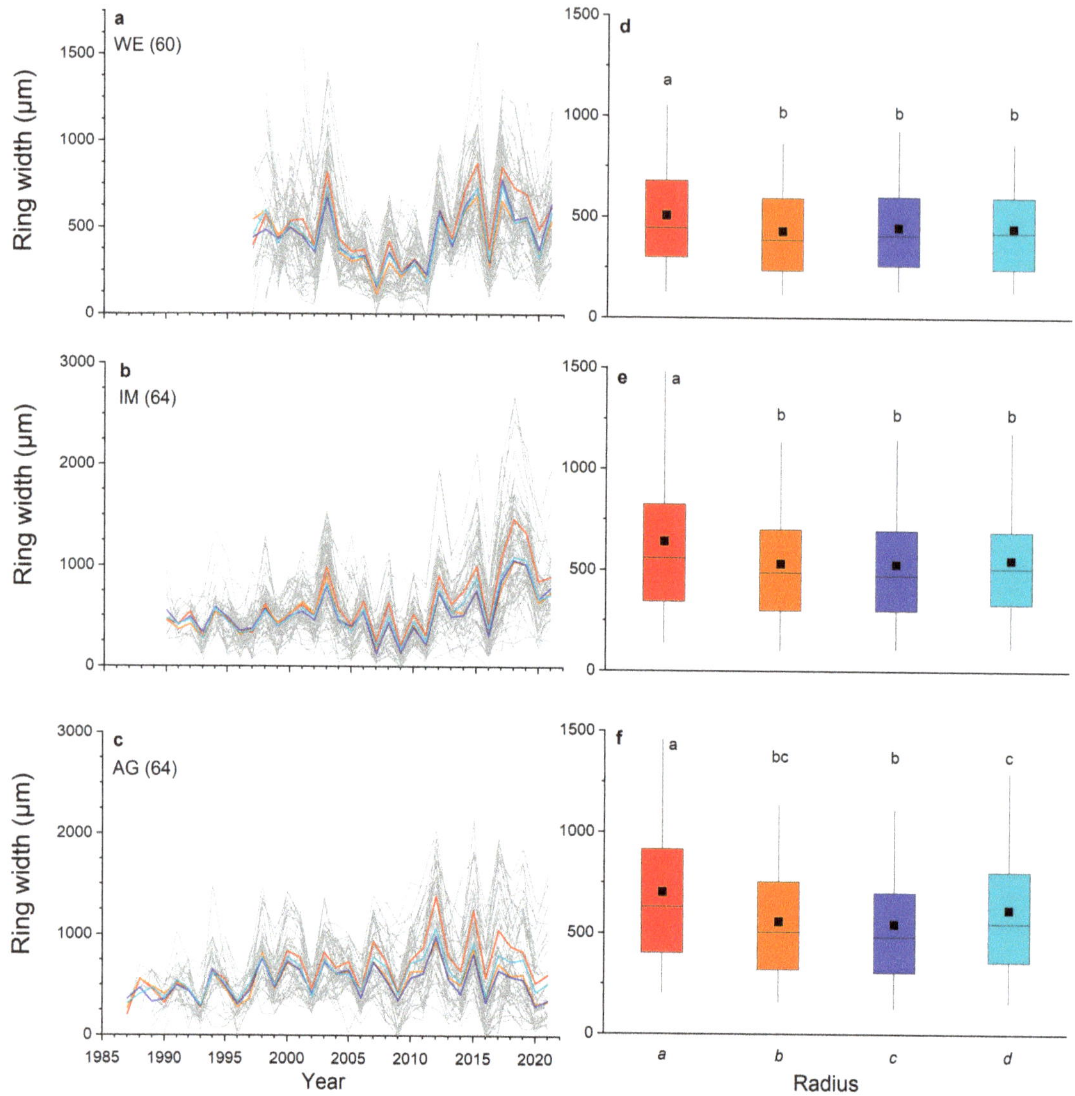

Figure 2. Ring-width series: (**a**–**c**) of all measured radii (grey lines; number of radii measured per stock are given in brackets) at a wind-exposed ridge (site WE), an intermediate site (IM) and a north-facing avalanche gully (site AG). Mean values of radii are shown in red (radius *a*), orange (radius *b*), dark blue (radius *c*) and light blue (radius *d*); (**d**–**f**) box-plots of ring width of all radii of each stock. The filled square and line indicate the mean and median ring width, respectively, the upper extent of the box indicates the 75th percentile, the lower extent indicates the 25th percentile and whiskers represent the 5th and 95th percentile. Statistically significant differences of mean values between radii are indicated by different letters ($p \leq 0.01$; Student's *t*-test).

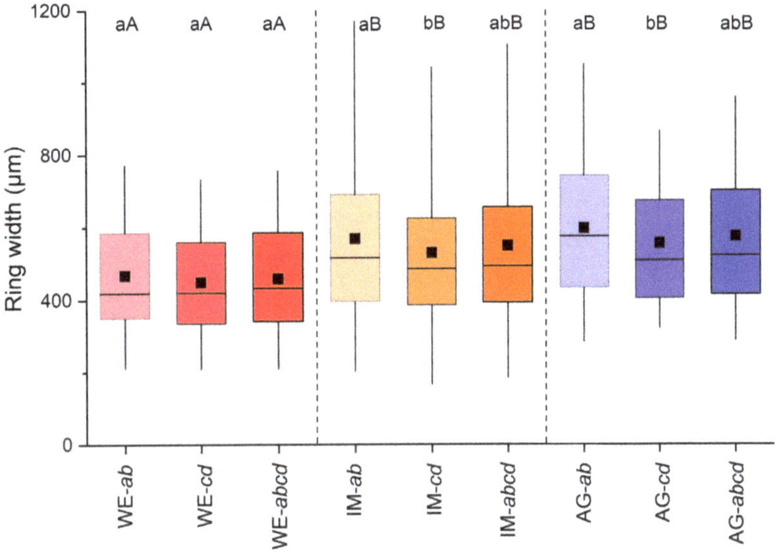

Figure 3. Box-plots showing variability in ring width among mean values of radii *a* and *b* (=*ab*), mean of radii measured perpendicular to these radii (=*cd*; cf. Figure 1) and all radii (=*abcd*) of all shoots taken from one stock at a wind-exposed ridge (site WE), an intermediate site (IM) and a north-facing avalanche gully (site AG). Statistically significant differences of mean values between and within stocks are indicated by different capital and small letters, respectively ($p \leq 0.05$; Student's *t*-test). For details on box-plots, see Figure 2.

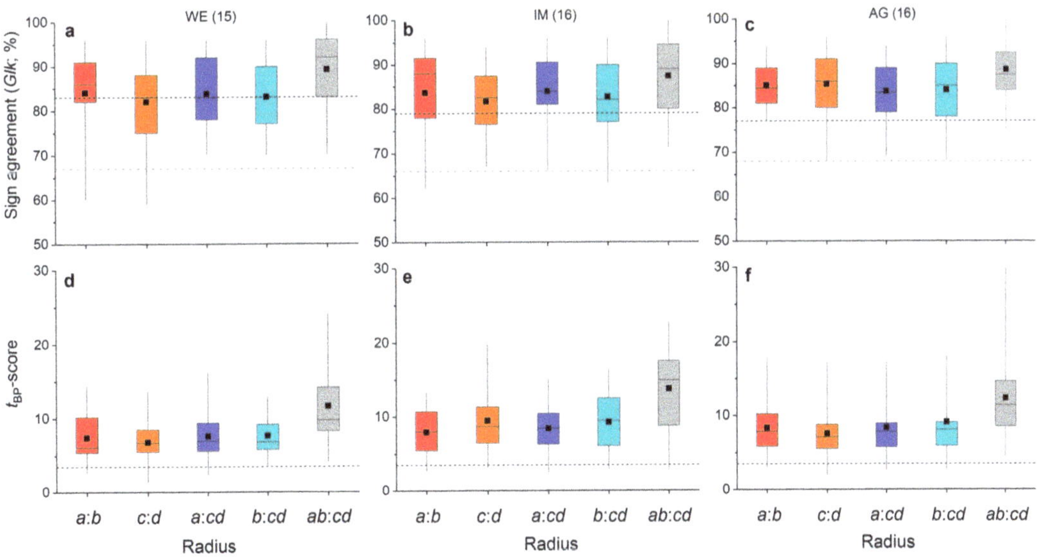

Figure 4. Box-plots showing variability in percentage of sign agreement ((**a–c**); *Glk* = Gleichläufigkeit; Eckstein and Bauch 1973) and the t_{BP}-score ((**d–f**); Baillie and Pilcher 1990) among opposite radii (i.e., *a*:*b* and *c*:*d*), radii taken at the lower and upper side of the stem to the mean of radii measured perpendicular to these radii (i.e., *a*:*cd* and *b*:*cd*, respectively) and mean of radii *a* and *b* vs. mean of radii *c* and *d* (i.e., *ab*:*cd*) of all shoots taken from one stock at a wind-exposed ridge (site WE), an intermediate

site (IM) and a north-facing avalanche gully (site AG). Number of shoots measured per stock are given in brackets. In (**a–c**), horizontal dashed black and grey lines indicate the significance level of $p \leq 0.001$ and $p \leq 0.05$, respectively. In (**d–f**), the horizontal dashed black line indicates a *t*-value of 3.5 (see Section 4 for details). For details on box-plots, see Figure 2.

Agreement of inter-annual radial growth among single radii and combinations of radii (=*ab*, *cd* and *abcd*) of shoots belonging to the same stock are depicted in Figure 5a–f. It is evident that radius *a* achieved the highest mean percentage of sign agreement (*Glk*; Figure 5a–c) and mean t_{BP}-score (Figure 5d–f) of all single radii in all individuals. Mean *Glk* for radius *a* of all shoots belonging to the same stock were statistically significant at $p \leq 0.01$ (stock WE) and at $p \leq 0.001$ (stocks IM and AG). Mean t_{BP}-score for radius *a* was close to (stock WE) or higher than (stocks IM and AG) 3.5. Time series of mean ring width calculated from measurement of all radii (=*abcd*) show high agreement in year-to-year variation in radial growth in shoots taken from one stock (Figures 5a–f and 6a–c). Larger radial growth variability in favorable years (e.g., 2003, 2015, 2017) than in years having poor growing conditions (e.g., 2002, 2004, 2016) are reflected by a wider and narrower interquartile range, respectively. There is also a pronounced coincidence in radial growth increase and reduction in 2003 and 2016, respectively, among shoots and WE and IM individuals. Ring width in 2003 is not significantly different from the mean in individual AG. However, in these extreme growth years, mean ring widths in all individuals were significantly different ($p \leq 0.001$) from the preceding year. Mean ring widths of all four radii measured (Figure 6d–f; "all years") are significantly ($p \leq 0.001$) lower at site WE compared to mean ring widths recorded at other sites, i.e., IM and AG, which are not significantly different ($p = 0.062$).

Table 2. Statistics of ring-width series measured along different radii (*a–d*) and mean values of combinations of radii (*ab, cd, abcd*) of all shoots taken from one stock at a wind-exposed ridge (site WE), an intermediate site (IM) and a north-facing avalanche gully (site AG). Statistics are calculated based on detrended ring-width series for the common interval 2001–2021 (AC = autocorrelation; EPS = expressed population signal; EV = variance 1st eigenvector; MS = mean sensitivity; Rbar = mean inter-series correlation among radii).

Site	Radius	MS (%)	AC	EPS	S/N-Ratio	Rbar	EV (%)
WE	*a*	41.9	0.255	0.961	24.9	0.657	68.8
	b	41.5	0.109	0.950	19.0	0.593	62.6
	c	42.2	0.112	0.943	16.5	0.559	60.2
	d	41.9	0.124	0.951	19.5	0.600	64.6
	ab	40.4	0.226	0.964	27.0	0.675	70.2
	cd	40.7	0.117	0.956	21.8	0.627	66.4
	abcd	40.5	0.171	0.966	28.1	0.684	71.2
IM	*a*	47.4	0.430	0.966	28.8	0.643	68.9
	b	48.9	0.326	0.959	23.3	0.593	63.5
	c	47.9	0.349	0.954	20.5	0.562	62.5
	d	45.5	0.333	0.970	32.3	0.669	69.4
	ab	47.7	0.389	0.970	32.4	0.669	70.7
	cd	45.4	0.332	0.971	34.0	0.680	70.7
	abcd	45.8	0.365	0.975	38.7	0.708	73.5
AG	*a*	41.2	0.208	0.919	11.3	0.465	52.6
	b	41.6	−0.058	0.904	9.4	0.419	49.2
	c	40.0	−0.026	0.894	8.4	0.392	47.2
	d	36.7	0.074	0.861	6.2	0.324	41.5
	ab	41.2	0.099	0.928	12.9	0.497	55.4
	cd	38.0	0.019	0.898	8.8	0.405	48.2
	abcd	39.4	0.072	0.926	12.5	0.489	55.4

2.3. Variability in Radial Growth among Different Sites

Agreement among ring-width series (single radius and combinations of radii) of different individuals are given in Table 3. Stocks WE and IM show high *Glk* ($p \leq 0.001$) and t_{BP}-scores (close to or higher than 7) for all single radii and calculated means (=*ab*, *cd*, *abcd*). *Glk* values are also significant ($p \leq 0.05$) for all radii among ring-width series at sites WE and AG (mean t_{BP}-score = 2.4). The lowest level of agreement was found among ring-width series of single radii of stock IM and AG (mean t_{BP}-score = 3.6). Radial growth of combinations of radii (i.e., *ab*, *cd*, *abcd*) are statistically significantly different ($p \leq 0.05$) between stock WE and stocks IM and AG, but not between the latter two (Figure 3). Although trends in ring-width series at different sites show a general consistency, they differ strongly with respect to the more recent growth trend. While at site AG there is a decreasing growth trend, constant growth is detectable at site WE, and at site IM *Alnus alnobetula* shows the most steadily increasing growth after the low-growth period occurring around 2007 (Figure 7).

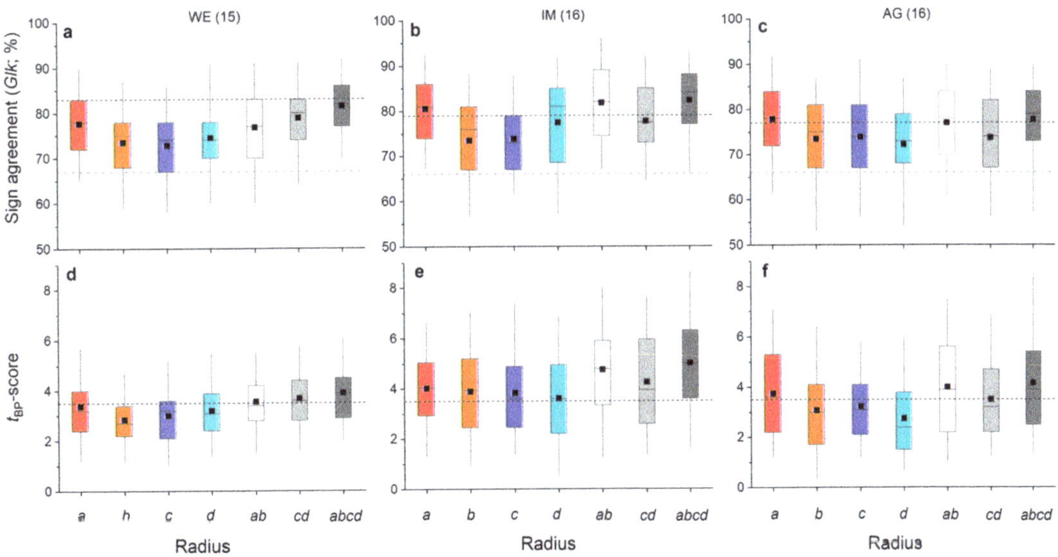

Figure 5. Box-plots showing variability in percentage of sign agreement ((**a–c**); *Glk* = Gleichläufigkeit; Eckstein and Bauch 1973) and t_{BP}-score ((**d–f**); Baillie and Pilcher 1990) among single radii (i.e., *a*, *b*, *c* and *d*), means of opposite radii (i.e., *ab* and *cd*) and mean of all radii (i.e., *abcd*) of shoots taken from one stock at a wind-exposed ridge (site WE), an intermediate site (IM) and a north-facing avalanche gully (site AG). Number of shoots measured per stock are given in brackets. In (**a–c**), horizontal dashed black and grey lines indicate the significance level of $p \leq 0.001$ and $p \leq 0.05$, respectively. In (**d–f**), the horizontal dashed black line indicates a *t*-value of 3.5 (see Section 4 for details). For details on box-plots, see Figure 2.

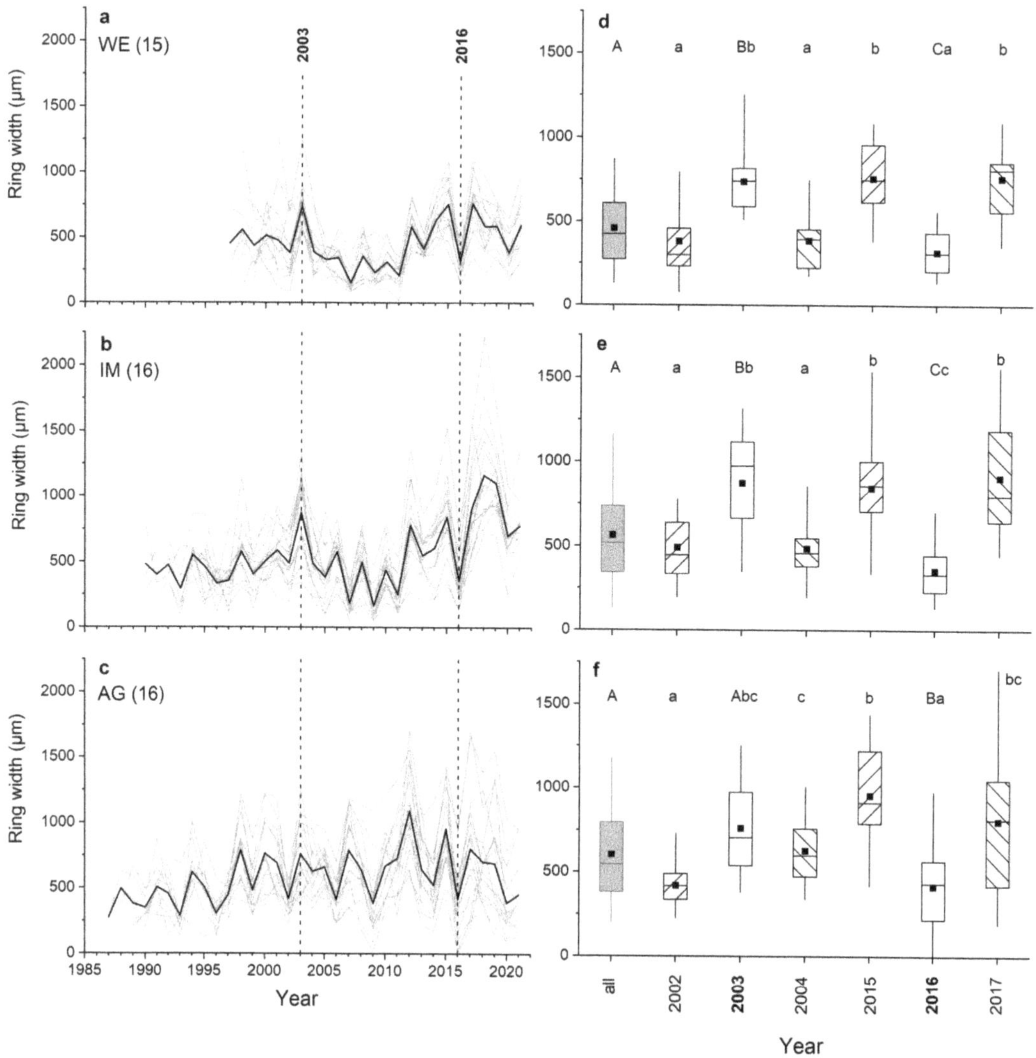

Figure 6. Ring-width series (**a**–**c**) of all radii (i.e., mean values of radii *a*, *b*, *c* and *d*) of all shoots taken from one individual (grey lines; number of shoots measured per stock are given in brackets) at a wind-exposed ridge (site WE), an intermediate site (IM) and a north-facing avalanche gully (site AG). Mean values of all shoots are given as black lines. Vertical dashed lines indicate exemplary extreme growth years: 2003 and 2016 show growth increase and decrease, respectively. Box-plots (**d**–**f**) showing ring-width variability in all years and in 2003 and 2016; the latter two cases include the previous and subsequent year. Statistically significant differences between (i) mean ring widths of all years and extreme growth years (2003, 2016) are indicated by upper case letters, and (ii) mean ring widths of extreme growth years and the preceding and subsequent year by lower case letters ($p \leq 0.001$; Student's t-test). For details on box-plots, see Figure 2.

Table 3. Percentage of sign agreement (*Glk*), statistical significance and t_{BP}-scores (given after) calculated among ring-width series of single radii (radius *a*, *b*, *c*, *d*), mean values of combinations of radii (*a* and *b* = *ab*; *c* and *d* = *cd*) and all radii (=*abcd*) among stocks sampled at a wind-exposed ridge (site WE), an intermediate site (IM) and a north-facing avalanche gully (site AG). Length of time series are given in brackets. * = $p \leq 0.05$; ** = $p \leq 0.01$; *** = $p \leq 0.001$.

	Radius *a*			**Radius *b***			**Radius *c***			**Radius *d***			**Mean *ab***			**Mean *cd***			**Mean *cd***		
	WE	**IM**	**AG**	**WE**	**IM**	**AG**	**WE**	**IM**	**AG**	**WE**	**IM**	**AG**	**WE**	**IM**	**AG**	**WE**	**IM**	**AG**	**WE**	**IM**	**AG**
WE (25)	-	96 ***/8.4	71 */2.3	-	83 ***/7.2	75 **/1.9	-	88 ***/6.7	71 */2.5	-	83 ***/7	81 ***/2.9	-	92 ***/8.2	75 **/2.2	-	88 ***/7	73 */2.7	-	88 ***/7.6	71 */2.4
IM (32)		-	65/3.6		-	68 */3.6		-	68 */3.4		-	66 */3.9		-	68 */3.6		-	66 */3.6		-	68 */3.6
AG (35)			-			-			-			-			-			-			-

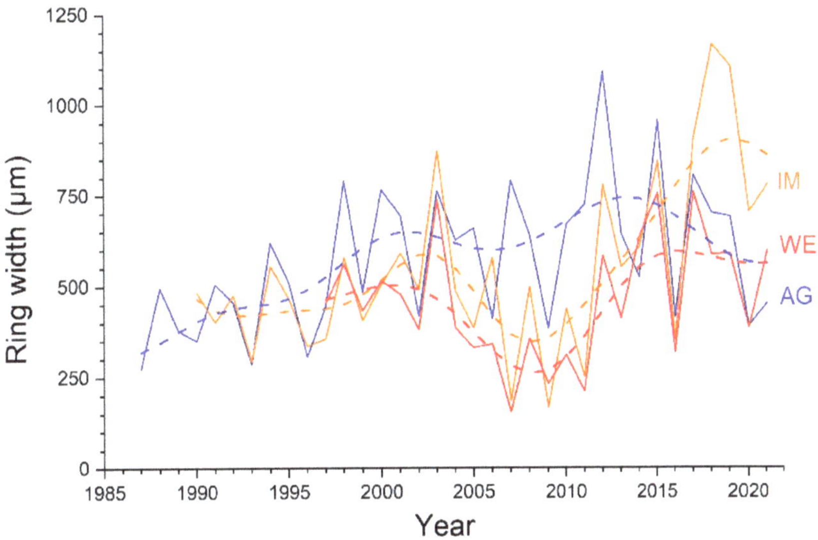

Figure 7. Long-term trend in ring-width series (mean of all radii, i.e., *abcd*) of all shoots of *Alnus alnobetula* sampled at a wind-exposed ridge (site WE), an intermediate site (IM) and a north-facing avalanche gully (site AG). Long-term trends (dashed lines) were calculated by fast Fourier transform low-pass filter (3 yrs).

3. Discussion

The multi-stemmed shrub *Alnus alnobetula* shows clonal growth by layering and root sprouts (e.g., [31]). The vegetative horizontal spread can be triggered by mechanical disturbances, e.g., avalanches, high snow load, wind exposure or livestock grazing. Shoots from one rootstock therefore emerge at intervals and have varying ages (Table 4). The highest variability in cambial age, ranging from 19 to 35 years, was found for shoots sampled in the avalanche track (site AG), indicating that this stock was exposed to the most frequent occurrences of disturbances, which most likely have induced successive sprouting. Extreme environmental conditions are most likely responsible for the significantly lower elongation and diameter growth at the site WE compared to other stocks (Table 4). To be specific, shallow soil depth and strong wind exposure on the south-facing ridge may impair water availability and make shoots of the individual WE also highly susceptible to frost drought during winter [26].

Low autocorrelation in ring-width series of all radii and radius combinations (Table 2) indicates that current climate conditions primarily determine radial growth of this deciduous shrub, which is consistent with a conceptual tree model put forward by Zweifel and Sterck [32] that shows that legacy effects are related to lifetimes of organs and reserves.

3.1. *Variability in Radial Growth and Synchronicity among Radii of One Shoot*

We expected that due to strong bending of shoots causing asymmetric radial growth and eccentric pith, mean ring width would differ between several radii. However, we found that in two of three individuals, only mean ring width of the longest radius (radius *a*) significantly differed from mean ring width of all other radii (Figure 2). In stock AG occurring within an avalanche track, growth was significantly different between several radii, which indicates that under a high disturbance regime (i.e., high and long duration of snow cover), cambial activity along different radii is more strongly influenced by irregular tension and compression load. Because radial growth only takes place during the snow-free period [27], we assume that due to heavy and prolonged snow load, shoots do not fully return to their original position, which accounts for the differences in mean ring width among radii.

Glk and t_{BP}-scores among radii belonging to the same shoot of one individual show very high agreement in year-to-year variability of radial growth independent of radius measured (Figure 4) and thus contradict our hypothesis (H1). Our results suggest that although irregular tension and compression load can affect amount of total annual increment (i.e., ring width of radius *a* is significantly wider than that of other radii), year-to-year variability in ring width of all radii is determined by environmental forcing, most likely climate forcing. Climate factors most strongly related to radial growth of *Alnus alnobetula* within the study area are summer temperature and winter precipitation [26]. Hence, results corroborate our recent finding that climate forcing is the primary determinant of cambial activity in *Alnus alnobetula*, irrespective of differences in physical strain acting on different radial directions, e.g., of the upper and lower side of the stem.

High values of MS (primarily > 40%) as well as statistics of chronology homogeneity of all single radii and combinations of radii found at all sites (Table 2) also indicate a strong and homogenous response of cambial activity to environmental forcing. It is worth noting that ring-width series of radius *a* having the widest ring widths in all individuals and also show the predominantly highest homogeneity statistics compared to other single radii (Table 2). This finding supports the use of the fastest growing radius, mainly occurring on the lower side of the stem, for recording intra-annual stem radius variations by applying point dendrometers. Averaging four radial measurements in the crosswise direction increases statistics of chronology homogeneity compared to radius *a* only slightly. Somewhat lower but still acceptable statistics were found in the individual located in the avalanche track (site AG), indicating that at a high disturbance regime, a decrease in agreement of inter-annual radial growth occurs among samples (here, shoots belonging to the same stock). This finding is well-known in dendrochronological studies by taking care of tree and site selection, i.e., avoidance of undesirable disturbance factors as a principle of sampling (e.g., [33,34]).

3.2. *Variability in Radial Growth and Synchronicity among Shoots Belonging to One Stock*

A high level of agreement among ring-width time series (single and combinations of radii) of shoots belonging to the same stock (Figure 5a–f) indicates strong climate forcing of radial growth, confirming our hypothesis (H2). Ring-width series of the longest radius (=radius *a*) achieve the highest agreement (synchronicity) among shoots in all individuals. That the growth rate affects agreement in year-to-year growth variability is also supported by higher values of *Glk* and t_{BP}-scores at sites IM and AG, showing significantly wider annual increments compared to stock WE. We suggest that under generally growth limiting environmental conditions prevailing at site WE (Table 4), disturbances limited to individual shoots can override climate forcing of radial growth, especially in "average" climate years, leading to lower synchrony in ring-width series.

Although based on an average of four radii measured per stem disc, high variability in radial growth among shoots of one individual shows itself through a large interquartile range, which is especially pronounced in favorable years (Figure 6a–f), confirming our hypothesis (H3). Because in *Alnus alnobetula* a closed canopy growth form is not generally developed, growth variability among shoots of one individual might be explained by influence of small-scale differences in microclimate on cambial activity. That a strong influence of canopy cover on microclimate might occur at the treeline has been demonstrated for deciduous beech (*Fagus sylvatica*) by Rita et al. [35].

During the heat wave that occurred in the extraordinary summer of 2003, when mean summer temperatures exceeded the 1961–1990 mean by c. 3 °C in Central Europe [36], a significant increase in radial growth was detected, which corroborates strong temperature forcing of radial growth (cf. Supplementary Material, Figure S1). In contrast, an exceptional late frost event (daily mean temperatures fell below −10 °C in late April to early May 2016) combined with a cool summer may have resulted in a distinct reduction of radial growth for all individuals in that year (cf. [26]). A significant increase and decrease in radial growth in 2003 and 2016, respectively ($p \leq 0.001$), compared to the previous year are found in all three stocks independent of site conditions (Figure 6d–f). Results indicate that extreme temperature forcing overlays microclimatic conditions and disturbances, which are present on a small scale and cause high growth variability and partially reduced agreement in year-to-year variability of radial growth at selected sites (Table 3).

The first few ring widths near the pith occasionally show reduced agreement among shoots (Figure 6a–c), and growth trends in ring-width series (even when developed from four radii) can be very pronounced and run in opposite directions in ring-width series developed from shoots belonging to the same individual, i.e., wide rings at low cambial age are becoming narrower with increasing age and vice versa. Most likely, the influence of environmental conditions (e.g., duration of snow cover, wind exposure, frost drought) and microtopography differently affect not only growth of young shoots, which are particularly sensitive to mechanical stress (e.g., snow load), but also growth of shoots over longer periods of time. A change in resource allocation over time between aboveground and belowground growth [37] and among shoots belonging to the same stock [21] has also to be considered.

3.3. Variability in Radial Growth and Synchronicity among Stocks

Significantly lower annual increments at site WE compared to sites IM and AG were found (Table 1 and Figure 6d–f). Because the elevational differences among sites are rather small (at most 35 m; Table 4), growth variability can be explained by differences in water and nutrient availability during the growing season. Shallow soil depth and significantly lower soil moisture at site WE (Table 4) compared to other sites supports this assumption. Although the long-term trend in radial growth over the past 30 years is largely consistent among individuals (Figure 7), striking differences in growth rates have occurred in recent years, i.e., growth rates are decreasing at site AG, constant at site WE or increasing at site IM. Changes in disturbance intensity, e.g., duration of snow cover or site-specific impacts of climate warming that affect the growth of single individuals, are likely the cause of temporarily occurring differences in growth trends. Although the basal area increment (BAI) reduces the effects of age and stem size on growth trends and is more closely related to biomass increment than ring width (e.g., [38,39]), BAI could not be adequately determined due to the frequent occurrence of pronounced stem eccentricity. It should also be noted that determining long-term growth trends from tree-ring series is generally a difficult task [40,41].

Highly synchronized radial growth between individuals at sites WE and IM (Table 3) suggests that similar climate variables control radial growth of these individuals. In contrast, inter-annual radial growth among individuals exposed to different aspects shows lower synchronicity, i.e., lower values of statistical parameters (*Glk*, t_{BP}-scores; Table 3). These findings corroborate results of a previous study within the study area, which required the elimination of about a quarter of the measured ring-width series to create a composite chronology out of eight study plots [26]. A high sensitivity of radial growth to site-specific environmental conditions (MS amounts to *c.* 40%; Table 2), which can change rapidly over small spatial scales within the treeline ecotone [27], is most likely responsible for lower agreement in ring-width variations among some individuals. For example, in the period 2005–2007, stock AG shows opposite growth variations compared to stocks WE and IM occurring on a SSE facing slope (Figure 7), which suggests the modulation of regional climate factors through site-specific conditions, e.g., water or nutrient availability, or disturbances such as avalanches (cf. [15]). Hence, our hypothesis (H3) that radial growth (absolute and trend) and year-to-year variations are influenced by prevailing site-specific environmental conditions is largely confirmed by results of this study.

3.4. Recommended Sampling Design for Determining Climate Forcing and Radial Growth Rates of *Alnus alnobetula*

The high agreement (=synchronous growth) among radii of one shoot and among shoots of one individual, which are also exposed to quite different environmental conditions (Table 4), underline the pronounced limitation of radial growth of *Alnus alnobetula* at treeline by climatic factors. Recently, we found that summer temperature and winter precipitation are the main climate drivers of radial stem growth of this tall shrub within the study area [26]. On the other hand, site-specific environmental conditions (e.g., water and nutrient availability) and/or disturbance factors (e.g., high snow load, wind exposure) cause a decrease in synchronicity among ring-width series developed at different sites ([26]; this study). Therefore, site-specific variability in climate-growth relationships of *Alnus alnobetula* must be considered in topographically heterogenous environments, which led us to the following recommendations: measuring 1 radius per shoot (preferably the one with the longest radius), 3 shoots per individual, and c. 10 individuals within 1 stand, in which individuals are exposed to the same (i.e., homogeneous) environmental conditions (total c. 30 radii per study plot). Note that the number of individuals to be sampled within one plot depends on the variability in radial growth among individuals, which we recommend determining in advance.

However, the determination of absolute growth rates as well as growth trends (e.g., with respect to recent climate warming) is strongly hampered by the pronounced variability in ring width (i) among radii of one shoot, (ii) among shoots of one individual and (iii) among individuals exposed to different environmental conditions and/or disturbance regimes. To be able to take this growth variability into account, the following recommendation is made: measuring 4 radii per shoot (preferably in crosswise-oriented radii along the longest and shortest diameter and excluding stem discs showing extreme abnormal growth), $\geq$5 shoots per individual, $\geq$10 individuals within 1 stand, in which individuals are exposed to the same (i.e., homogeneous) environmental conditions (total c. 200 radii per study plot). At sites characterized by a high frequency of disturbances (e.g., in highly active avalanche tracks or grazed areas), a higher number of shoots and individuals is required to account for anomalous growth patterns, discontinuous rings and extreme asymmetric growth.

3.5. Recommendations for Determining Intra-Annual Radial Growth

Our results are also relevant for studies focusing on intra-annual dynamics of radial growth of tall shrubs by applying dendrometers, which allow determination of (i) key phenological stages of cambial activity (i.e., onset, end, time of maximum growth and growth duration) and (ii) influence of seasonal and short-term extreme weather conditions on radial growth (e.g., [42–44]). Due to pronounced variability in growth rates among shoots sprouting from one individual, $\geq$5 shoots per individual should be equipped with appropriate measuring devices. Preliminary point dendrometer records gathered from

Alnus alnobetula within the study area confirmed the high growth variability among shoots found in this study, and also revealed pronounced temporal differences (up to three weeks) in key phenological dates of radial growth, demonstrating the high sensitivity of cambial activity to microclimate in this species. Thus, in future studies we suggest the use of band dendrometers instead of point dendrometers, as the former integrate growth over the entire stem circumference. On the other hand, we suggest that point dendrometers be mounted on the radius showing the highest growth rate (usually at the lower side of the stem), because inter-annual agreement in radial growth is related to growth rate.

4. Materials and Methods

4.1. Study Area and Site Selection

The study area is within the treeline ecotone of Mt. Patscherkofel (2246 m asl) in the Central European Alps (Tyrol, Austria; 47°12′ N, 11°27′ E) where *Alnus alnobetula* stands are developed between c. 1950 and 2200 m asl primarily in leeward avalanche tracks but also on south- to southeast-facing wind-exposed slopes. The geology of the study area is dominated by gneisses and schists [45]. According to the World Base for Soil Resources [46], the soil type within the study area is classified as haplic podzol [47]. The local climate is characterized by the frequent occurrence of strong southerly foehn winds [48]. During the period 1991–2020, the mean annual temperature was 0.8 ± 0.7 °C and the coldest and warmest months were February (−6.6 °C) and July (8.9 °C), respectively (Supplementary Material, Figure S1). The mean annual precipitation during this time period was 889 mm, with the majority falling during summer (June–August; 371 mm), while the winter (December–February) was the driest season (134 mm). Snow depth shows pronounced spatial variation due to irregular distribution caused by strong winds. While at south-facing wind-exposed slopes snow depth hardly is >1m and only small patches of snow exist till early April, on leeward north-facing slopes (particularly in avalanche tracks) snow height can reach up to 3 m and a permanent snow cover may persist till late spring and into early June.

4.2. Sampling and Growth Ring Measurements

In order to cover a large range of environmental conditions, we selected three *Alnus alnobetula* stocks exposed to different site conditions and disturbance regimes between 2115 to 2150 m asl (Table 4, Supplementary Material Figure S2): a north-facing avalanche gully (hereafter referred to as the AG site or AG stock), a wind-exposed ridge (hereafter WE site or WE stock) and an intermediate site (hereafter IM site or IM stock). At each site, soil moisture at 10 cm soil depth was recorded every 30 min during July and August 2022 using four ThetaProbe ML2 devices (Delta-T Devices Ltd., Burwell, UK). Snow load and water availability were at maximum and minimum at the AG site and at the WE site, respectively, and were intermediate at the IM site (Table 4). Accordingly, stem bending and growth were highest at the AG site and lowest at the WE site.

Table 4. Description of selected study sites (WE = wind-exposed ridge; IM = intermediate site; AG = avalanche gully; asl = above sea level). Statistically significant differences of mean values among sites are indicated by different letters ($p \leq 0.01$; Student's *t*-test).

Site	Elevation (m asl)	Aspect	Topography	Slope (°)	Soil Depth (cm)	Soil Moisture (Vol.%)
WE	2150	SSE	Ridge	5	5–10	11.5 ± 3.0^{a}
IM	2140	SSE	Slope	10	10–15	20.1 ± 2.7^{b}
AG	2115	N	Avalanche gully	20	NA [1]	23.0 ± 3.2^{c}

[1] Boulder debris impaired the correct determination of soil depth.

Special care was taken to ensure that selected shoots belonged to one stock, i.e., one individual, and that no vegetative links existed to other stocks nearby (cf. [31]). Around the rootstock, i.e., from all cardinal directions, we sampled 20 shoots from each stock (i.e., about half of all shoots sprouting from one stock). All shoots of one stock were taken within a radius of no more than 50 cm from the center. Because the lifespan of shoots is less than 50 years within the study area [26], stem discs were cut using a folding saw from the basal part of the shoot to allow the development of the longest possible ring-width series. After an initial visual inspection of stem cross sections, 4–5 stem discs sampled from each individual were discarded due to expression of extreme anomalous growth patterns, e.g., multisectoral eccentricity, lobed growth or high number of wedging rings. The remaining stem discs were air-dried, and the surface was prepared with a sharp razor blade and treated with wood stain to make the annual ring boundaries more clearly visible.

Particularly at the stem base, the arched ascending growth form of shoots of *Alnus alnobetula* cause eccentricity of the pith and elliptical-shaped annual increments. Therefore, radial growth was measured along four radii from the pith towards the cambium (Figure 1) including the longest and shortest radius, which were most frequently occurring at the lower (radius *a*) and upper (radius *b*) stem side, respectively. Additionally, radial growth along two radii (radius *c* and *d*) oriented orthogonal to the longest stem diameter were measured. Annual increments were measured along each radius to 1 µm precision using the LINTAB measuring system (Rinn, Heidelberg, Germany). A total of 188 radii were surveyed. Correct dating of ring-width time series (identifying missing or false rings) was checked by visually cross-dating all radii of one shoot, whereby special attention was given to the presence of extreme growth years, and finally using COFECHA software [49]. In a few cases it was necessary to insert a missing ring into the ring-width series (value of zero). Based on cross-dated ring-width time series measured from one stem disc, three radial combinations, i.e., averages of $\geq$2 radii, were calculated: mean of radii *a* and *b* (=*ab*), mean of radii *c* and *d* (=*cd*) and mean of all four radii (=*abcd*; Figure 1).

4.3. Statistical Analysis

Two different cross-dating methods were applied to quantify the agreement of the year-to-year ring-width changes among time series. Changes in ring width from year to year can be simplified to a binary variable (increase or decrease in ring width from one year to the next) and agreement can be quantified non-parametrically by counting the number of agreements and disagreements. Expressed as percentage of agreement, this is referred to as percentage of parallel variation [50] or "Gleichläufigkeit" (*Glk*; [51]). Agreement among ring-width values was also quantified parametrically using the product–moment correlation coefficient *r*, which in turn was adjusted for the amount of overlap between ring-width series using the standard *t*-statistic [52]. The value of *t* is defined as:

$$t = \frac{r \cdot \sqrt{N-2}}{\sqrt{1-r^2}}$$

where *r* is the product–moment correlation coefficient and *N* is the amount of overlap between ring-width series.

A t_{BP}-score of 3.5 is regarded as an acceptable agreement among ring-width series [50,52,53], and this t_{BP}-score is also employed as a minimum for dating control in the software COFECHA. The value of *t* cannot be converted to a corresponding significance level, because autocorrelation is not considered [54] and cross-dating involves a high degree of multiplicity [55].

To determine the high-frequency (i.e., annual) agreement among ring-width series, age/size-related growth trends and other transient disturbance pulses had to be eliminated [55]. All ring-width series (i.e., ring-width series developed from single radii and all combinations of radii) were standardized to dimensionless radial growth indices by the ARSTAN program [56]. A conservative detrending method, i.e., a linear regression of negative slope or a negative exponential decline, was fitted to each ring-width series. Dimensionless radial growth indices were formed by dividing the observed ring-width value by the predicted ring-width value. Several statistics were then calculated for the standardized ring-width series. Mean sensitivity (MS) is a measure of the mean relative change between adjacent ring widths [33]. A high MS indicates that radial growth is highly responsive to the environment. The first-order autocorrelation (AC) assesses relationships with prior growth. The strength and confidence of ring-width series were estimated by calculating the signal-to-noise ratio (S/N-ratio) and the expressed population signal (EPS). The S/N-ratio is an expression of the strength of the observed common signal [57] and is defined as:

$$S/N = N * r/(1 - r)$$

where r is the average correlation between radii and N is the number of radii. The mean correlation technique was applied to estimate the chronology signal strength (EPS), which quantifies the degree to which a given sample chronology reflects the hypothetically perfect chronology. Though a specific range of EPS values, which constitute acceptable statistical quality, cannot be given, Wigley et al. [57] suggested a threshold value of 0.85 as reasonable. Correlation matrices were calculated for the maximum period common to a specific radius and the mean inter-series correlation coefficient (Rbar) estimated according to formulae given in Briffa and Jones [58]. ARSTAN also calculates the common variance among ring-width series using principal component analysis [59]. Higher common variance accounted for by the first principal component (or eigenvector, EV) indicates a greater climatic influence on growth [33,58].

Student's independent sample t-test was applied to determine significant differences among growth characteristics (shoot length, diameter and cambial age) of different stocks, and ring-width time series of single radii or combinations of radii. All the statistical analyses were performed using the STATISTICA software package (version 13.5.0.17; TIBCO Software Inc., Palo Alto, CA, USA).

5. Conclusions

We can summarize that at the alpine treeline, climate factors are the primary drivers of radial growth in the tall multi-stemmed shrub *Alnus alnobetula* leading to a high agreement among ring-width series developed from different radii, shoots and individuals, while differences in compressive and tensile forces and particularly variation in microsite conditions determine absolute growth rates and long-term growth trends. Thus, a high agreement of inter-annual radial growth contrasts with a high variability of growth rates, indicating that the research focus is also of central importance for the development of a suitable sampling design for dendroclimatological and ecological studies of tall multi-stemmed shrubs (*cf.* [21,60]).

Supplementary Materials: The following supporting information can be downloaded at: https://www.mdpi.com/article/10.3390/plants12081708/s1. Figure S1: Mean air temperature and precipitation during June–August and January–December recorded during 1991–2020 at Mt. Patscherkofel (2246 m asl). Figure S2: Green alder (*Alnus alnobetula*) individual located within an avalanche gully.

Author Contributions: Conceptualization, W.O., G.W. and A.G.; methodology, W.O. and A.G.; investigation, A.-L.D., T.H. and F.S.; data curation, W.O.; writing—original draft preparation, W.O.; writing—review and editing, W.O. and G.W.; supervision, W.O.; project administration, W.O.; funding acquisition, W.O., G.W. and A.G. All authors have read and agreed to the published version of the manuscript.

Funding: This research was funded by the Austrian Science Fund (FWF) [P34706-B]. For the purpose of open access, the authors have applied a CC BY public copyright licence to any Author Accepted Manuscript version arising from this submission.

Data Availability Statement: The ring-width data presented in this study are openly available in Zenodo at 10.5281/zenodo.7708713.

Acknowledgments: Open Access Funding by the Austrian Science Fund (FWF).

Conflicts of Interest: The authors declare no conflict of interest.

References

1. Hallinger, M.; Manthey, M.; Wilmking, M. Establishing a missing link: Warm summers and winter snow cover promote shrub expansion into alpine tundra in Scandinavia. *New Phytol.* **2010**, *186*, 890–899. [CrossRef] [PubMed]
2. Myers-Smith, I.H.; Forbes, B.C.; Wilmking, M.; Hallinger, M.; Lantz, T.; Blok, D.; Tape, K.D.; Macias-Fauria, M.; Sass-Klaassen, U.; Lévesque, E.; et al. Shrub expansion in tundra ecosystems: Dynamics, impacts and research priorities. *Environ. Res. Lett.* **2011**, *6*, 045509. [CrossRef]
3. Frost, G.V.; Epstein, H.E. Tall shrub and tree expansion in Siberian tundra ecotones since the 1960s. *Glob. Chang. Biol.* **2014**, *20*, 1264–1277. [CrossRef] [PubMed]
4. Anthelme, F.; Michalet, R.; Barbaro, L.; Brun, J.-J. Environmental and spatial influences of shrub cover (*Alnus viridis* DC.) on vegetation diversity at the upper treeline in the inner Western Alps. *Arct. Antarct. Alp. Res.* **2003**, *35*, 48–55. [CrossRef]
5. Sturm, M.; Douglas, T.; Racine, C.; Liston, G.E. Changing snow and shrub conditions affect albedo with global implications. *J. Geophys. Res. Biogeosci.* **2005**, *110*, G01004. [CrossRef]
6. Hiltbrunner, E.; Aerts, R.; Bühlmann, T.; Huss-Danell, K.; Magnusson, B.; Myrold, D.D.; Reed, S.C.; Sigurdsson, B.D.; Körner, C. Ecological consequences of the expansion of N_2-fixing plants in cold biomes. *Oecologia* **2014**, *176*, 11–24. [CrossRef] [PubMed]
7. Bühlmann, T.; Körner, C.; Hiltbrunner, E. Shrub expansion of *Alnus viridis* drives former montane grassland into nitrogen saturation. *Ecosystems* **2016**, *19*, 968–985. [CrossRef]
8. Caviezel, C.; Hunziker, M.; Kuhn, N.J. Green alder encroachment in the European Alps: The need for analyzing the spread of a native-invasive species across spatial data. *Catena* **2017**, *159*, 149–158. [CrossRef]
9. Anthelme, F.; Villaret, J.; Brun, J.-J. Shrub encroachment in the Alps gives rise to the convergence of sub-alpine communities on a regional scale. *J. Veg. Sci.* **2007**, *18*, 355–362. [CrossRef]
10. David, F. Expansion of green alder (*Alnus alnobetula* Ehrh K. Koch) in the northern French Alps: A palaeoecological point of view. *Comptes Rendus Biol.* **2007**, *333*, 424–428. [CrossRef]
11. Huber, B.; Frehner, M. Die Verbreitung und Entwicklung der Grünerlenbestände in der Ostschweiz. *Schweizerische Zeitschrift für Forstwesen* **2013**, *164*, 87–94. [CrossRef]
12. Wettstein, S. Der Einfluss abiotischer Faktoren auf die Morphologie der Grünerle. *Bot. Helv.* **2001**, *111*, 31–44.
13. Brüchert, F.; Gallenmüller, F.; Bogenrieder, A.; Speck, T. Stem mechanics, functional anatomy and ecology of *Alnus viridis* and *Alnus glutinosa*. *Feddes Repertorium* **2003**, *114*, 181–197. [CrossRef]
14. Lantz, T.C.; Gerge, S.E.; Henry, G.H.R. Response of green alder (*Alnus viridis* subsp. *fruticosa*) patch dynamics and plant community composition to fire and regional temperature in north-western Canada. *J. Biogeogr.* **2010**, *37*, 1597–1610. [CrossRef]
15. Tape, K.D.; Hallinger, M.; Welker, J.M.; Ruess, R.W. Landscape heterogeneity of shrub expansion in arctic Alaska. *Ecosystems* **2012**, *15*, 711–724. [CrossRef]
16. Salmon, V.G.; Breen, A.L.; Kumar, J.; Lara, M.J.; Thornton, P.E.; Wullschleger, S.D.; Iversen, C.M. Alder distribution and expansion across a tundra hillslope: Implications for local N cycling. *Front. Plant. Sci.* **2019**, *10*, 1099. [CrossRef]
17. Wiedmer, E.; Senn-Irlet, B. Biomass and primary productivity of an *Alnus viridis* stand—A case study from the Schächental valley, Switzerland. *Bot. Helv.* **2006**, *116*, 55–64. [CrossRef]
18. Gehrig-Fasel, J.; Guisan, A.; Zimmermann, N.E. Tree line shifts in the Swiss Alps: Climate change or land abandonment? *J. Veg. Sci.* **2007**, *18*, 571–582. [CrossRef]
19. Bühlmann, T.; Hiltbrunner, E.; Körner, C. *Alnus viridis* expansion contributes to excess reactive nitrogen release, reduces biodiversity and constrains forest succession in the Alps. *Alp. Bot.* **2014**, *124*, 187–191. [CrossRef]
20. Mitka, J.; Kucharzyk, S.; Capelo, J.; Stachurska-Swakoń, A. Subalpine woody vegetation in the Eastern Carpathians after release from agropastoral pressure. *Sci. Rep.* **2022**, *12*, 17897. [CrossRef]
21. Myers-Smith, I.H.; Hallinger, M.; Blok, D.; Sass-Klaassen, U.; Rayback, S.A.; Weijers, S.; Trant, A.J.; Tape, K.D.; Naito, A.T.; Wipf, S.; et al. Methods for measuring arctic and alpine shrub growth: A review. *Earth Sci. Rev.* **2015**, *140*, 1–13. [CrossRef]
22. Forbes, B.C.; Macias-Fauria, M.; Zetterberg, P. Russian Arctic warming and 'greening' are closely tracked by tundra shrub willows. *Glob. Change Biol.* **2010**, *16*, 1542–1554. [CrossRef]
23. Macias-Fauria, M.; Forbes, B.C.; Zetterberg, P.; Kumpula, T. Eurasian Arctic greening reveals teleconnections and the potential for structurally novel ecosystems. *Nat. Clim. Chang.* **2012**, *2*, 613–618. [CrossRef]
24. Jørgensen, R.H.; Hallinger, M.; Ahlgrimm, S.; Friemel, J.; Kollmann, J.; Meilby, H. Growth response to climatic change over 120 years for *Alnus viridis* and *Salix glauca* in West Greenland. *J. Veg. Sci.* **2015**, *26*, 155–165. [CrossRef]
25. Dearborn, K.D.; Danby, R.K. Topographic influences on ring widths of trees and shrubs across alpine treelines in southwest Yukon. *Arct. Antarct. Alp. Res.* **2018**, *50*, e1495445. [CrossRef]
26. Oberhuber, W.; Wieser, G.; Bernich, F.; Gruber, A. Radial stem growth of the clonal shrub *Alnus alnobetula* at treeline is constrained by summer temperature and winter desiccation and differs in carbon allocation strategy compared to co-occurring *Pinus cembra*. *Forests* **2022**, *13*, 440. [CrossRef]
27. Körner, C. *Alpine Plant Life. Functional Plant Ecology of High Mountain Ecosystems*; Springer: Heidelberg, Germany, 2021.
28. Kolishchuk, V.G. Dendroclimatological study of prostrate woody plants. In *Methods of Dendrochronology. Applications in the Environmental Sciences*; Cook, E.R., Kairiukstis, L.A., Eds.; Kluwer Academic: Dordrecht, The Netherlands, 1990; pp. 51–55.
29. Bär, A.; Bräuning, A.; Löffler, J. Dendroecology of dwarf shrubs in the high mountains of Norway—A methodological approach. *Dendrochronologia* **2006**, *24*, 17–27. [CrossRef]

30. Buras, A.; Wilmking, M. Straight lines or eccentric eggs? A comparison of radial and spatial ring width measurements and its implications for climate transfer functions. *Dendrochronologia* **2014**, *32*, 313–326. [CrossRef]
31. Anthelme, F.; Cornillon, L.; Brun, J.-J. Secondary succession of *Alnus viridis* (Chaix) DC. in Vanoise National Park, France: Coexistence of sexual and vegetative strategies. *Ann. For. Sci.* **2002**, *59*, 419–428. [CrossRef]
32. Zweifel, R.; Sterck, F. A conceptual tree model explaining legacy effects on stem growth. *Front. For. Glob. Change* **2018**, *1*, 9. [CrossRef]
33. Fritts, H.C. *Tree Rings and Climate*; Academic Press: London, UK, 1976.
34. Speer, J.H. *Fundamentals of Tree-Ring Research*; University of Arizona Press: Tucson, AZ, USA, 2010.
35. Rita, A.; Bonanomi, G.; Allevato, E.; Borghetti, M.; Cesarano, G.; Mogavero, V.; Rossi, S.; Saulino, L.; Zotti, M.; Saracino, A. Topography modulates near-ground microclimate in the Mediterranean Fagus sylvatica treeline. *Sci. Rep.* **2021**, *11*, 8122. [CrossRef] [PubMed]
36. Beniston, M. The 2003 heat wave in Europe: A shape of things to come? An analysis based on Swiss climatological data and model simulations. *Geophys. Res. Lett.* **2004**, *31*, 2022–2026. [CrossRef]
37. Buchwal, A.; Rachlewicz, G.; Fonti, P.; Cherubini, P.; Gärtner, H. Temperature modulates intra-plant growth of *Salix polaris* from a high Arctic site (Svalbard). *Polar Biol.* **2013**, *36*, 1305–1318. [CrossRef]
38. LeBlanc, D.C. Relationships between breast-height and whole-stem growth indices for red spruce on Whiteface mountain, New York. *Can. J. For. Res.* **1990**, *20*, 1399–1407. [CrossRef]
39. Biondi, F.; Qeadan, F. A theory-driven approach to tree-ring standardization: Defining the biological trend from expected basal area increment. *Tree Ring Res.* **2008**, *64*, 81–96. [CrossRef]
40. Bowman, D.M.; Brienen, R.J.M.; Gloor, E.; Phillips, O.L.; Prior, L.D. Detecting trends in tree growth: Not so simple. *Trends Plant Sci.* **2013**, *18*, 11–17. [CrossRef]
41. Peters, R.L.; Groenendijk, P.; Vlam, M.; Zuidema, P.A. Detecting long-term growth trends using tree-rings: A critical evaluation of methods. *Glob. Change Biol.* **2015**, *21*, 2040–2054. [CrossRef] [PubMed]
42. Rossi, S.; Deslauriers, A.; Anfodillo, T.; Morin, H.; Saracino, A.; Motta, R.; Borghetti, M. Conifers in cold environments synchronize maximum growth rate of tree-ring formation with day length. *New Phytol.* **2006**, *170*, 301–310. [CrossRef]
43. Deslauriers, A.; Rossi, S.; Anfodillo, T. Dendrometer and intra-annual tree growth: What kind of information can be inferred? *Dendrochronologia* **2007**, *25*, 113–124. [CrossRef]
44. Gruber, A.; Zimmermann, J.; Wieser, G.; Oberhuber, W. Effects of climate variables on intra-annual stem radial increment in *Pinus cembra* (L.) along the alpine treeline ecotone. *Ann. For. Sci.* **2009**, *66*, 503. [CrossRef]
45. Tollmann, A. *Geologie von Österreich, Band 1. Die Zentralalpen*; Deuticke: Wien, Austria, 1977.
46. FAO; ISRIC; ISSS. *World Reference Base for Soil Resources*; FAO: Rome, Italy, 1998.
47. Neuwinger, I. Böden der subalpinen und alpinen Stufe in den Tiroler Alpen. *Mitteilungen der Ostalpin-Dinarischen Gesellschaft für Vegetationskunde* **1970**, *11*, 135–150.
48. Richner, H.; Hächler, P. Understanding and forecasting alpine foehn. In *Mountain Weather Research and Forecastings*; Chow, F.K., De Wekker, S.F.J., Snyder, B.J., Eds.; Springer Atmospheric Sciences, Springer: Dordrecht, The Netherlands, 2013; pp. 219–260.
49. Holmes, R.L. Computer-assisted quality control in tree-ring dating and measurement. *Tree Ring Bull.* **1983**, *43*, 69–78.
50. Baillie, M.G.L. *Tree-Ring Dating and Archaeology*; Routledge: London, UK, 1982.
51. Eckstein, D.; Bauch, J. Beitrag zur Rationalisierung eines dendrochronologischen Verfahrens und zur Analyse seiner Aussagesicherheit. *Forstwissenschaftliches Centralblatt* **1969**, *88*, 230–250. [CrossRef]
52. Baillie, M.G.L.; Pilcher, J.R. A simple cross-dating program for tree-ring research. *Tree Ring Bull.* **1973**, *33*, 7–14.
53. Wigley, T.M.L.; Jones, P.D.; Briffa, K.R. Cross-dating methods in dendrochronology. *J. Archaeol. Sci.* **1987**, *14*, 51–64. [CrossRef]
54. Pilcher, J.R.; Baillie, M.G.L.; Schmidt, B.; Becker, B. A 7272-year tree-ring chronology from Western Europe. *Nature* **1984**, *312*, 150–152. [CrossRef]
55. Cook, E.; Kairiukstis, L. (Eds.) *Methods of Dendrochronology: Applications in the Environmental Sciences*; Kluwer Academic: Dordrecht, The Netherlands, 1990.
56. Cook, E.R.; Holmes, R.L. *User Manual for Program ARSTAN*; University of Arizona, Laboratory of Tree Ring Research: Tucson, AZ, USA, 1984.
57. Wigley, T.M.L.; Briffa, K.R.; Jones, P.D. On the average value of correlated time series with applications in dendroclimatology and hydrometeorology. *J. Clim. Appl. Meteor.* **1984**, *23*, 201–213. [CrossRef]
58. Briffa, K.R.; Jones, P.D. Basic chronology statistics and assessment. In *Methods of Dendrochronology: Applications in the Environmental Sciences*; Cook, E.R., Kairiukstis, L.A., Eds.; Kluwer Academic: Dordrecht, The Netherlands, 1990; pp. 137–152.
59. Jolliffee, I.T. *Principal Component Analysis*; Springer Series in Statistics; Springer: New York, NY, USA, 2002.
60. Nehrbass-Ahles, C.; Babst, F.; Klesse, S.; Nötzli, M.; Bouriaud, O.; Neukom, R.; Dobbertin, M.; Frank, D. The influence of sampling design on tree-ring-based quantification of forest growth. *Glob. Change Biol.* **2014**, *20*, 2867–2885. [CrossRef]

Article

Characterization of *Polylepis tarapacana* Life Forms in the Highest-Elevation Altiplano in South America: Influence of the Topography, Climate and Human Uses

Victoria Lien López [1,2,*], Lucia Bottan [3], Guillermo Martínez Pastur [3,*], María Vanessa Lencinas [3], Griet An Erica Cuyckens [4] and Juan Manuel Cellini [1]

1 Laboratorio de Investigaciones en Maderas (LIMAD), Facultad de Ciencias Agrarias y Forestales (FCAyF), Universidad Nacional de La Plata (UNLP), La Plata 1900, Argentina; jmc@agro.unlp.edu.ar
2 CCT-La Plata (CONICET), La Plata 1900, Argentina
3 Centro Austral de Investigaciones Científicas (CADIC), Consejo Nacional de Investigaciones Científicas y Te\u0301cnicas (CONICET), Houssay 200, Ushuaia 9410, Argentina; lucia.bottan@agro.unlp.edu.ar (L.B.)
4 Instituto de Ecorregiones Andinas (INECOA-CONICET-UNJu) y Centro de Estudios Territoriales Ambientales y Sociales (CETAS-UNJu), Alberdi 47, San Salvador de Jujuy 4600, Argentina
* Correspondence: victoria.lopez@agro.unlp.edu.ar (V.L.L.); gpastur@conicet.gov.ar (G.M.P.); Tel.: +54-(29941)-539503 (V.L.L.)

Abstract: In the upper vegetation limit of the Andes, trees change to shrub forms or other life forms, such as low scrubs. The diversity of life forms decreases with elevation; tree life forms generally decrease, and communities of shrubs and herbs increase in the Andean highlands. Most of treeline populations in the northwestern Argentina Altiplano are monospecific stands of *Polylepis tarapacana*, a cold-tolerant evergreen species that is able to withstand harsh climatic conditions under different life forms. There are no studies for *P. tarapacana* that analyze life forms across environmental and human impact gradients relating them with environmental factors. This study aims to determine the influence of topographic, climatic, geographic and proxies to human uses on the occurrence of life forms in *P. tarapacana* trees. We worked with 70 plots, and a new proposal of tree life form classification was presented for *P. tarapacana* (arborescent, dwarf trees, shrubs and brousse tigrée). We describe the forest biometry of each life form and evaluate the frequency of these life forms in relation to the environmental factors and human uses. The results show a consistency in the changes in the different life forms across the studied environmental gradients, where the main changes were related to elevation, slope and temperature.

Keywords: tree growth form; stem; multi-stemmed; scrublands; scrub; tree; Argentinean highlands; high Andean vegetation

Citation: López, V.L.; Bottan, L.; Martínez Pastur, G.; Lencinas, M.V.; Cuyckens, G.A.E.; Cellini, J.M. Characterization of *Polylepis tarapacana* Life Forms in the Highest-Elevation Altiplano in South America: Influence of the Topography, Climate and Human Uses. *Plants* **2023**, *12*, 1806. https://doi.org/10.3390/plants12091806

Academic Editor: Gregor Kozlowski

Received: 31 March 2023
Revised: 25 April 2023
Accepted: 26 April 2023
Published: 28 April 2023

1. Introduction

Life forms are functional types that have been used to describe the adaptation of vegetative structures of plants to certain ecological conditions. They refer to the general physiognomy of the tree or the common habit of the individual [1,2], and are used to interpret the functionality of vegetation [3,4] as well as to group individuals with similar morphologies [5], which allows comparisons between floras from different regions [4]. Solbrig [6] described life forms as functional groups based on a single character and Raunkiær expressed that, in response to harsh environments, plants can develop adaptations that allow the protection of renewing shoots [7].

In the upper limit of vegetation, trees change their growth strategy to shrub or other life forms, such as low scrubs [8]. These ecosystems are characterized by a marked thermal seasonality defined by long winters and short growing seasons, which greatly influence plant growth. Trees are subjected to the synergistic actions of stress, due to extreme fluctuations in temperature, drought, nutrient limitations and high levels of radiation [9,10].

Tree distribution boundaries are rarely sharp, and the transition from tree to shrub-only stages may be fragmented and stretched over a few meters in steep terrains or over many kilometers in flat terrains [8].

The topography, soils and the degree of human disturbance can modify the distribution and structure of arid and semi-arid vegetation [11]. The diversity of life forms and tree growth forms decreases with elevation; tree life forms generally decrease while shrubs and herbs increase in the Andean highlands [4]. Elevation represents the best example of an environmental gradient, where resources change due to a complex combination of climatic factors (e.g., temperature, rainfall, soils and substrate stability) [12], and it is a decisive factor that shapes the spatial patterns of plant life forms [13]. Therefore, the identification and estimation of different tree life forms are relevant for the evaluation of ecosystem structure and function [14,15].

In the Andean regions at the tree line, trees show a series of morphological and physiological responses to face extreme low temperatures and other environmental stresses, such as drought (caused by greater levels of evaporation), lower rainfall and higher solar radiation [16,17]. The response to this type of environmental stress is a reduction in the aboveground biomass, manifested as a decrease in tree height [18]. In this area, some individuals with unbranched or slightly branched woody stems have different tree life forms [19], standing out in places with little vegetation that does not exceed 50 cm in height [20]. These observations have generated greater interest in the adaptive importance of tree forms in the highlands [21]. In the Andean forests, there are individuals that growth in prostate or creeping forms, due to the effect of wind and snow, generating crooked trunks [19]. This type of growth is described for *Nothofagus pumilio* (Poepp. *et* Endl.) Krasser and *Austrocedrus chilensis* (D. Don) Pic. Serm. and Bizzarri, where low temperatures and drought are the limiting factors that condition the growth and tree development [22], as well as for *Polylepis tarapacana* Phil. forests in the Chilean highlands [19].

In the highlands of the South American Andes, *P. tarapacana* forests present differences in forest structure at different elevations and climate [23]. This species is sensitive to soil moisture content and, in response to drought, it can distribute its biomass in multiple individual stems with smaller diameters and heights to conserve the available moisture of the plant [19,21].

Rios [24] and Saavedra [19] described two life forms for *P. tarapacana* in Chilean forests: (i) single stem trees and (ii) multi-stemmed trees. These authors calculated the frequency of each tree life form, but they did not assess its relationship with environmental gradients or past human uses. Due to the correlation between the environment and the structure of these forests [21,23], it is expected that there is some influence of environmental factors on tree life forms [13]. Therefore, there is a need to review the classification developed by Ríos [24] and Saavedra [19], and propose a broader classification of tree life forms based on those described for different tree-line life forms [25–32].

Tree life form classifications, which can better present the tree line [26], are: (i) "*Treelets*", or trees shorter than 5–10 m, and (ii) certain scrubby but very tree-like forms of 1–5 m, called "dwarf-trees". The noun 'arborescent' is used to refer to woody plants that branch near the ground level. (iii) "*Arborescents*" are thus an intermediate form between trees and shrubs. (iv) "*Krummholz*" is a wind-stunted woody scrub occurring primarily at the tree-line and other exposed sites in mountains. (v) "*Shrubs*" are woody plants with multiple stems arising from the ground level. (vi) "*Dwarf shrubs*" are woody, generally small-leaved chamaephytes, not taller than 0.50 m, and more commonly 0.30 m. (vii) "*Cushion shrubs*" are small krummholz forms occurring near to constantly desiccating, often cold winds, with a completely limited vertical branch extension, resulting in an extremely dense mass of short branches forming a flat or rounded shape. Likewise, (viii) "*brousse tigrée*" (tiger bush) appear on slope gradient bands of vegetation formed perpendicular to this direction [26,29–32].

There are no studies that relate *P. tarapacana* life forms with environmental and human factors along elevation gradients [33], which is a topic of great ecological importance as a species adapted to very restrict environmental conditions. Moreover, it is considered a Near Threatened species [34] due to human impact, which mainly led to habitat degradation caused especially by the extraction of wood for fuel and construction in the area [35]. Some studies have attempted to make comparisons among plant life forms in the Altiplano region, but they always study different species in a community [13,36] and are not focused on a single species that might present different life forms. These studies are needed to assess the variability in tree life form composition within the region. There are also some intriguing differences in this region that deserve attention. For example, where are the tree life forms with only one trunk found? Is it more frequent to find multi-stemmed individuals? Does the extraction and use of firewood and poles modify the frequency of single-trunk life forms in the landscape? This and other similar questions are based on more or less anecdotal evidence of distribution patterns, which generally relate to forest structure data rather than form description.

In this paper, we elucidate how tree life forms change along different environmental and human use gradients. This study aims to determine the influence of topographic, climatic, geographic and human use factors on the occurrence of *P. tarapacana* life forms. In particular, we aim to answer the following questions: (i) Which tree life forms does this species have and what tree-specific biometric characteristics do the different tree life-forms have? (ii) How does the tree life forms and its distribution change with topography (elevation, slope and aspect), climate (temperature and precipitation), life zones and proxies of human uses (human footprint and distance to towns)?

2. Results

The 70 plots presented a large heterogeneity in topography (e.g., location, elevation, slope and aspect), where the N–S range was 156 km and W–E was 61 km. The elevation gradient was 789.1 m, where four life zones occur: Tropical alpine moist tundra (TAMT), Tropical alpine wet tundra (TAWT), Tropical subalpine dry scrub (TSDS) and Tropical subalpine moist forest (TSMF). Likewise, the average distance to towns was 9.7 ± 4.7 km (average ± standard deviation). During the surveys, 1801 trees were recorded, with an average area of 1.29 m^2 of tree crown, 7.8 cm of DBT and 86.1 cm of H.

2.1. *Life Forms of P. tarapacana*

We found four different life forms in the studied plots: arborescent (Ar), dwarf trees (Dt), shrubs (Sh) and brousse tigrée (Bt). The Ar life form has a unique base and branches that cover the entirety of the trunk, protecting buds from wind damage on the bark (Figure 1). The Ar life form presents a highly variable size between individuals (DAB 8.3 ± 7.0 cm, H 73.8 ± 51.6 cm). Its tree crown was 0.42 ± 0.57 m^2 and showed a CsR circular to oval shape (1.3 ± 0.3). In Dt, the crown is observed at the end of the main axis of the trunk, which is not covered by branches and has frequent signs of crown dieback (Figure 1). The DAB was 10.9 ± 7.4 cm and H was 101.0 ± 55.3 cm. Its TC and CsR were similar to those of Ar, with values of 0.35 ± 0.41 m^2 for TC and a similar CsR (1.3 ± 0.3). In the shrub life form, we observed multiple trunks, where the shortest ones were located at the periphery of the bush and the tallest ones in the center. A high mortality of the central trunks was observed in some shrub life forms (Figure 1). The size was 9.1 ± 6.3 cm for DAB and 100.7 ± 49.4 cm for H. A broad and highly variable tree crown (size and shape) was observed, measuring 2.0 ± 2.4 m^2, while the CsR value was similar to that of Ar and Dt (1.3 ± 0.3) (Figure 1). Brousse tigrée (Bt) occurs in *P. tarapacana* in bands, with different sizes of trunks, where the smallest were in the direction of the slope and the tallest were upslope (Figure 1). In the brousse tigree life form, some trunks located upslope were dead or with a low presence of live leaves. The size was 4.3 ± 3.0 cm for DAB and 57.6 ± 27.1 cm for H, where the tree crown was 1.4 ± 1.6 m^2 with an elongated shape (2.74 ± 1.36).

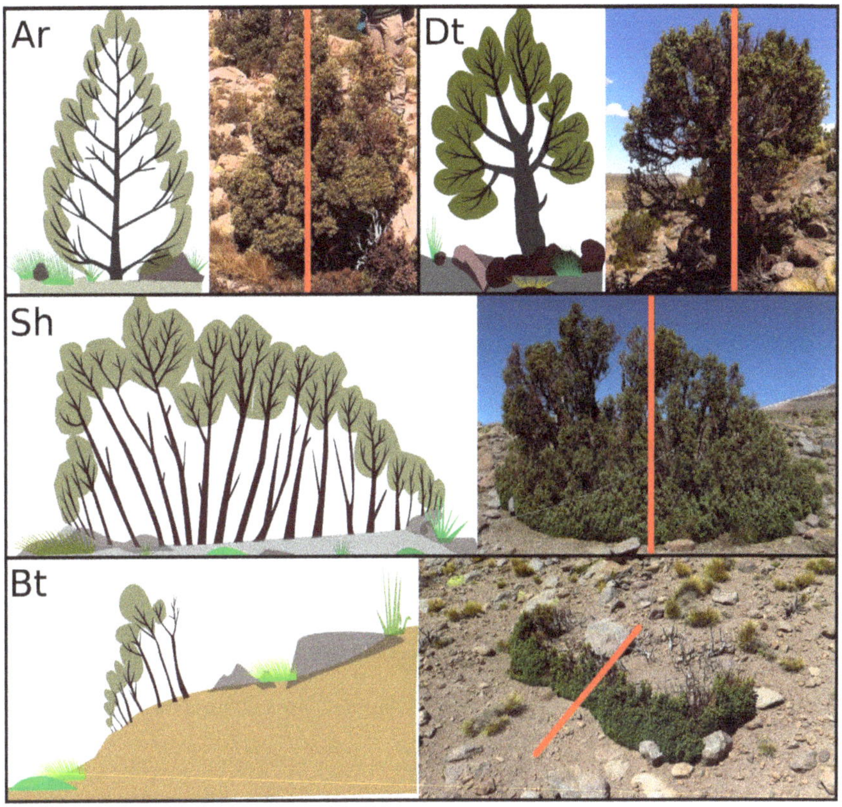

Figure 1. Classification of life forms in *P. tarapacana*. Ar: Arborescent; Dt: Dwarf tree; Sh: Shrubs; Bt: Brousse tigrée. The red line in the photo indicates the vertical cut that is observed in the graph on the left.

The size of the individuals among the different life forms presented significant differences (DAB H_{Test}: 77.1, $p < 0.0001$, and H = H_{Test}: 94.8, $p < 0.0001$), where the smallest values were observed in Bt, followed by Ar, Sh and finally Dt (Figure 2). A higher frequency (H_{Test}: 305.0, $p < 0.0001$) of small crowns was observed in Ar and Dt, intermediate values were observed in Bt, and the highest values were observed in Sh. The crown spread ratio did not present differences among Ar, Dt and Sh, while Bt was always more elongated, with an average value of 2.74 and a maximum value of 9.33 (H_{Test}: 185.6, $p < 0.0001$). The vitality of the individuals, classified by life form, presented significant differences (H_{Test}: 41.3, $p < 0.0001$), following the gradient of Dt < Sh < Ar = Bt, with Dt values close to intermediate vitality and Ar and Bt close to healthy. We provide complementary information of the biometric characteristics of each life form in Table A1 in Appendix A.

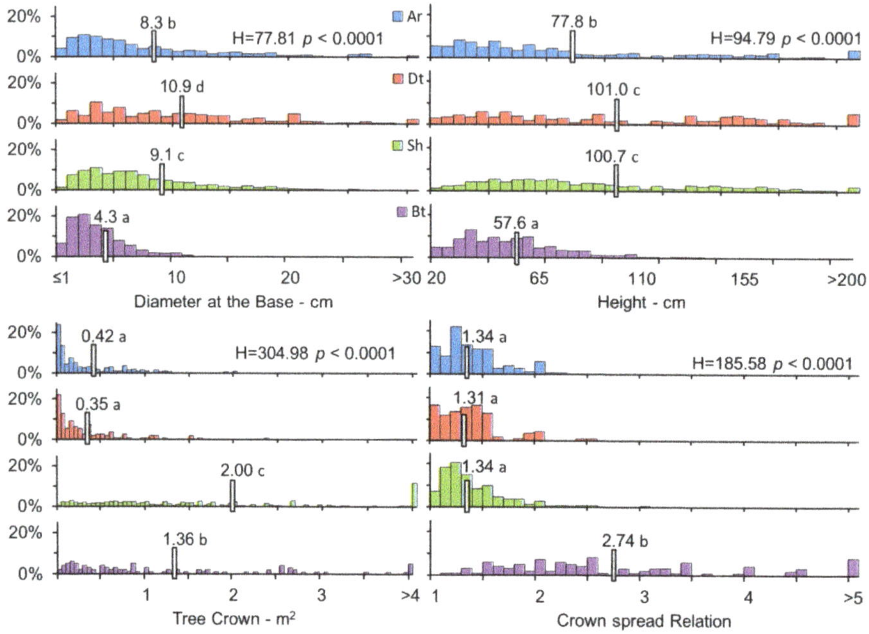

Figure 2. Kruskal–Wallis test for the diameter, height, tree crown and crown spread ratio of *P. tarapacana* life forms. Different letters indicate significant differences ($p < 0.05$) by Conover–Iman test.

2.2. Changes in the Frequency of Life Forms According to Topographic, Climatic, Geographic and Human Use Factors

The frequency of each of the four life forms was analyzed by plot, and showed a higher value for Sh (50%), followed by Ar (24%), Bt (16%) and Dt (9%). Table 1 shows that topographic factors (elevation, slope and aspect) significantly affect the frequency of the defined life forms. There were significant differences for all *P. tarapacana* life form frequency in elevation, while slope only showing differences in Ar and Bt. At higher elevations, the frequency of Ar and Dt was higher (>30% in Ar and >13% in Dt for >4700 m a.s.l.), while for Sh and Bt, the frequency decreased (from 59% in <4400 m a.s.l. to 40 % at >4700 m a.s.l. in Sh and from 19% to 12% at the same elevations for Bt). In the same way, there is an inverse relationship between the slope and frequency of Ar, and a direct one in Bt. For the different aspects (north and east aspects), we did not detect significant differences among the studied life forms.

Differences in the frequency of *P. tarapacana* life forms were observed in relation to the following factors: life zone, annual mean temperature and distance to towns (Table 2). AMT showed a similar pattern to those observed in the elevation gradient analysis (Table 1), where the frequencies of the lowest temperatures are those located at the higher elevations. Additionally, the frequency of all life forms presented significant differences among the different life zones, with a greater number of Ar and Dt in the alpine life zone (TAMT and TAWT) and Sh and Bt in the subalpine areas (TSMF and TSDS). The alpine life zones are located at higher elevations compared with the subalpine zones, which corroborates the change in the structure described in the elevation analysis. For temperature, it was observed that the frequency of Ar and Dt increased in colder zones, and the opposite trend occurred for Sh. Likewise, neither AP nor HF showed significant differences. However, it was observed that the frequency of Ar and Bt increased and decreased, respectively, with the distance to towns (DTT).

Table 1. Kruskal–Wallis test for the relative frequency of *Polylepis tarapacana* life forms compared with topographic factors (elevation in m a.s.l., slope in degrees and two aspects). Different letters indicate significant differences ($p < 0.05$) by the Conover–Iman test.

Variable	Range	n	Ar	Dt	Sh	Bt
Elevation	<4400	14	17 a	5 a	59 b	19 b
	4400–4500	17	17 a	8 ab	51 ab	24 b
	4500–4600	19	26 ab	10 abc	53 ab	11 a
	4600–4700	6	30 b	16 c	39 a	14 ab
	>4700	14	35 b	13 bc	40 a	12 a
	p		0.0082	0.0349	0.0123	0.0017
	H_{Test}		13.7	10.2	12.7	17.1
Slope	<15	18	30 b	11	55	8 a
	15–25	25	27 ab	11	48	14 a
	>25	27	18 a	7	50	24 b
	p		0.0268	0.2320	0.5501	0.0002
	H_{Test}		7.2	2.8	1.2	17.2
NA	N	27	22	8	52	18
	Rest	43	25	10	49	15
	p		0.3250	0.3049	0.5542	0.2000
	H_{Test}		0.9	1.0	0.3	2.7
EA	W	13	28	10	50	12
	Rest	34	22	10	51	17
	E	23	24	9	50	17
	p		0.6889	0.9677	0.9456	0.5978
	H_{Test}		0.7	0.1	1.1	1.0

Elevation in m a.s.l.; slope in degrees; NA: North aspect (N: from 315° to 45°); EA: East aspect (E: from 45° to 135°, W: from 225° to 315°). The aspects were calculated as sine and cosine functions, where sine values range from −1 (west) to 1 (east), while cosine values range from −1 (south) to 1 (north). Ar: Arborescents; Sh: Shrubs; Dt: Dwarf trees; Bt: Brousse tigrée. Different letters indicate significant differences determined by comparisons of means (Conover–Iman test, $p < 0.05$).

Table 2. Kruskal–Wallis of the frequency of *Polylepis tarapacana* life forms compared with climatic (temperature and precipitation), geographical (life zones), human footprint and distance to towns factors.

Variable	Range	n	Ar	Dt	Sh	Bt
Life Zone	TAMT	8	42 b	17 b	36 a	5 a
	TAWT	2	39 ab	18 ab	39 ab	5 ab
	TSDS	43	18 a	8 a	52 ab	22 b
	TSMF	17	29 b	8 a	55 b	8 a
	p		0.0006	0.0446	0.0426	0.0001
	H_{Test}		17.3	7.9	8.2	21.8
AMT	<5	15	34 b	15 b	39 a	11
	5–6	33	22 a	9 a	52 b	17
	>6	22	20 a	7 a	55 b	18
	p		0.0340	0.0232	0.0199	0.0629
	H_{Test}		6.8	7.5	7.8	5.5
AP	<150	26	24	9	50	17
	>150	44	24	10	51	16
	p		0.9399	0.6171	0.9890	0.3618
	H_{Test}		<0.01	0.3	<0.01	0.9
HF	0	53	25	10	50	15
	>0	17	20	8	50	21
	p		0.2758	0.5673	0.9891	0.1700
	H_{Test}		1.2	0.3	<0.01	1.9

Table 2. *Cont.*

Variable	Range	n	Ar	Dt	Sh	Bt
DTT	<5	13	14 a	6	58	23 b
	5–10	32	23 a	9	51	17 ab
	>10	25	31 b	11	45	12 a
	p		0.0039	0.2661	0.1225	0.0436
	H_{Test}		11.0	2.6	4.2	6.2

Life zones: Tropical alpine moist tundra (TAMT), Tropical alpine wet tundra (TAWT), Tropical subalpine dry scrub (TSDS) and Tropical subalpine moist forest (TSMF). AMT: Annual mean temperature in °C; AP: Annual precipitation in mm.yr^{-1}; HF: Human footprint; DTT: Distance to towns in km. A: Arborescents; Sh: Shrubs; Dt: Dwarf trees; Bt: Brousse tigrée. Different letters indicate significant differences determined by comparisons of means (Conover–Iman test, $p < 0.05$).

Topographic (elevation, slope and aspects), climatic (temperature and precipitation), and human use (human footprint and distance to towns) factors were used to define the frequency of life forms per plot (Figure 3). The plots were categorized into one simple (Sh n = 12), three doubles (Sh | Ar n = 22, Sh | Bt n = 18 and Sh | Dt n = 4), one triple (Ar | Bt | Dt n= 5) and one multiple (M n = 9) in relation to the abundance of life forms. Elevation and slope presented two differentiated groups, where, in the plots with multiple forms, Sh | Ar and Sh | Dt occurred in high- and low-slope areas, shrubs occurred at low-elevation and -slope areas, and finally the formations with Bt occurred on high slopes. The aspect factor presented less marked differences, with one single group located in the north aspect and dwarf trees (Dt) occurred in the west aspect, while shrubs with Bt and Ar tended to occupy the east aspects. Regarding human footprint (HF) and DTT, the Dt life form was observed far away from towns, while shrubs occurred closer to towns. The HF showed a gradient in which Dt and Ar were found in values close to 0, and as this value increased, Sh and Bt were more frequently observed.

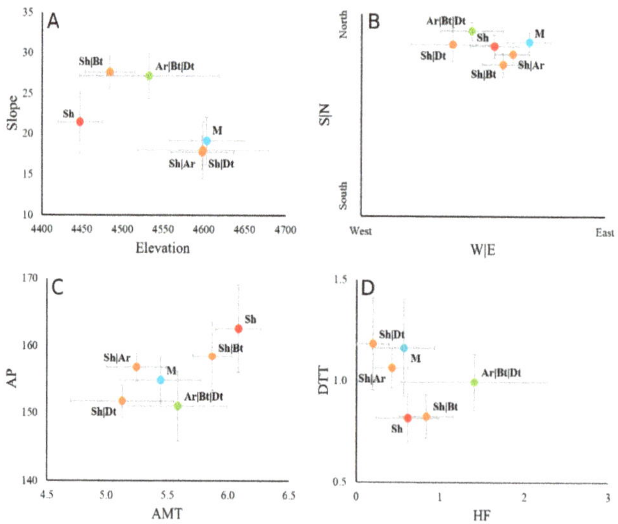

Figure 3. Topographic, climatic and human use factors classified by the frequency of life forms. Relationship among topographic variables ((**A**) Elevation and slope, (**B**) Aspects), climatic ((**C**) temperature and precipitation) and human use ((**D**) human footprint and distance to towns). Bars indicate the standard deviation of each axis. Elevation in m a.s.l.; Slope in degree; S | N: North Aspect; W | E: East aspect. The aspects factors were calculated as sine and cosine functions, where sine values range from −1 (west) to 1 (east), while cosine values range from −1 (south) to 1 (north). AMT: Annual mean temperature in °C; AP: Annual precipitation in mm.yr^{-1}; HF: Human footprint; DTT: Distance to towns in km. Ar: Arborescents; Sh: Shrubs; Dt: Dwarf trees; Bt: Brousse tigrée; M: Multiple forms.

3. Discussion

In this study, a new proposal to classify life forms of *P. tarapacana* was presented. The new life forms considered were arborescent, dwarf trees, shrubs and brousse tigrée. This proposal differs from that of Ríos [24] and Saavedra [19] as it divides the single stem category into two (Ar and Dt) and the multi-stem into another two (Sh and Bt). The justification for this new classification lies in the marked differences in their specific biometric characteristics as well as in the described influence of topographic, climatic and human uses factors, mainly on their frequency. The inclusion of the additional life form categories (Ar and Bt, novel for this species) help to achieve a more comprehensive overview of the life form composition for *P. tarapacana*.

3.1. Frequency of P. tarapacana Life Forms in the Argentinean Altiplano

Several authors agree in the classification of the life forms of *P. tarapacana* into single trunk and multi-trunk [19,24], concluding that the highest proportion of individuals found correspond to multi-trunk. These observations were made in small and specific areas of the distribution of *P. tarapacana*, and the present work offered results in a larger study area, covering most of the heterogeneity of its distribution (e.g., location, elevation, slope and aspects).

According to Mooney [37], there should be, for a given combination of climate and community succession, an optimal dominant life form. With this, a high degree of similarity in life forms would be expected at each of the sampled stands in this study because a large environment condition was covered. We found that the highest frequency of life form corresponded to Sh (50.3%), where the highest proportion was related to the extreme climatic conditions of the study area. The Altiplano plateau of the central Andes represents one of the harshest places on Earth for plant growth [8], where shrubs are associated with disturbed and stressful environments [38,39], and due to their lower height, plants take advantage of the relatively favorable climate near the surface [40,41]. A relatively smooth surface reduces the turbulence and, thus, minimizes heat loss, implying that, in calm and clear conditions, the daytime temperature of shrubs is generally higher than the air temperature [42]. The highest proportion of Sh was found by other authors in the Chilean *P. tarapacana* [19], where 64% of the individuals have a multi-stem tree life form and only 36% present a single stem. Ríos [24] found, further south in the Province of Iquique (Chile), that the proportion of life forms changes according to different topographic conditions. Likewise to other species that inhabit extreme ecosystems [43], the different life forms of *P. tarapacana* show a great phenotypic plasticity (i.e., an ability to respond to changes in their architecture to adapt to different environmental variations throughout their distribution).

3.2. Biometric Characteristics of Different P. tarapacana Life Forms

The size and shape of the life forms was determined by the growth restrictions that affect the lateral shoots and apical buds as well as the ability of plants to produce shoots from the root [4,7,13]. For example, in our study, Shrubs (Sh) presented the largest tree crown. Epicormic shoots under the bark facilitate the horizontal growth of shrubs [44], so they can expand horizontally as their length and mass increase to capture more light than a small tree that tends to grow mainly upwards [45–47]. In cold and alpine areas, low vegetation survives extreme weather and strong winds better, due to a better aerodynamic resistance [48–50]. In *P. tarapacana* shrubs, the highest trunk mortality occurs in the central zone due to the risk of cavitation caused by drought and freezing [51,52]. This could be a beneficial adaptation to hostile environments, such as the highlands, where the maximum height of the trees is determined in part by the problem of exposure to water in the upper crown. A lack of water can cause xylem embolism [53] and the risk of cavitation increases with stem height due to gravity. Related to this, both the brousse tigrée (Bt) and shrub (Sh) life forms, by developing multiple stems, have a continuous horizontal growth, close to the ground, allowing new roots and vertical shoots to develop [44,51].

In the case of *P. tarapacana*, our results show that the dwarf trees (Dt) life form had a lower vitality than the arborescents (Ar). This could be, as in the case of *Pinus aristata*, due to the exposure of the trunk in this tree life form to wind-induced desiccation and wind-driven cambial dieback [54,55]. The tree life form Ar had a greater general vitality, since its trunk is covered with branches. In the case of brousse tigrée (Bt) and Sh, multiple trunks protect one another, generating intermediate values of vitality. This could explain the increase in the proportion of Ar as elevation increases, because of the extreme characteristics of the climate [23].

In *P. tarapacana* forests, brousse tigrée occurs perpendicular to the slope line, where the smallest are down to the slope and the tallest, with a low presence of live leaves, are upslope. The formation of banded vegetation patterns on the slopes responds to a deposition of sediments by the interception of plants [56,57] that affects the properties and structure of the soil, leading to the deterioration of the environment [58] and to mortality in the upper slope vegetation bands. The form and function of different life forms of the alpine plant communities reflect various avoidance, tolerance, or resistance strategies to the interactions of cold temperature, radiation, wind, and desiccation stresses that prevail in the short growing season [59]. In this way, it can be considered that each *P. tarapacana* life form fills a particular niche, so each life form has different adaptations in response to different environmental conditions. The concept of life form as a morphological expression of belonging to a group can help to understand the functioning of this species.

3.3. *Influence of Topographic, Climatic, Life Zones and Human Use Factors in the Frequency of Life Forms*

A life form is the morphological outcome of a number of selection pressures, both abiotic (e.g., climate and elevation) and biotic (e.g., competitive interactions and human uses). It is a structural and functional compromise that allows for the optimization of cost–benefit relationships [60]. In *P. tarapacana*, Hoch and Körner [21] found that the greater the abundance of shrubby forms, the higher the elevation. However, not only elevation influences the general behavior of tree life forms, but also slope, temperature, life zone and resource use, as observed in this study.

Elevation represents a complex combination of climatic variables to which species have to adjust and has been considered an important environmental factor affecting community structure and organization [61,62]. Two variables closely related to elevation are temperature and precipitation [63] and this is the best example of a complex gradient where resources change [12]. Although elevation is the most important variable explaining differences in tree shape, this factor indisputably exercises an indirect influence through interactions with temperature, humidity and topography [17,64]. The development of *P. tarapacana* in higher elevation areas is limited by temperature drops [21,65], where forms that protect the cambium dieback are necessary. According to an ecophysiological study in the Chilean Altiplano, the photosynthetic processes and carbon assimilation of this species are well adapted to withstand cold temperatures [66], which allows them to thrive at elevations higher than 4700 m a.s.l. This was observed in this study, highlighting the differences found in the frequencies of Ar, Dt and Sh with changes in temperature as well as the different life zones [67].

The structure of the studied areas presented smaller trees and a higher crown cover when the elevation decreased [23], where there was a higher proportion of Sh and Bt. This response could be the result of the combined effect of the decrease in temperature and the increase in precipitation as elevation increases [68], increasing the proportion of the Ar form over the others.

Moderate to steep slopes are environments where this species presents its greatest development [19,23], because these landscapes provide safe and suitable sites for seed survival, germination and development of individuals, provided by, among other factors, the higher soil moisture and proportion of daytime with shade offered by rocks that reduce soil evaporation, as well as the smaller temperature fluctuations moderating competition for water with herbaceous vegetation [69]. We observed that the frequency of Ar decreases with the slope, while the Bt form increases its frequency considerably, mainly due to the instability of the soil, which produces the banded shape for this species, and the damages caused by this movement of gravel and soil reduces the frequency of Ar forms. We observed that the frequency of single-trunk forms (Ar and Dt) increases in alpine life zones (TAMT and TAWT), while in subalpine zones (TSDS and TSMF), the Sh form increases in proportion. This could be due to alpine areas presenting low temperatures and intermediate precipitation values, optimal factors for the development of this species [23], with the Sh form being more adapted to higher temperatures and low rainfall [26,37,70]. We observed the largest number of Sh and Bt specimens farther from the towns. This could be related to human uses and access to these areas, since individuals at lower elevations are the most accessible to the local communities, who select the Dt form that has larger dimensions [23,24] for fuel and construction purposes [35]. This is due to the fact that the trunks of *P. tarapacana* in the Argentinean highlands were used as beams for the construction of the roofs of houses, requiring trunks of at least 2–3 m in length [35]. However, the high durability of the wood of this species allows people to not need to replace the cut pieces frequently, generating a lower impact on forests. Likewise, in lower elevation regions (below 3500 m a.s.l.), there are forests of other tree species (*Polylepis tomentella* Wedd. and *Strombocarpa ferox* (Griseb.) C.E. Hughes and G.P. Lewis) that are greatly preferred compared with *P. tarapacana*, reducing its use in those regions [35].

The results of this study indicate the possible life forms that can be managed sustainably. In the case of multi-trunk life forms (Bt and Sh), it is possible to extract those that have a low vitality or are totally dry for their use for fuel or construction. In this context, the specimen is not completely eliminated, maintaining the conservation of genetic diversity [71]. In addition, it is important to consider the different life forms of *P. tarapacana* when conducting conservation studies and management plans, since each life form occupies a particular environmental situation and conserving these forests without differentiating its life forms is to lose part of its ecological niche.

The results of this investigation are preliminary, due to the short period of research and being concerned with an arboreal species with a very slow growth and long life, and the lack of information about how these life forms are subject to succession. It is necessary to continue with studies on the structure of these forests and their relationship with environmental characteristics, taking into account *P. tarapacana* life forms. Likewise, we suggest to carry out studies that evaluate the combined effect of biotic and abiotic factors on various forms of tree life to identify the important factors that have a major influence on their distribution and biometric characteristics.

4. Materials and Methods

4.1. Study Species

Polylepis tarapacana (*Rosaceae*: *Sanguisorbeae*) is a species that constitute rare, monospecific evergreen forests distributed in the sem-iarid high beds of the western Altiplano from southern Peru to south-western Bolivia, northern Chile and adjacent northwest Argentina (16°–23° S) [72] at 3400–5013 m a.s.l. [18,23,73,74]. This species is adapted to the Altiplano, which is capable of withstanding harsh climatic conditions. Under different life forms, *P. tarapacana* ranges from small shrubs to trees up to 7 m height [75]; however, specimens usually oscillate between 1 to 5 m high [73]. *Polylepis tarapacana* is a unique tree because it lives at a higher elevation compared to any other tree species and comprises the highest elevation tree line on Earth [74].

4.2. Study Area

We worked in shrublands and forests of *P. tarapacana* distributed in the Andean cordillera in the Province of Jujuy, Argentina (22°04′–23°40′ SL at 66°46′–65°49′ WL; Figure 4), located in the high peaks of the Argentinean Andes mountain range from 4160 to 4952 m a.s.l. [23]. The climate is cold and dry with strong winds, characterized by a reduced seasonality of temperature, but marked seasonality in precipitation [76,77] with 135 to 165 mm.yr^{-1} concentrated in summer, and with 4.2 to 6.5 °C of annual mean temperature [23]. The vegetation in this area is composed of many species with traits linked to extremely low temperatures, wind and xerophytism [77], and specially with dwarf shrubs and cushion plants [20,78,79].

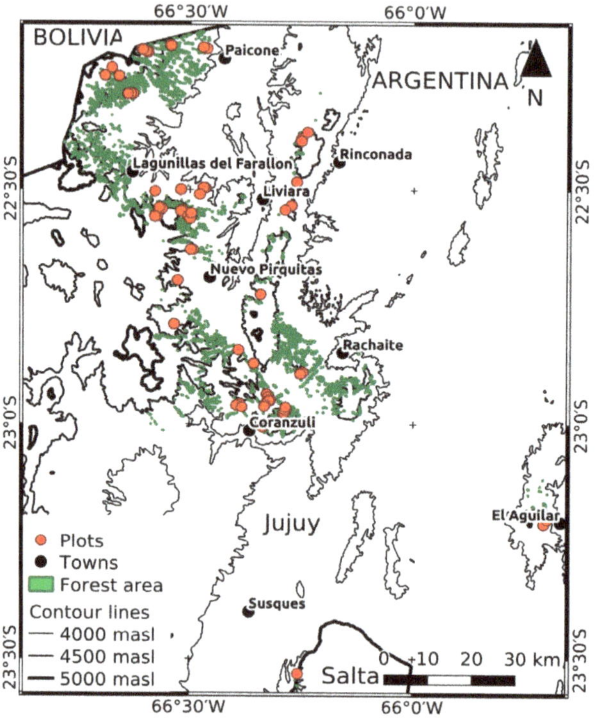

Figure 4. Distribution of *Polylepis tarapacana* forests (green) in the study area (Altiplano, Argentina), showing plots (red circles), towns (black circles) and contour lines, at 4000 m a.s.l. (narrow line), 4500 m a.s.l. (medium line) and 5000 m a.s.l. (tick line). Modified from López et al. [23].

4.3. Classification of Life Forms of P. tarapacana

The considered life forms were arborescents (Ar), dwarf trees (Dt), shrubs (Sh) and brousse tigrée (Bt), where Ar is an intermediate form between trees and shrubs, with a single base and branches that arise from the base of the trunk and along the main axis of the tree. Dwarf trees have a single main stem shorter than 5 m, where the trunk is not covered by branches, and well-developed lateral branches forming a crown. Shrubs are multi-stemmed short woody plants, branching at the ground with vegetative buds to form new shoots. The brousse tigrée tree life form occurs in bands perpendicular to the slope line (Figure 1).

4.4. Data Obtained

We worked in 70 *P. tarapacana* shrublands and forests patches throughout the distribution range [23]. The patches were selected according to: (i) homogeneous cover, where the distance between individuals was nearly constant; (ii) accessibility; and (iii) the patch size being >1 ha. In the center of each patch, we established one plot (20 × 50 m) to describe the vegetation structure, maintaining the elevation level. We corrected the areas according to the slope of the terrain using the following formula: Corrected area = Area × cosine (slope in degrees) [80]. In this way, the starting point and the ending point of the plot had the same elevation. We measured all live plants ≥0.20 m height, recording: (i) tree life form of each individual; (ii) diameter at the base (DAB, cm), corresponding to the trunk or the tallest trunk of multi-stem plants (Figure 5); (iii) height (H, cm) of the trunk or tallest trunk of multi-stem trees (Figure 5); (iv) diameter of the maximum axis and of the axis at 90 degrees from the each crown to calculate the area of the tree crown (TC, m^2) with the ellipse formula and the crown spread ratio (CsR) as the relation between the two measurements; and (v) vitality or health status of each individual to test the influence of wind damage for the different tree life forms into 3 types: (a) low vigor (more than 50 % of the foliage light green, more than 50% of dead branches and more than 50% of dead trunk section), (b) intermediate vigor (less than 50% of light green foliage, less than 50% of dead branches and less than 50% of dead trunk section) and (c) healthy (deep green foliage, no dead branches and no damage in the trunk). The measurements were made in March–April and October–November, coinciding with the periods of less rain and intermediate temperatures.

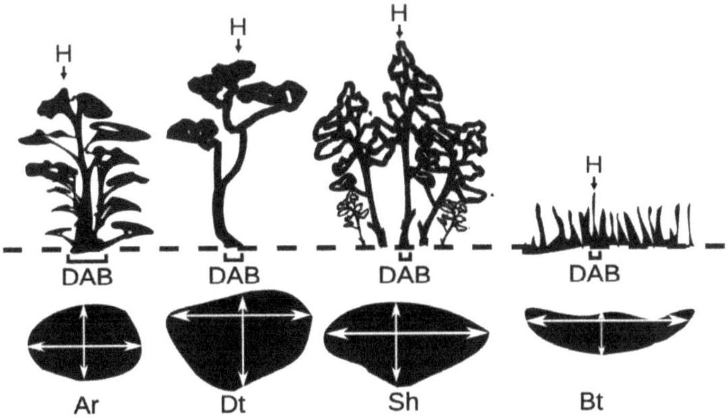

Figure 5. Determination of the height, diameter at the base, crown diameter of the maximum axis and of the axis at 90 degrees for each life forms in *P. tarapacana*. H: height; DAB: diameter at the base of the tree; Ar: Arborescent; Dt: Dwarf tree; Sh: Shrubs; Bt: Brousse tigrée.

4.5. Environmental Characterization

In all the plots, we registered longitude, latitude and elevation (m a.s.l.) with a global geopositioning device (GPS), slope using a clinometer (°) and the aspect with compass as sine and cosine functions of the north magnetic direction. Sine values ranged from −1 (west) to 1 (east), while cosine values ranged from −1 (south) to 1 (north) [80]. The environmental characteristics of each plot were registered through Holdridge life zones following Derguy et al. [67]: (i) Tropical alpine moist tundra (TAMT), (ii) Tropical alpine wet tundra (TAWT), (iii) Tropical subalpine dry scrub (TSDS) and (iv) Tropical subalpine moist forest (TSMF). The climatic factors, annual mean temperature (AMT) (°C) and annual precipitation (AP) ($mm.yr^{-1}$), for the period of 1970–2000 were obtained from WorldClim [81], the values of the human footprint (HF) were extracted from Lizárraga and Monguillot [70], and the distance to the nearest town (DTT) was obtained using QGIS software.

4.6. Data Analysis

We conducted non-parametric Kruskal–Wallis tests to compare forest structure variables (DAB, H, TC, CsR and vitality) for the different life form categories determined for *P. tarapacana*. Additionally, we also performed non-parametric Kruskal–Wallis tests using topographic, climate, geographical and human use variables as the main factors to analyze the frequency of each life form. Differences were determined by comparisons of the means (Conover–Iman test at $p < 0.05$).

To characterize the life forms, we used the maximum relative frequency of the plots in relation to topographic, climatic and human uses, where each plot was categorized into single, double, triple and multiple categories in relation to the abundance of the different life forms. Single categories were considered when a frequency was >70% for a certain tree form (e.g., 74% life form Ar, 12% Dt, 8% Sh, 6% Bt = Category Ar) and double categories when the single categories do not reach 70% but the sum of two higher categories reaches >70% (e.g., 54% life form Ar, 22% Dt, 16% Sh, 8% Bt = Category Ar | Dt). The multiple (M) category is when three or more categories are needed to reach >70% frequency (e.g., 36% life form Ar, 24% Dt, 24% Sh, 16% Bt = Category M). When the categories were represented by less than 5% of the plots (<4 plots), these categories were grouped creating a new category (e.g., Ar | Dt = 3 plots; Ar | Sh = 2 plots: Ar | Dt | Sh = 5 plots). With these categorizations, the means and the standard deviations were calculated for each topographic, climatic and human use. Additionally, the variability of the plots was determined in terms of the simultaneous occurrence of the different forms life forms depending on the topography, climate and human use.

5. Conclusions

A life form is the morphological result of selection pressures, both abiotic (e.g., climate and altitude) and biotic (e.g., competitive interactions and human uses). In this study, a new proposal for the classification of *P. tarapacana* life forms was presented, and we demonstrated the influence of elevation, slope, life zone, AMT and DTT in the life form frequency. Each life form of *P. tarapacana* occupies a particular niche, and this concept can help us to understand the ways in which these Altiplano communities function.

In this study, the relative success of each life form was measured in terms of relative frequency, but other measures, such as the biomass of each life form, may be useful and provide additional information. This study showed a general consistency in the changes in life forms of the high-altitude Andean vegetation, and the changes generated by environmental gradients in this species were explained. This consistency provides a framework for broader comparisons with species of the genus *Polylepis* in other parts of South America. These comparisons will provide information on the distribution of these and other life forms and may help us to understand the mechanisms that determine the structure of these high-elevation forest communities and beyond.

Author Contributions: V.L.L. and J.M.C. designed the research; V.L.L., L.B., G.A.E.C. and J.M.C. collected the field data; V.L.L., L.B., G.M.P., M.V.L., G.A.E.C. and J.M.C. analyzed the data; V.L.L., L.B., G.M.P., M.V.L., G.A.E.C. and J.M.C. wrote the paper. All authors have read and agreed to the published version of the manuscript.

Funding: This research received no external funding.

Data Availability Statement: Not applicable.

Conflicts of Interest: The authors declare no conflict of interest.

Abbreviations

DAB	Diameter at the base of the tree
H	Height
TC	Tree crown
CsR	Crown spread ratio
GPS	Global geopositioning device
TAMT	Tropical alpine moist tundra
TAWT	Tropical alpine wet tundra
TSDS	Tropical subalpine dry scrub
TSMF	Tropical subalpine moist forest
AMT	Annual mean temperature
AP	Annual precipitation
HF	Human footprint
DTT	Distance to town
M	Multiple category
Ar	Arborescents
Dt	Dwarf tree
Sh	Shrubs
Bt	Brousse tigrée
NA	North aspect
EA	East aspect

Appendix A

Table A1. Biometric characteristics mean (± standard deviation) (DAB, height, tree crown and crown spread ratio) of different life forms (Ar: Arborescents; Sh: Shrubs; Dt: Dwarf trees; Bt: Brousse tigrée) in relation to topography (elevation, slope and N and E aspect), climate (annual mean temperature, AMT; annual precipitation, AP), life zones (tropical alpine moist tundra, TAMT; tropical alpine wet tundra, TAWT; tropical subalpine dry scrub, TSDS; tropical subalpine moist forest, TSMF) and human uses (human footprint, HF; distance to towns in km, DTT).

		Diameter at Base (cm)				Height (cm)				Tree Crown (m^2)				Crown Spread Ratio			
Variable		AR	Dt	Sh	Bt	AR	Dt	Sh	Bt	AR	Dt	Sh	Bt	AR	Dt	Sh	Bt
Elevation	<4400	6.3 ± 3.5	5.4 ± 3.3	6.3 ± 4.0	4.1 ± 2.5	56.2 ± 20.1	59.3 ± 24.7	71.9 ± 29.4	59.4 ± 23.5	0.3 ± 0.4	0.1 ± 0.2	1.8 ± 2.3	1.3 ± 1.4	1.4 ± 0.3	1.2 ± 0.2	1.3 ± 0.3	2.5 ± 0.9
	4400–4500	6.7 ± 4.0	6.8 ± 6.7	5.9 ± 3.8	4.1 ± 2.7	63.9 ± 26.6	57.4 ± 30.5	76.8 ± 36.7	55.9 ± 18.8	0.3 ± 0.3	0.1 ± 0.1	1.2 ± 2.3	1.0 ± 0.5	1.3 ± 0.2	1.2 ± 0.1	1.3 ± 0.3	2.2 ± 1.6
	4500–4600	8.6 ± 7.3	11.1 ± 7.6	8.1 ± 5.7	3.7 ± 2.2	87.4 ± 61.0	108.5 ± 61.1	99.1 ± 46.5	52.9 ± 17.4	0.5 ± 0.5	0.4 ± 0.4	1.3 ± 1.5	1.1 ± 0.7	1.5 ± 0.4	1.3 ± 0.2	1.4 ± 0.4	2.9 ± 1.3
	4600–4700	8.0 ± 6.8	11.9 ± 8.1	9.7 ± 7.8	4.1 ± 2.6	72.6 ± 54.9	108.2 ± 65.2	100.2 ± 62.1	53.9 ± 22.0	0.5 ± 0.7	0.4 ± 0.5	1.3 ± 1.3	0.9 ± 0.8	1.2 ± 0.2	1.4 ± 0.4	1.3 ± 0.4	2.9 ± 0.9
	>4700	9.7 ± 8.3	12.6 ± 7.6	10.8 ± 6.7	3.7 ± 3.0	92.5 ± 59.0	118.6 ± 49.1	124.3 ± 54.5	53.6 ± 30.0	0.5 ± 0.6	0.4 ± 0.4	1.6 ± 1.5	0.9 ± 0.9	1.3 ± 0.3	1.3 ± 0.3	1.3 ± 0.3	3.4 ± 1.4
Slope	<15	9.1 ± 7.8	12.5 ± 8.1	10.3 ± 6.7	5.0 ± 3.3	85.5 ± 59.6	118.5 ± 58.4	113.1 ± 50.6	62.4 ± 23.5	0.5 ± 0.6	0.4 ± 0.5	1.5 ± 1.3	1.2 ± 1.4	1.4 ± 0.4	1.3 ± 0.2	1.4 ± 0.4	3.3 ± 1.2
	15–25	8.9 ± 7.3	11.5 ± 7.7	8.4 ± 5.6	4.2 ± 2.7	86.4 ± 56.3	105.8 ± 56.5	101.7 ± 50.7	57.0 ± 25.6	0.5 ± 0.5	0.4 ± 0.4	1.7 ± 2.3	1.0 ± 0.9	1.3 ± 0.3	1.3 ± 0.3	1.3 ± 0.3	2.8 ± 1.5
	>25	6.4 ± 4.5	5.8 ± 3.8	5.2 ± 3.3	3.5 ± 2.1	64.5 ± 37.2	60.8 ± 30.3	67.0 ± 28.8	52.7 ± 18.9	0.3 ± 0.3	0.1 ± 0.2	1.1 ± 1.6	1.0 ± 0.7	1.4 ± 0.3	1.3 ± 0.3	1.3 ± 0.3	2.5 ± 1.3
N aspect	N	9.0 ± 7.3	10.4 ± 7.5	8.2 ± 6.0	4.2 ± 2.7	84.4 ± 56.4	98.8 ± 54.1	93.8 ± 47.9	58.8 ± 22.4	0.5 ± 0.6	0.3 ± 0.4	1.5 ± 2.1	1.1 ± 0.8	1.3 ± 0.3	1.3 ± 0.3	1.4 ± 0.3	2.5 ± 1.4
	Rest	7.4 ± 6.5	11.3 ± 7.9	7.5 ± 5.4	3.7 ± 2.5	75.9 ± 51.2	106.3 ± 62.5	94.6 ± 50.0	51.2 ± 21.4	0.4 ± 0.4	0.4 ± 0.4	1.4 ± 1.4	1.0 ± 1.1	1.3 ± 0.3	1.3 ± 0.3	1.3 ± 0.3	3.0 ± 1.3
E aspect	W	9.0 ± 7.0	9.9 ± 7.6	8.8 ± 5.1	4.0 ± 2.2	88.6 ± 51.9	91.5 ± 55.6	103.2 ± 46.8	54.0 ± 21.3	0.5 ± 0.5	0.3 ± 0.3	1.5 ± 1.5	1.0 ± 0.8	1.3 ± 0.3	1.1 ± 0.1	1.3 ± 0.3	1.9 ± 0.4
	Rest	8.9 ± 7.3	10.2 ± 7.4	7.5 ± 5.7	4.2 ± 2.8	82.5 ± 58.6	97.7 ± 54.0	88.6 ± 45.3	58.8 ± 23.2	0.5 ± 0.6	0.3 ± 0.4	1.4 ± 2.2	1.1 ± 0.9	1.4 ± 0.3	1.3 ± 0.3	1.4 ± 0.3	2.9 ± 1.3
	E	7.4 ± 6.8	11.9 ± 8.0	8.1 ± 6.2	3.6 ± 2.5	74.4 ± 50.6	112.1 ± 60.3	96.6 ± 53.6	51.5 ± 20.6	0.4 ± 0.4	0.4 ± 0.5	1.4 ± 1.5	0.9 ± 1.1	1.3 ± 0.3	1.4 ± 0.3	1.3 ± 0.3	2.7 ± 1.5
AMT	<5	9.5 ± 8.2	12.6 ± 8.4	11.1 ± 7.2	4.1 ± 3.2	89.2 ± 60.7	116.4 ± 57.5	120.9 ± 55.4	58.9 ± 30.5	0.5 ± 0.7	0.5 ± 0.4	1.6 ± 1.5	1.2 ± 1.2	1.3 ± 0.3	1.3 ± 0.3	1.4 ± 0.3	3.2 ± 1.3
	5–6	8.2 ± 6.9	9.3 ± 6.8	7.3 ± 5.3	3.6 ± 2.2	81.3 ± 54.5	91.3 ± 52.9	92.0 ± 46.6	51.3 ± 18.2	0.4 ± 0.5	0.3 ± 0.3	1.3 ± 1.9	0.9 ± 0.6	1.4 ± 0.4	1.3 ± 0.3	1.3 ± 0.4	2.4 ± 1.5
	>6	7.2 ± 4.8	9.8 ± 7.0	6.1 ± 4.3	3.0 ± 2.8	69.5 ± 41.8	89.5 ± 57.3	78.8 ± 38.5	59.9 ± 21.5	0.4 ± 0.4	0.3 ± 0.3	1.5 ± 1.9	1.2 ± 1.2	1.4 ± 0.3	1.2 ± 0.2	1.3 ± 0.3	2.5 ± 0.9
AP	<150	7.3 ± 6.1	11.3 ± 7.5	8.6 ± 5.8	5.0 ± 3.0	73.0 ± 47.5	96.6 ± 52.9	96.5 ± 44.4	64.5 ± 25.4	0.4 ± 0.5	0.3 ± 0.4	1.8 ± 2.3	1.3 ± 1.2	1.3 ± 0.3	1.3 ± 0.3	1.4 ± 0.3	2.9 ± 1.5
	>150	9.0 ± 7.5	10.4 ± 7.7	7.6 ± 5.7	3.3 ± 2.1	85.8 ± 57.7	103.7 ± 59.0	92.8 ± 50.9	50.0 ± 18.0	0.5 ± 0.5	0.3 ± 0.4	1.2 ± 1.5	0.8 ± 0.6	1.4 ± 0.4	1.3 ± 0.3	1.3 ± 0.3	2.6 ± 1.1
Life Zone	TAMT	10.3 ± 8.7	12.7 ± 8.2	11.3 ± 7.1	5.5 ± 4.0	95.7 ± 62.3	119.2 ± 51.6	128.2 ± 53.4	71.0 ± 32.4	0.5 ± 0.7	0.5 ± 0.4	1.8 ± 1.7	1.2 ± 1.1	1.3 ± 0.3	1.3 ± 0.3	1.3 ± 0.3	3.3 ± 1.1
	TAWT	9.5 ± 7.8	11.8 ± 4.5	13.2 ± 5.0	2.9 ± 1.2	97.8 ± 56.1	123.8 ± 42.4	157.7 ± 47.4	42.5 ± 7.2	0.5 ± 0.4	0.4 ± 0.2	2.3 ± 1.2	0.6 ± 0.1	1.3 ± 0.3	1.3 ± 0.3	1.3 ± 0.3	3.3 ± 1.1
	TSDS	7.1 ± 5.9	9.5 ± 7.6	7.0 ± 5.3	3.7 ± 2.4	69.3 ± 46.9	88.0 ± 58.2	85.7 ± 44.6	53.3 ± 20.6	0.4 ± 0.5	0.3 ± 0.4	1.3 ± 1.9	1.0 ± 0.8	1.3 ± 0.3	1.3 ± 0.3	1.3 ± 0.3	2.6 ± 1.5
	TSMF	8.5 ± 6.5	10.6 ± 7.6	7.9 ± 5.3	5.0 ± 2.3	82.5 ± 53.7	101.7 ± 56.4	90.1 ± 43.2	64.4 ± 22.7	0.4 ± 0.5	0.4 ± 0.4	1.5 ± 1.8	1.4 ± 1.6	1.4 ± 0.4	1.3 ± 0.2	1.3 ± 0.3	2.9 ± 0.9
HF	0	8.4 ± 6.8	11.3 ± 7.4	8.5 ± 6.0	4.2 ± 2.8	80.3 ± 52.0	104.5 ± 54.9	98.6 ± 49.5	57.0 ± 23.6	0.4 ± 0.5	0.4 ± 0.4	1.6 ± 2.0	1.1 ± 1.1	1.3 ± 0.3	1.3 ± 0.3	1.3 ± 0.3	2.9 ± 1.4
	>0	8.4 ± 8.1	8.4 ± 8.3	6.1 ± 4.5	3.4 ± 2.0	85.8 ± 64.9	88.3 ± 62.9	79.5 ± 43.3	52.3 ± 18.7	0.5 ± 0.6	0.3 ± 0.4	0.8 ± 0.7	0.9 ± 0.4	1.3 ± 0.2	1.1 ± 0.0	1.3 ± 0.3	1.8 ± 0.3
DTT	<5	5.4 ± 3.7	4.8 ± 3.4	5.6 ± 3.9	3.4 ± 2.5	51.1 ± 20.8	49.4 ± 26.2	68.0 ± 30.9	50.5 ± 22.3	0.3 ± 0.4	0.1 ± 0.2	1.9 ± 2.2	1.3 ± 1.7	1.4 ± 0.3	1.1 ± 0.2	1.4 ± 0.3	2.5 ± 1.1
	5–10	9.0 ± 7.4	10.8 ± 6.8	7.6 ± 5.4	3.6 ± 2.1	87.3 ± 57.9	107.4 ± 59.0	96.1 ± 48.3	52.3 ± 18.6	0.5 ± 0.6	0.3 ± 0.4	1.4 ± 2.0	0.9 ± 0.6	1.3 ± 0.3	1.3 ± 0.3	1.3 ± 0.4	2.8 ± 1.0
	>10	8.3 ± 7.1	11.6 ± 8.5	9.6 ± 6.4	5.2 ± 3.0	80.2 ± 53.6	103.3 ± 54.3	103.5 ± 51.9	65.6 ± 25.1	0.4 ± 0.5	0.4 ± 0.4	1.3 ± 1.3	1.1 ± 0.9	1.3 ± 0.3	1.3 ± 0.3	1.3 ± 0.3	3.0 ± 1.7

Elevation in m a.s.l.; Slope in degrees; North Aspect: N: from 315° to 45°; East aspect: E: from 45° to 135°; W: from 225° to 315°.

References

1. Körner, C. Scaling from species to vegetation: The usefulness of functional groups. In *Biodiversity and Ecosystem Function*; Schulze, E.D., Mooney, H.A., Eds.; Springer: Berlin, Germany, 1994; pp. 117–140.
2. Odland, A. Interpretation of altitudinal gradients in South Central Norway based on vascular plants as environmental indicators. *Ecol. Indic.* **2009**, *9*, 409–421. [CrossRef]
3. Hadar, L.; Noy-Meir, I.; Perevolotsky, A. The effect of shrub clearing and grazing on the composition of a Mediterranean plant community: Functional group versus species. *J. Veg. Sci.* **1999**, *10*, 673–682. [CrossRef]
4. Pavón, N.P.; Hernández-Trejo, H.; Rico-Gray, V. Distribution of plant life forms along an altitudinal gradient in the semi arid valley of Zapotitlán, México. *J. Veg. Sci.* **2000**, *11*, 39–42. [CrossRef]
5. Mueller-Dombois, D.; Ellenberg, H. *Aims and Methods of Vegetation Ecology*; John Wiley & Sons Inc.: New York, NY, USA, 1974; p. 547.
6. Solbrig, O.T. Plant traits and adaptive strategies: Their role in ecosystem function. In *Biodiversity and Ecosystem Function*; Schulze, E.D., Mooney, H.A., Eds.; Springer: Berlin/Heidelberg, Germany, 1993; pp. 97–116.
7. Klimes, L. Life-forms and clonality of vascular plants along an altitudinal gradient in E Ladakh (NW Himalayas). *Basic Appl. Ecol.* **2003**, *4*, 317–328. [CrossRef]
8. Körner, C. Climatic treelines: Conventions, global patterns, causes (Klimatische Baumgrenzen: Konventionen, globale Muster, Ursachen). *Erdkunde* **2007**, *61*, 316–324. [CrossRef]
9. Arroyo, M.T.K.; Cavieres, L.A. High-Elevation Andean Ecosystems. In *Encyclopedia of Biodiversity*, 2nd ed.; Levin, S.A., Ed.; Academic Press: Cambridge, MA, USA, 2013; pp. 96–110. [CrossRef]
10. Jaramillo Aguilar, D.A. Fotosíntesis en los Bosques a Mayor Elevación en el Planeta: *Polylepis tarapacana* en un Gradiente de Elevación en los Andes de Arica y Parinacota, Chile. Ph.D. Thesis, Universidad Austral de Chile, Valdivia, Chile, 2015.
11. Cowling, R.M.; Esler, K.J.; Midgley, G.F.; Honing, M.A. Plant functional diversity, species diversity and climate in arid and semi-arid southern Africa. *J. Arid Environ.* **1994**, *27*, 141–158. [CrossRef]
12. Huston, M.A. *Biological Diversity: The Coexistence of Species*; Cambridge University Press: Cambridge, UK, 1994; p. 681.
13. Ramsay, P.M.; Oxley, E.R.B. The Growth Form Composition of Plant Communities in the Ecuadorian Páramos. *Plant Ecol.* **1997**, *131*, 173–192. [CrossRef]
14. Diaz, S.; Cabido, M. Plant functional types and ecosystem function in relation to global change. *J. Veg. Sci.* **1997**, *8*, 463–473. [CrossRef]
15. Mahdavi, P.; Akhani, H.; Van der Maarel, E. Species Diversity and Life-Form Patterns in Steppe Vegetation along a 3000 m Altitudinal Gradient in the Alborz Mountains, Iran. *Folia Geobot.* **2013**, *48*, 7–22. [CrossRef]
16. Leuschner, C. Are high elevations in tropical mountains arid environments for plants? *Ecology* **2000**, *81*, 1425–1436. [CrossRef]
17. Körner, C. *Alpine Plant Life: Functional Ecology of High Mountain Ecosystems*, 2nd ed.; Springer: Berlin, Germany, 2003; p. 349.
18. Kessler, M.; Schmidt-Lebuhn, A.N. Taxonomical and distributional notes on *Polylepis* (*Rosaceae*). *Org. Divers. Evol.* **2006**, *6*, 67–70. [CrossRef]
19. Saavedra, J. Caracterización de la Estructura Poblacional de *Polylepis tarapacana* en Sector Cerro Chiguana, Cuenca del Salar de Surire, Para Establecer Los Lineamientos de un Programa de Restauración Ecológica. Ph.D. Thesis, Universidad de Tarapacá, Arica, Chile, 2013.
20. López, V.L.; Cellini, J.M. Plantas medicinales asociadas a bosques de *Polylepis tarapacana* del altiplano jujeño: Riqueza y cobertura en gradientes geográficos, topográficos y de estructura forestal. *Ecol. Austral.* **2022**, *32*, 894–907. [CrossRef]
21. Hoch, G.; Körner, C. Growth, demography and carbon relations of *Polylepis* trees at the world\u2019s highest treeline. *Funct. Ecol.* **2005**, *19*, 941–951. [CrossRef]
22. Donoso, C. *Bosques Templados de Chile y Argentina: Variación, Estructura y Dinámica*; Editorial Universitaria: Santiago, Chile, 1993; p. 483.
23. López, V.L.; Huertas Herrera, A.; Rosas, J.M.; Cellini, J.M. Optimal environmental drivers of high-mountains forest: *Polylepis tarapacana* cover evaluation in their southernmost distribution range of the Andes. *Trees For. People* **2022**, *9*, 100321. [CrossRef]
24. Ríos, S. Estudio de la Distribución y Caracterización Ecológica de las Poblaciones Locales de Queñoa de Altura (*Polylepis tarapacana* Phil.) en el Sector de la Provincia de Iquique. Ph.D. Thesis, Universidad de Chile, Tarapacá, Chile, 1998.
25. De Laubenfels, D.J. *Mapping the World's Vegetation. Regionalization of Formations and Flora*, 1st ed.; Syracuse University Press: New York, NY, USA, 1975; p. 246.
26. Box, E.O. *Macroclimate and Plant Forms: An Introduction to Predictive Modeling in Phytogeography*; Tasks for Vegetation Science; Antiquariaat Junk: The Hague, The Netherlands, 1981; Volume 1, p. 58.
27. Rundel, P.W. Shrub life forms. In *Response of Plants to Multiple Stress*; Mooney, H.A., Winner, W.E., Pell, E.J., Eds.; Academic Press: San Diego, CA, USA, 1991; pp. 345–371.
28. Wilson, B.F. Shrub stems: Form and function. In *Plant Stems: Physiology and Functional Morphology*; Gartner, B.L., Ed.; Academic Press: New York, NY, USA, 1995; pp. 91–102.
29. Valentin, C.; d'Herbès, J.M. Niger tiger bush as a natural water harvesting system. *Catena* **1999**, *37*, 231–256. [CrossRef]
30. Rietkerk, M.; Boerlijst, M.C.; van Langevelde, F.; Hille-Ris-Lambers, R.; van de Koppel, J.; Kumar, L.; Prins, H.H.T.; de Roos, A.M. Self-organization of vegetation in arid ecosystems. *Am. Nat.* **2002**, *160*, 524–530. [CrossRef]

31. Ludwig, J.A.; Wilcox, B.P.; Breshears, D.D.; Tongway, D.J.; Imeson, A.C. Vegetation patches and runoff–erosion as interacting ecohydrological processes in semiarid landscapes. *Ecology* **2005**, *86*, 288–297. [CrossRef]
32. Lefever, R.; Barbier, N.; Couteron, P.; Lejeune, O. Deeply gapped vegetation patterns: On crown/root allometry, criticality and desertification. *J. Theor. Biol.* **2009**, *261*, 194–209. [CrossRef]
33. Renison, D.; Morales, L.; Cuyckens, G.A.E.; Sevillano, C.S.; Cabrera Amaya, D.M. Ecología y conservación de los bosques y arbustales de *Polylepis*: ¿qué sabemos y qué ignoramos? *Ecol. Austral* **2018**, *28*, 163–174. [CrossRef]
34. IUCN. Red List of Threatened Species. Version 2020-1. 2020. Available online: https://www.iucnredlist.org (accessed on 5 March 2023).
35. Renison, D.; Cuyckens, G.A.E.; Pacheco, S.; Guzmán, G.F.; Grau, H.R.; Marcora, P.; Robledo, G.; Cingolani, A.M.; Dominguez, J.; Landi, M.; et al. Distribución y estado de conservación de las poblaciones de árboles y arbustos del género *Polylepis* (*Rosaceae*) en las montañas de Argentina. *Ecol. Austral.* **2013**, *23*, 27–36. [CrossRef]
36. Montana, C.; Valientebanuet, A. Floristic and life-form diversity along an altitudinal gradient in an intertropical semiarid Mexican region. *Southwest Nat.* **1998**, *43*, 25–39.
37. Mooney, H.A. Plant forms in relation to environment. In *Handbook of Vegetation Science*; Strain, B.R., Billings, W.D., Eds.; Vegetation and Environment: The Hague, The Netherlands, 1974; Volume 6, pp. 111–122.
38. Givnish, T.J. Plant stems: Biomechanical adaptation for energy capture and influence on species distributions. In *Plant Stems: Physiology and Functional Morphology*; Gartner, B.L., Ed.; Academic Press: San Diego, CA, USA, 1995; pp. 3–49.
39. Sheffer, M.; Vergnon, R.; Cornelissen, H.C.; Hantson, S.; Holmgren, M.; van Nes, E.H.; Xu, C. Why trees and shrubs but rarely trubs? *Trends Ecol. Evol.* **2014**, *29*, 433–434. [CrossRef] [PubMed]
40. Raunkiær, C. Types biologiques pour la géographie botanique. *Overs. Kongel. Dan. Vidensk. Selsk. Forh. Medl. Arb.* **1905**, *5*, 347–437.
41. Körner, C. A re-assessment of high-elevation treeline positions and their explanation. *Oecologia* **1998**, *115*, 445–459. [CrossRef]
42. Wilson, C.; Grace, J.; Allen, S.; Slack, F. Temperature and stature: A study of temperatures in montane vegetation. *Funct. Ecol.* **1987**, *1*, 405–413. [CrossRef]
43. Pereg, D.; Payette, S. Development of black spruce growth forms at treeline. *Plant Ecol.* **1998**, *138*, 137–147. [CrossRef]
44. Pickett, S.T.A.; Kempf, J.S. Branching patterns in forest shrubs and understory trees in relation to habitat. *New Phytol.* **1980**, *86*, 219–228. [CrossRef]
45. Givnish, T.J. Leaf and canopy adaptations in tropical forests. In *Physiological Ecology of Plants in the Wet Tropics*; Tasks for Vegetation Science; Antiquariaat Junk: The Hague, The Netherlands, 1984; pp. 51–84.
46. Küppers, M. Ecological significance of above-ground architectural patterns in woody plants: A question of cost-benefit relationships. *Trends Ecol. Evol.* **1989**, *4*, 375–379. [CrossRef]
47. Sun, F.; Kuang, Y.; Wen, D.; Xu, Z.; Li, J.; Zuo, W.; Hou, E. Long-term tree growth rate, water use efficiency, and tree ring nitrogen isotope composition of *Pinus massoniana* L. in response to global climate change and local nitrogen deposition in Southern China. *J. Soils Sediments* **2010**, *10*, 1453–1465. [CrossRef]
48. Grace, J. Plant response to wind. *Agric. Ecosyst. Environ.* **1988**, *22–23*, 71–88. [CrossRef]
49. Hallinger, M.; Manthey, M.; Wilmking, M. Establishing a missing link: Warm summers and winter snow cover promote shrub expansion into alpine tundra in Scandinavia. *New Phytol.* **2010**, *186*, 890–899. [CrossRef]
50. Neuner, G. Frost resistance in alpine woody plants. *Front. Plant Sci.* **2014**, *5*, 654. [CrossRef] [PubMed]
51. Ryan, M.G.; Yoder, B.J. Hydraulic Limits to Tree Height and Tree Growth. *BioScience* **1997**, *47*, 235–242. [CrossRef]
52. Götmark, F.; Götmark, E.; Jensen, A.M. Why Be a Shrub? A Basic Model and Hypotheses for the Adaptive Values of a Common Growth Form. Front. *Plant Sci.* **2016**, *7*, 1095. [CrossRef]
53. Tyree, M.; Sperry, J. Vulnerability of xylem to cavitation and embolism. *Ann. Rev. Plant Biol.* **1989**, *40*, 19–36. [CrossRef]
54. LaMarche, V.C. Environment in Relation to Age of Bristlecone Pines. *Ecology* **1969**, *50*, 53–59. [CrossRef]
55. Schauer, A.J.; Schoettle, A.W.; Boyce, R.L. Partial cambial mortality in high-elevation Pinus aristata (Pinaceae). *J. Bot.* **2001**, *88*, 646–652. [CrossRef]
56. Beuselinck, L.; Steegen, A.; Govers, G.; Nachtergaele, J.; Takken, I.; Poesen, J. Characteristics of sediment deposits formed by intense rainfall events in small catchments in the Belgian Loam Belt. *Geomorphology* **2000**, *32*, 69–82. [CrossRef]
57. Pan, C.; Ma, L.; Shangguan, Z.; Ding, A. Determining the sediment trapping capacity of grass filter strips. *J. Hydrol.* **2011**, *405*, 209–216. [CrossRef]
58. Gallart, F.; Puigdefa'bregas, J.; del Barrio, G. Computer simulation of high mountain terracettes as interaction between vegetation growth and sediment movement. *Catena* **1993**, *20*, 529–542. [CrossRef]
59. Germino, M.J. Plants in Alpine Environments. In *Ecology and the Environment*; Monson, R., Ed.; Springer: New York, NY, USA, 2014; pp. 1–30. [CrossRef]
60. Baruch, Z. Patterns of energy content in plants from the Venezuelan pa\u0301ramos. *Oecologia* **1982**, *55*, 47–52. [CrossRef] [PubMed]
61. Stevens, G.C. The elevational gradient in altitudinal range: An extension of Rapoport's latitudinal rule to altitude. *Am. Nat.* **1992**, *140*, 893–911. [CrossRef]
62. Heitz, P.; Heitz-Seifert, U. Intra and interspecific relations within an epiphyte community in a Mexican humid montane forest. *Selbyana* **1995**, *16*, 135–140.

63. Austin, M.P.; Cunningham, R.B.; Fleming, P.M. New approaches to direct gradient analysis using environmental scalars and statistical curve-fitting procedures. *Vegetatio* **1984**, *55*, 11–27. [CrossRef]
64. Holten, J.J. Vascular plant species richness in relation to altitudinal and slope gradients in the mountain landscape of central Norway. In *The Impacts of Climate Variability on Forests*; Beniston, M., Innes, J.L., Eds.; Springer: Berlin, Germany, 1998; pp. 231–239.
65. Macek, P.; Macková, J.; de Bello, F. Morphological and ecophysiological traits shaping altitudinal distribution of three Polylepis treeline species in the dry tropical Andes. *Acta Oecol.* **2009**, *35*, 778–785. [CrossRef]
66. García-Plazaola, J.I.; Rojas, R.; Christie, D.A.; Coopman, R.E. Photosynthetic responses of trees in high-elevation forests: Comparing evergreen species along an elevation gradient in the Central Andes. *AoB Plants* **2015**, *7*, plv058. [CrossRef] [PubMed]
67. Derguy, M.R.; Frangi, J.L.; Drozd, A.A.; Arturi, M.F.; Martinuzzi, S. Holdridge Life Zone Map Republic of Argentina. *Gen. Tech. Rep. Dep. Agric.* **2019**, *51*, 1–48.
68. Luebert, F.; Gajardo, R. Vegetacion alto andina de Parinacota (norte de Chile) y una sinopsis de la vegetacion de la Puna meridional. *Phytocoenologia* **2005**, *35*, 79–128. [CrossRef]
69. López, V.L.; Cellini, J.M.; Cuyckens, G.A.E. Influencia del micrositio y el ambiente en la instalación de *Polylepis tarapacana* en los Altos Andes. *Neotrop. Biodivers.* **2021**, *7*, 135–145. [CrossRef]
70. Lizárraga, L.; Monguillot, J. Huella Humana en Argentina. Dirección Regional Noroeste, Administración de Parques Nacionales. Salta, Argentina. 2018. Available online: http://geoportal.idesa.gob.ar/layers/geonode%3Ah_humana_v01 (accessed on 15 March 2023).
71. Peng, Y.; Lachmuth, S.; Gallegos, S.C.; Kessler, M.; Ramsay, P.M.; Renison, D.; Suarez, R.; Hensen, I. Pleistocene climatic oscillations rather than recent human disturbance influence genetic diversity in one of the world's highest treeline species. *Am. J. Bot.* **2015**, *102*, 1676–1684. [CrossRef]
72. Cuyckens, G.; Christie, D.; Domic, A.; Malizia, L.; Renison, D. Climate change and the distribution and conservation of the world's highest elevation woodlands in the South American Altiplano. *Glob. Planet. Chang.* **2015**, *137*, 79–87. [CrossRef]
73. Boza Espinoza, T.E.; Kessler, M. A monograph of the genus *Polylepis* (*Rosaceae*). *PhytoKeys* **2022**, *203*, 1–274. [CrossRef]
74. Rojas, R.; Flexas, J.; Coopman, R.E. Particularities of the highest elevation treeline in the world: *Polylepis tarapacana* Phil. as a model to study ecophysiological adaptations to extreme environments. *Flora* **2022**, *292*, 152076. [CrossRef]
75. Araya-Presa, J.; Squeo, F.A.; Barrientos, L.; Belmonte, E.; Mamani, M.; Arancio, G. *Manual de Plantas y Canciones Aymará*. Proyecto ExploraCONYCIT ED7/027085: Etnobotánica y Etnomusicología Aymará: Divulgación de La Sabiduría Ancestral Sobre Las Plantas Nativas Del Altiplano de Arica. Universidad de La Serena. Chile. 2003. Available online: http://www.biouls.cl/aymara (accessed on 3 March 2023).
76. Garreaud, R.; Vuille, M.; Clement, A.C. The climate of the Altiplano: Observed current conditions and mechanisms of past changes. *Palaeogeogr. Palaeoclim. Palaeoecol.* **2003**, *194*, 5–22. [CrossRef]
77. Morello, J.; Matteucci, S.D.; Rodríguez, A.F.; Silva, M.; De Haron., J.C. *Ecorregiones y Complejos Ecosistémicos Argentinos*, 1st ed.; Orientación Gráfica Editora: Buenos Aires, Argentina, 2012; p. 773.
78. Cabrera, A.L. Ecología vegetal de la Puna. *Colloq. Geogr.* **1968**, *9*, 91–116.
79. Oyarzabal, M.; Clavijo, J.; Oakley, L.; Biganzoli, F.; Tognetti, P.; Barberis, I.; Maturo, H.M.; Aragón, R.; Campanello, P.I.; Prado, D.; et al. Unidades de vegetación de la Argentina. *Ecol. Austral* **2018**, *28*, 40–63. [CrossRef]
80. Jenness, J.S. Calculating landscape surface area from digital elevation models. *Wildl. Soc. Bull.* **2004**, *32*, 829–839. [CrossRef]
81. Fick, S.E.; Hijmans, R.J. WorldClim 2: New 1 km spatial resolution climate surfaces for global land areas. *Int. J. Climatol.* **2017**, *37*, 4302–4315. [CrossRef]

MDPI AG
Grosspeteranlage 5
4052 Basel
Switzerland
Tel.: +41 61 683 77 34
www.mdpi.com

Plants Editorial Office
E-mail: plants@mdpi.com
www.mdpi.com/journal/plants

www.ingramcontent.com/pod-product-compliance
Lightning Source LLC
Chambersburg PA
CBHW061606210926
51850CB00009B/2227
9783725815531